Behavioral Modeling of Nonlinear RF and Microwave Devices

For a recent listing of titles in the *Artech House Microwave Library*,
turn to the back of this book.

Behavioral Modeling of Nonlinear RF and Microwave Devices

Thomas R. Turlington

Artech House
Boston • London

Library of Congress Cataloging-in-Publication Data
Turlington, Thomas R.
Behavioral modeling of nonlinear RF and microwave devices / Thomas R. Turlington
p. cm. — (Artech House microwave library)
Includes bibliographical references and index.
ISBN 1-58053-014-1 (alk. paper)
1. Microwave devices—Mathematical models. I. Title. II. Series.
TK7876.T87 1999 99-045833
621.381'3—dc21 CIP

British Library Cataloguing in Publication Data
Turlington, Thomas R.
Behavioral modeling of nonlinear RF and microwave devices.
— (Artech House microwave library)
1.Microwave devices—Mathematical models
I. Title
621.3'8133

ISBN 1-58053-014-1

Cover design by Jennifer Stuart

685 Canton Street

Norwood, MA 02062

International Standard Book Number: 1-58053-014-1
Cataloging-In-Publication: 99-045833

10 9 8 7 6 5 4 3 2 1

To the Eternal, who planted the seeds of thought.
To my patient wife, Jeannette, who endured the times of growth.
To my encouraging children, Charles and Jeannine, who, I'm told, look up to their dad.
And to my adorable grandchildren, Rachel and Jordan, too young to understand.

Table of Contents

Preface

What Is Behavioral Modeling?

In the course of scientific investigation, there often arises a need to express an observed object's behavior with an equation such as a behavioral model. More often than not, the observed object's behavior is the result of a nonlinear process that does not lend itself to easy curve fitting, and the physics of the object are not completely known. Furthermore, the equation that models the observed behavior should provide acceptable results over the full range of all independent variable values. The equation should be capable of generating acceptable, smoothly flowing, continuous, dependent data at independent variable values between the observed, or measured, data points. The method used to formulate the curve fit equation is extremely important if the model's output data is to be operated on by a Fourier transform creating new data in a paired domain.

The methodology of behavioral modeling, developing equations that acceptably reproduce an observed object's behavior, be it linear or nonlinear, is the topic of this book. The objects of interest in this text are electronic in nature, and operate at radio and microwave frequencies. The new curve-fitting methods defined and described in Chapter 2 have the ability to generate equations that provide acceptably accurate fits to data that cannot otherwise be represented by a single equation. Polynomial expansions are sometimes acceptable where measured data flow is well behaved, and simple quadratic or cubic expressions suffice to model the measured data. Interpolation errors always exist between known data values when higher order polynomials are used to capture sharp inflections in data flow. Third-order, fourth-order, and high-order spline curve fit techniques can be made to fit data flow with acceptable accuracy, but no single equation results from the process. For that reason high-order spline curve fits are difficult to explain and publish in the literature, and equation coefficients have little or no direct relationship to measured physical parameters. The curve fit methods used here are readily classified as additive functions but involve entirely new and unique right-hand functions and left-hand functions. They work well through sharp inflections because they involve summations of logarithmic functions which are infinite series in disguise. Coefficients used in these new curve fit equations are measurable parameters such as power, voltage, current, gain, phase, noise figure, and compression coefficient. The equations can easily be applied to behavior of objects other than electronic devices with equal ease and flexibility of coefficient definitions.

Development of Curve Fit Methodology

An arbitrary set of data is proposed and polynomial curve-fit methods are applied to show the difficulties and errors that arise in practice. Portions of the data set are modeled with low-order polynomials, giving reasonable results but not one single equation. The magnitude of error becomes unacceptable when a higher order polynomial is used to model the entire data set. The new approach to curve fitting is developed using the same example that demonstrated failure of polynomial and spline techniques. New functions needed to complete the new curve-fit method are defined and characteristics of those functions are explained. The first example is then used to demonstrate the new method's effectiveness.

Application to RF and Microwave Transistors

The new curve-fit method is applied to microwave and RF transistor devices. Nonlinear models are developed for bipolar and MESFET devices used in amplifiers. Those models are used to generate characteristic amplifier data as a function of bias current level and drive level. Amplifier characteristic data includes plots of device output waveforms as the device is driven into saturation and cut off. Output waveforms are analyzed for harmonic content by performing discrete fourier transforms on the generated output data. Harmonic content is then studied as a function of bias condition and power input level.

Class A and Class AB Amplifiers

Data generated by behavioral models of transistors at different bias conditions and varying input power levels, at saturation and cutoff, is the basis for developing nonlinear behavioral models for class A and class AB amplifiers. New modeling techniques are used to characterize class A and class AB amplifiers, and develop relationships between the P_{1dB} point, the *3rd OIP*, P_{sat}, small signal gain (G_{ss}), compression characteristic (K), compression depth, bias current, drain to source voltage, and power input (P_{in}). Compression characteristic (K) is defined, and methods for determining the value K for an amplifier are discussed. The nonlinear amplifier behavioral model equations developed have coefficients of P_{in}, P_{sat}, G_{ss}, K, i_{dss}, v_{ds}, v_k, *3rd OIP*, P_{1dB}, and *NF*. They relate directly to measurable parameters.

The two-tone method of determining amplifier third-order intercept is discussed, and the two-tone intermodulation sideband level for any amplifier at 1 dB compression is calculated. A strong argument is made that the only two-tone input level useful for *3rd OIP* calculation and characterization is when output power is compressed 1 dB (the P_{1dB} point). Two-tone, third-order intermodulation sideband growth rate of 3:1 is not always observed in practice. Some so-called "high dynamic range" amplifiers exhibit 6:1 and even 8:1 growth rate of two-tone test third-order *IM* sidebands. An amplifier third-order intercept point determined at output power levels other than the P_{1dB} point for these amplifiers is not correct. Such "high dynamic range" amplifiers benefit from large clear dynamic range for small signals, but suffer the same cross modulation as any other amplifier when high power signals drive the first amplifier stages into compression. Behavioral model equations predict these abnormal third-order *IM* sideband growth rates as a function of compression characteristic (K) value.

A rule of thumb based on years of observation of devices and circuits is proposed for estimating amplifier noise figure. The relationships between current, noise figure, *3rd OIP*, P_{sat}, G_{ss}, and P_{1dB} are then used to determine optimum cascaded amplifier configurations for low-noise amplification and power generation.

Optimizing Cascaded Amplifier Design

The systems engineer and the circuit designer are often faced with the need to know and understand the trades between noise figure, current consumption, and amplifier input *3rd OIP*. The unique amplifier stage behavioral model equations developed in the first part of the book are applied to cascaded amplifiers, and trade spaces are developed showing the relationship between the three parameters as functions of amplifier stage gain distribution and compression characteristic (K). Third-order intercept value for cascaded stages is shown to be a strong driver of minimum obtainable noise figure and minimum obtainable current consumption.

Nonlinear behavioral model equations of cascaded stages are combined to develop a direct relationship between saturated power output of each cascaded stage, compression characteristic (K), small signal gain per stage, and maximum saturated power output available in a power amplifier design. The relationship developed is unique and extremely useful to the systems designer as well as the MMIC circuit developer. Power-added efficiency of cascaded power amplifier stages is shown to be a dynamic function of power input.

Accounting for Statistical Variation in Nonlinear Amplifier Parameters

Probability distributions of small signal gain, saturated power output, bias current, and noise figure are discussed. Methods are developed showing how to account for these parameter variations lot to lot as large quantities of subsystems are assembled. Use of the spreadsheet add-in CRYSTAL BALL (developed by Decisioneering, Denver, Colorado) to determine the number of stages required to realize a minimum gain over frequency, at maximum operating temperature, and over a large population of assemblies, is illustrated. A spreadsheet that determines the gain trim range needed to tune a large population of nonlinear power amplifiers that have random small signal gain values for 2 dB compression is developed and demonstrated.

Trades are developed that show the noise figure and current consumption benefits that result when designing with fewer high-gain stages rather than more low-gain stages. A trade method is developed to determine the optimum number of gain stages per MMIC to use when designing cascaded stage amplifiers where standard deviation of small signal gain is additive for stages on the same MMIC and is root sum squared for MMICs randomly selected from a wafer population.

Expanding Behavioral Modeling Into the Frequency Domain

Parameters vary as a function of frequency. Behavioral model results will also vary as key parameters change with frequency. Amplifier compression depth and phase shift will change as small signal gain and saturated power output vary differently with respect to frequency. Compression characteristic (K) will change as a function of frequency, thereby changing the softness or sharpness of amplifier compression and the intermodulation and cross modulation components that result when overdriven. Power input may change as a function of frequency if an amplitude weighted waveform is used as a modulation technique. Techniques for modeling measured and simulated frequency responses are discussed and developed.

Adding Temperature Sensitivity to Behavioral Models

Techniques for adding temperature sensitivity to behavioral models are described. Temperature sensitivity of parameters such as small signal gain, saturated power output, and noise figure are discussed. Variable temperature sensitivity of parameters as a function of frequency is also discussed and methods of modeling this additional complexity are explained.

The Sum of All Models

All of the modeling techniques are brought to bear on system and subsystem design and analysis and on the flow down of system parameters to component and device levels. A receiver system, and a transmitter amplifier subsystem are studied with the object of optimizing parameter values for mass production to obtain highest production yield to specified requirements. The use of risk analysis add-in software for spreadsheets is described and results are shown.

Odds and Ends

The new curve fit technique is used to develop equations for S-parameters where families of S-parameter files exist as a function of frequency at different bias levels. A single equation is developed that generates S-parameter data as a function of frequency and bias current. Eight equations fully characterize the entire family of eight S-parameters, magnitude, and phase.

The book closes with thought-provoking examples of simulating time-related functions, and modeling diode junction capacitance with a single equation. The purpose here is to encourage expansion of the modeling techniques into areas heretofore not considered. The author believes the applications of right-hand and left-hand functions are unlimited, and, once mastered, they can become a synthesis and design tool throughout the scientific community.

Thomas Turlington
Lecturer and Consultant
Linthicum, MD

1

Numbers and Traces

1.1 Common Curve-Fitting Techniques

Behavioral modeling is the science of accurately expressing measured behavior of an object, be it linear or nonlinear, with easily generated equations. The behavioral modeling objective is to formulate a single closed form equation representing a measured parameter. That measured parameter might be a function of multiple orthogonal independent variables that simultaneously control the object's behavior, as indicated in Equation (1.1).

$$P_{out} = f(P_{in}, f, T, V_{ds}, I_{ds}, V_{cmd}, B_{data}) \tag{1.1}$$

Figure 1.1 illustrates a typical object and some of the parameters associated with that object that define and control its behavior.

The object may be linear or nonlinear, may have simultaneous analog and digital inputs, and may be sensitive to environmental conditions. The analog inputs might be variable in magnitude, phase, and frequency, all occurring simultaneously. Digital inputs might be dynamic in nature, causing continuous discrete changes in the output signal. Internal design details of the object may not be known. Measured data or catalog data representing the object's behavior in terms of scattering parameters over frequency may be the only information available. That data may be in table or chart form.

The process of converting measured data into equations relies on curve-fitting techniques. Many of the common curve-fitting techniques are useful where the data trace is well behaved over a defined independent variable range and where behavior of an object is known to follow a specific mathematical model. Well-behaved data can be represented by linear, logarithmic, power

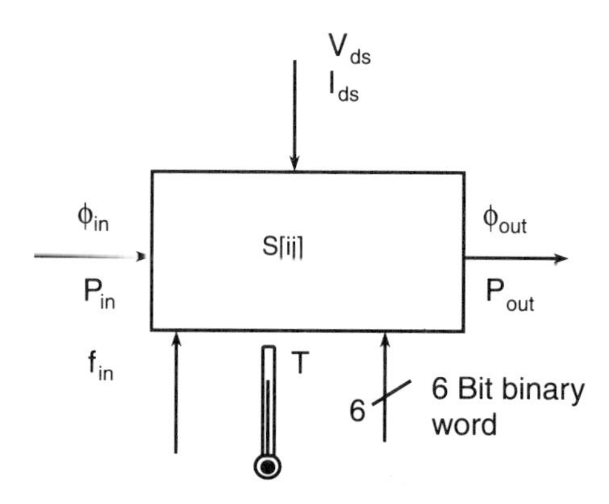

Figure 1.1 A typical object, its parameters, sensitivities, and characteristics.

function, exponential, or polynomial function curve fits. Problems arise when nonlinear characteristics such as saturation and cutoff abruptly modify the object's behavior. Additional problems arise when the object's complex internal parameters, represented by locations of poles and zeros in the complex plane, cause the performance data trace to exhibit sharp inflections as a function of frequency; then data ceases to be well behaved and common curve-fitting techniques become useless. Detailed discussions of contemporary curve-fitting techniques are published by Chapman & Hall in a continuing series of books, *Monographs on Statistics and Applied Probability.* Two volumes of interest are Volume 43, *Generalized Additive Models* [1], and Volume 58, *Nonparametric Regression and Generalized Linear Models* [2].

Fundamental common curve fit techniques and their usefulness to behavioral modeling are discussed in this chapter. Equations representing an object's behavior are often used in calculations where convolution and domain transformation are performed. For example, the discrete Fourier transform (DFT) is one operation that requires a large number of data points in order to obtain an acceptably detailed transformation into a paired domain. Fabrication of data points by interpolation between measured data points is required in order to obtain sufficient numbers of input data values for the DFT. Typical DFTs require 512 or 1024 input data points. The trace of fabricated data points must be smooth and acceptably realistic if the desired detail and accuracy are to be obtained in the domain transformation. The discussion of fundamental curve-fitting techniques in this chapter makes clear the need for a new curve-fit technique that provides smoothness and continuity through plotted trace sharp inflections. Real data describing an object's behavior when signal saturation and signal cutoff occur and behavior as a function of signal frequency are examples on which this book focuses on.

1.2 Linear Regression

Linear regression is used to represent data that traces a straight line as a function of the independent variable. Coefficients m and b for the equation

$$y = mx + b \tag{1.2}$$

can be valued such that a straight line defined by the equation falls on or near most of the data points. The first derivative of a linear regression equation is

$$\frac{\partial y}{\partial x} = m \tag{1.3}$$

and all higher derivatives are zero. Behavioral modeling methods use curve fitting of this type to express linear traces of data and to generate equations for asymptotes to curved traces of data. A thorough discussion of asymptotes is found in Chapter 2.

1.3 Linear Interpolation

As mentioned above, transformation between paired domains by DFT requires a large number of data points. Interpolation and fabrication of information between measured data points is required to obtain sufficient data to perform a DFT. Measured data representing behavior of a single object often does not include sufficient resolution to enable acceptably accurate interpolation of values between the measured points. Of the many tools at hand to perform curve fitting, the easiest to use is linear interpolation between measured data points. This technique is a simple "connect the dots" approach resulting in straight lines between the measured data points with no single, closed form, publishable equation that describes data over the entire independent variable range. Linear interpolation does not result in a smooth curve. Unacceptable errors result in domain transform calculations if data generated by linear interpolation are used in convolution and DFT operations.

1.4 Logarithmic Regression

Logarithmic curve fit is obtained by determining coefficient values c and b that cause the equation

$$y = c \ln x + b \tag{1.4}$$

to match given data over a range of independent variables. The logarithmic curve fit cannot be used where $x \leq 0$. Logarithmic regression generates a monotonic function that has an ever-decreasing first derivative

$$\frac{\partial y}{\partial x} = \frac{1}{x} \tag{1.5}$$

This curve-fit equation is difficult to accurately fit to a large number of data points unless the object's behavior is truly related to a natural logarithmic function. Figure 1.2 illustrates a logarithmic function where coefficients $c = 2$ and $b = 3$.

Logarithmic regression as defined by (1.4) is not used in its pure form in behavioral modeling. Modified versions of logarithmic regression are developed in Chapter 2 that are extremely useful and are the basis of the curve-fitting technique that enables the behavioral modeling method. The logarithm is a smoothly varying function that can be approximated by a power series and is therefore rich in high-order harmonic content when transformed into a paired domain through DFT.

1.5 Power Function Regression

The power function

$$y = cx^b \tag{1.6}$$

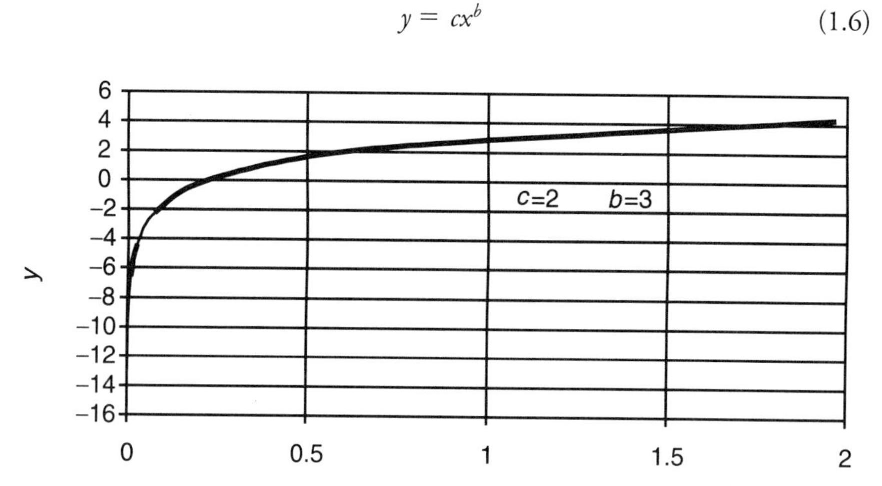

Figure 1.2 Logarithmic function plotted.

is useful in modeling objects that are known to obey a power law. The power function's first derivative

$$\frac{\partial y}{\partial x} = bcx^{b-1} \tag{1.7}$$

defines the slope of a tangent to the curve at any point x. The value of the first derivative is often used in behavioral modeling as the slope of an asymptote to the curve at point x. Figure 1.3 illustrates the square law power function where $c = 1$ and $b = 2$.

Vacuum tubes obey a 3/2 power law, and field effect transistors exhibit a square law characteristic behavior under small signal operation. The power function, (1.6), is limited in its range of usefulness as a model for most nonlinear objects when extreme conditions of nonlinearity are encountered. For instance, vacuum tubes and field effect transistors cutoff; no current flows for all negative input signal values less than cutoff. They also saturate at some large positive input signal value. The power function, (1.6), obtains zero value at $x = 0$ representing cutoff of the nonlinear object, but the power function value continues to increase or decrease as x becomes smaller than zero. This does not accurately model the object's behavior of cutoff where there is no output for input signal more negative than the cutoff point. As the power function variable x increases into the positive value range, y increases without bound and fails to reproduce that aspect of the object's behavior of saturation. Figure 1.4

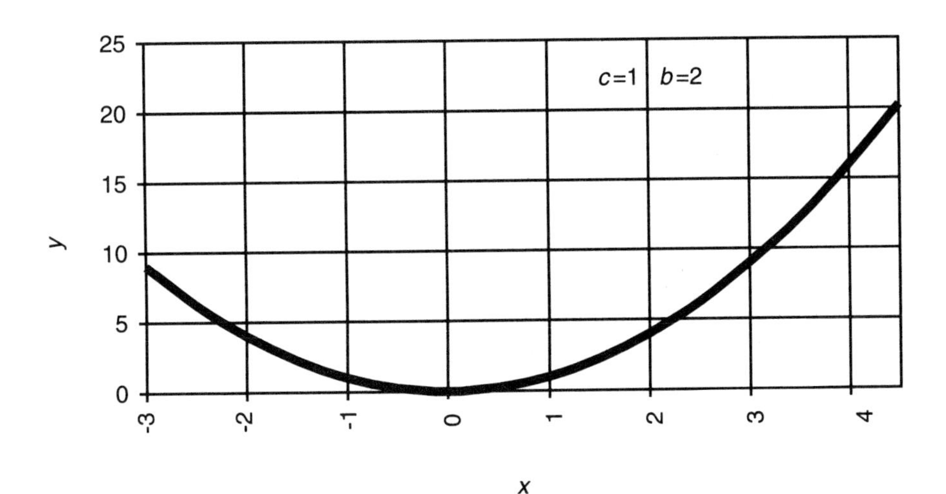

Figure 1.3 The power function plotted.

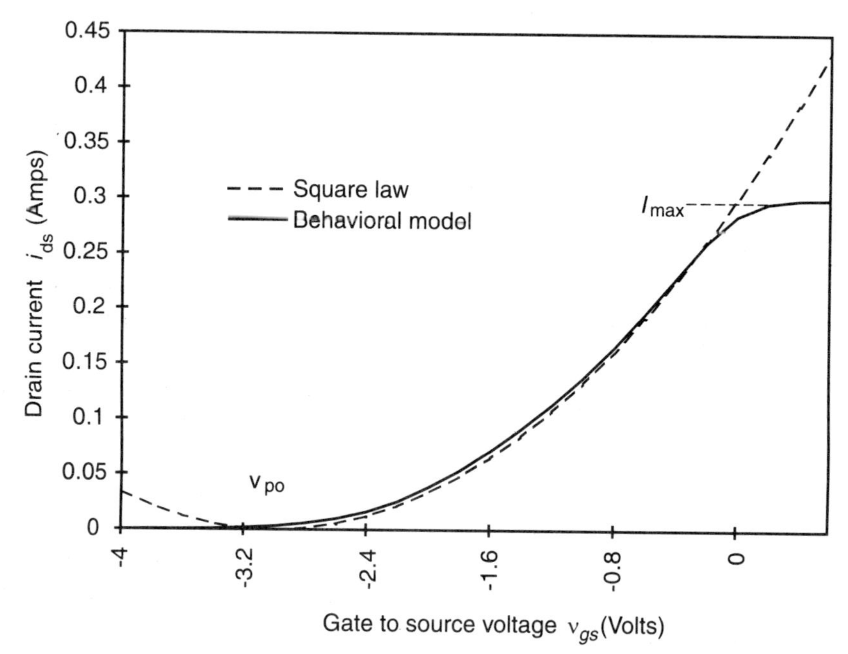

Figure 1.4 Comparison between square law function and a field effect transistor behavioral model.

shows a comparison between the square law power function and a behavioral model function that duplicates the object's saturation at I_{dss} and cutoff at V_{po}.

1.6 Exponential Regression

The exponential equation

$$y = ce^{bx} \tag{1.8}$$

can be used to model object characteristics that are known to behave in an exponential fashion. The exponential equation has a first derivative

$$\frac{\partial y}{\partial x} = bce^{bx} \tag{1.9}$$

which is monotonic; it doesn't change sign over the entire range of independent variable $-\infty < x < +\infty$. Figure 1.5 illustrates the form of an exponential equation where $c = 1$ and $b = 2$.

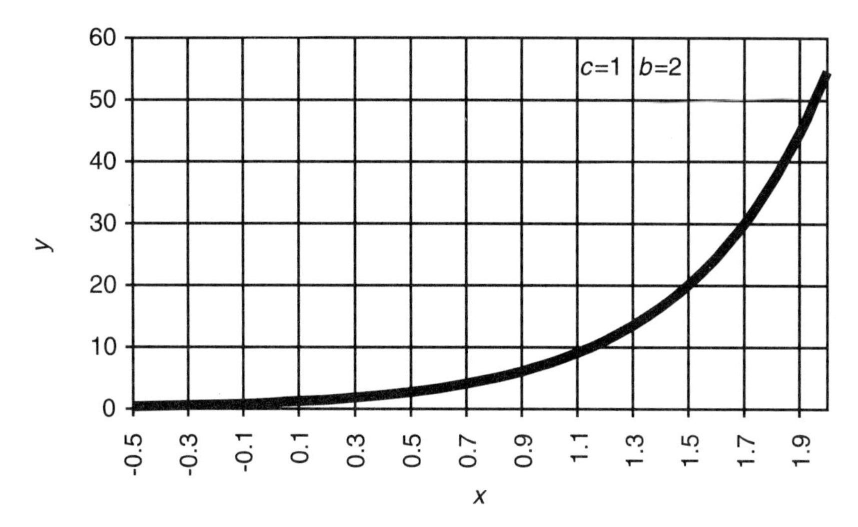

Figure 1.5 The exponential function.

Diode junctions and bipolar transistors are objects that behave in a manner described by an exponential equation. The bipolar transistor, like the field effect transistor, exhibits saturation and cutoff when overdriven. The exponential function does model pinchoff nicely. As the exponential function variable x decreases in value, the function approaches zero and remains there for all lesser values of x. However, the exponential function increases without bound as x increases. Like the power function, the exponential function fails to model the abrupt nonlinearity caused by saturation. Figure 1.6 shows a comparison between the exponential function and a behavioral model of the bipolar transistor that saturates.

Many time-dependent processes involving energy storage or energy dissipation are known to be exponential in nature and can be modeled by slight modification of the exponential function.

1.7 Polynomial Regression

An nth order polynomial equation

$$y = a_0 + a_1x + a_2x^2 + a_3x^3 + \cdots + a_nx^n \tag{1.10}$$

can be used to create a curvilinear regression or average fit to a data trace that exhibits changes in sign of first, second, and higher derivatives. This technique works nicely if the data trace is well behaved. Objects that lend themselves

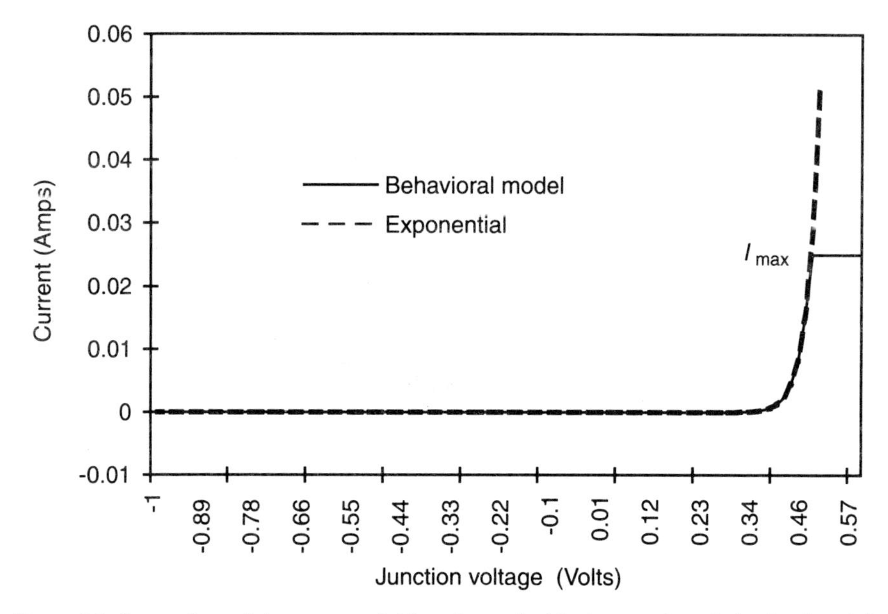

Figure 1.6 Comparison of the exponential function and a bipolar transistor behavioral model.

nicely to behavioral modeling with polynomial expressions are AT cut quartz resonators which exhibit resonant frequency value shifts as a function of temperature. These shifts in resonant frequency follow a third-order polynomial expression. An example of a third-order polynomial equation where $a_0 = 0$, $a_1 = -3$, $a_2 = 0$, and $a_3 = 2$ is shown in Figure 1.7.

The method of coefficient derivation for a polynomial equation curve fit depends on the independent variable reference point chosen. If the polynomial expansion is about $x = 0$, Maclaurin's procedure is used. If the expansion is about a point other than $x = 0$, Taylor's procedure is used. A detailed discussion of Maclaurin's and Taylor's polynomial series can be found in most books on college calculus, differential equations, and applied engineering mathematics. It is sufficient here to repeat Taylor's theorem, that the equation

$$f(z) = f(a) + f'(a)(z - a) + f''(a)\frac{(z - a)^2}{2!} + f'''(a)\frac{(z - a)^3}{3!} + \text{oooo} \quad (1.11)$$

is a valid representation of $f(z)$ at all points in the interior of any circle having its center at a and within which $f(z)$ is analytic. Maclaurin's expansion is the unique case where $a = 0$. A method for obtaining Taylor's series equation coefficient values is given in Example 1.1. Example 1.1 also illustrates the inaccura-

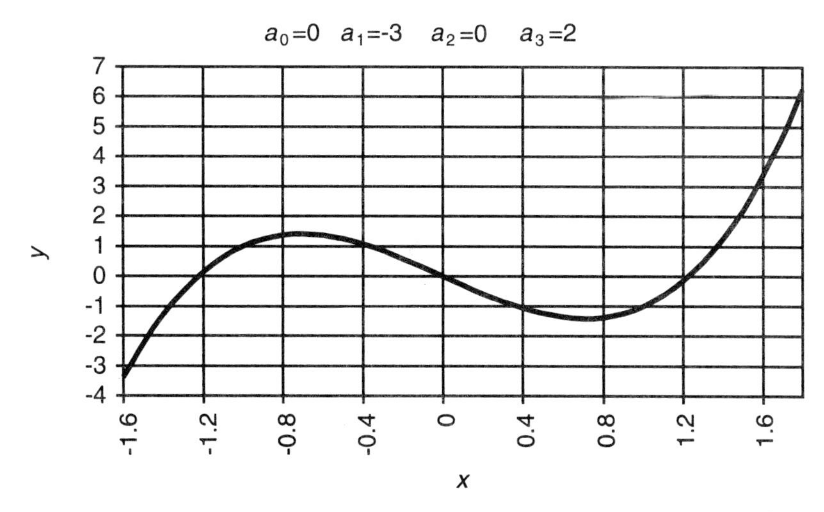

Figure 1.7 Third-order polynomial expansion example.

cies experienced when using Taylor's expansion about a data point to generate a single equation for the entire range of independent variables.

Example 1.1 Obtaining Polynomial Equation Coefficients

Taylor's expansion about a data point is a technique that results in a publishable equation which satisfies a requirement of the behavioral modeling method. The resulting polynomial expansion equation converges to each measured data point with very small error, and provides a smooth continuum of interpolated data between data points if the data flow is well behaved. However, unacceptably large errors can and do develop between measured data points if the data flow is not well behaved. Table 1.1 shows a set of data that has a flow that is typical in characteristic to input or output reflection coefficient of a complex microwave circuit. Data shown in Table 1.1 is plotted in Figure 1.8.

A region of this data ($6 \le x \le 7.2$) has a domain that can be modeled by a cubic equation. Another region of this data ($7.2 \le x \le 9$) is easily modeled by a fifth-order equation. Both of these regions by themselves are well behaved. Third-order and fifth-order equations are developed using Taylor's expansions to demonstrate how well the formulated equations fit data in these two independent regions. A fourth-order polynomial equation is then formulated with a Taylor' s expansion around the independent variable value $x = 7.1$ to obtain a curve fit for the region ($6.6 \le x \le 7.8$) around the sharp inflection point at $x = 7.2$. Finally, an eighth-order polynomial equation is formulated

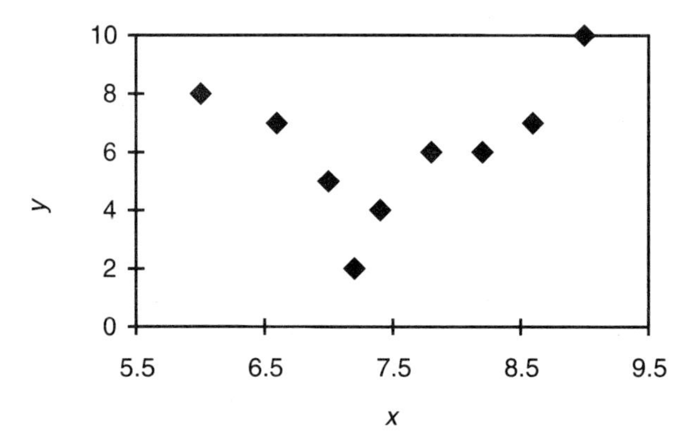

Figure 1.8 Example data points plotted with no connecting trace.

with a Taylor's expansion around the value $x = 7.1$ to demonstrate the large errors that develop between data points if the data flow is not well behaved.

Expand the first region of data around $x = 6.7$. Use matrix methods to solve four simultaneous equations to obtain coefficient values a_0, a_1, a_2, and a_3 in the polynomial expression

$$y = a_0 + a_1(x - x_0) + a_2(x - x_0)^2 + a_3(x - x_0)^3$$

$$8 = a_0 + a_1(6 - 6.7) + a_2(6 - 6.7)^2 + a_3(6 - 6.7)^3$$

$$7 = a_0 + a_1(6.6 - 6.7) + a_2(6.6 - 6.7)^2 + a_3(6.6 - 6.7)^3$$

$$5 = a_0 + a_1(7 - 6.7) + a_2(7 - 6.7)^2 + a_3(7 - 6.7)^3$$

$$2 = a_0 + a_1(7.2 - 6.7) + a_2(7.2 - 6.7)^2 + a_3(7.2 - 6.7)^3$$

$$\begin{bmatrix} a_0 \\ a_1 \\ a_2 \\ a_3 \end{bmatrix} = \begin{bmatrix} 8 \\ 7 \\ 5 \\ 2 \end{bmatrix} * \begin{bmatrix} 1 & -0.7 & 0.49 & -0.34 \\ 1 & -0.1 & 0.01 & -0.001 \\ 1 & 0.3 & 0.09 & 0.027 \\ 1 & 0.5 & 0.25 & 0.125 \end{bmatrix}^{-1}$$

$$\begin{bmatrix} 6.833 \\ -2.444 \\ -8.888 \\ -11.11 \end{bmatrix} \qquad \begin{bmatrix} a_0 \\ a_1 \\ a_2 \\ a_3 \end{bmatrix} =$$

Table 1.1
Example Data Showing Sharp Inflection

x	y
6	8
6.6	7
7	5
7.2	2
7.4	4
7.8	6
8.2	6
8.6	7
9	10

The cubic equation obtained by this process is

$$y = 6.833 - 2.444(x - 6.7) - 8.888(x - 6.7)^2 - 11.11(x - 6.7)^3 \quad (1.12)$$

which reduces to

$$y = 2965.703 - 1379.529x + 214.423x^2 - 11.11x^3 \quad (1.13)$$

Equation (1.13) defines the modeled domain over the region ($6 \leq x \leq 7.2$). Figure 1.9 shows the curve fit obtained with this equation where values between the given data points are calculated and plotted to illustrate the domain trace. Notice the curve fits the original data points precisely and flows smoothly between original data points. There is a disturbing third-order twist in the trace at the low end of the range.

Expand the second data region around independent variable value $x = 8$ and develop a fifth-order equation to fit data over the region ($7.2 \leq x \leq 9$). Use matrix methods again to solve six simultaneous equations.

$$2 = a_0 + a_1(7.2 - 8) + a_2(7.2 - 8)^2 + a_3(7.2 - 8)^3 + a_4(7.2 - 8)^4 + a_5(7.2 - 8)^5$$

$$4 = a_0 + a_1(7.4 - 8) + a_2(7.4 - 8)^2 + a_3(7.4 - 8)^3 +$$

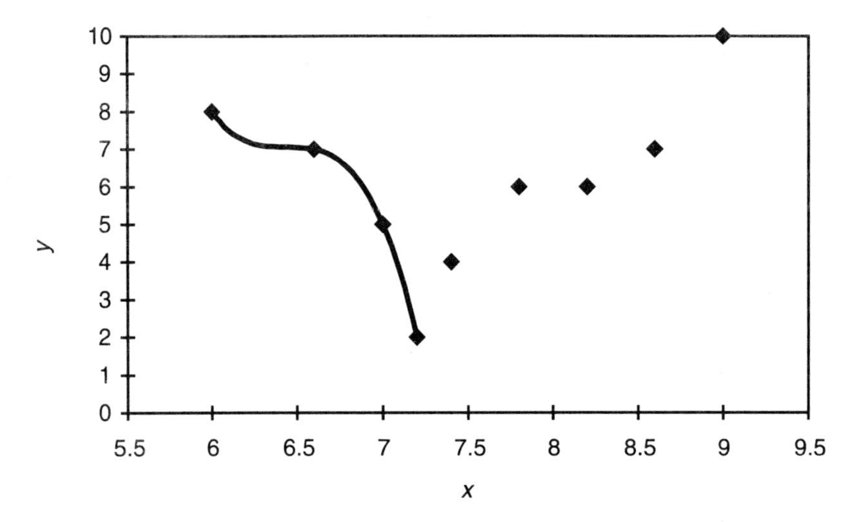

Figure 1.9 Cubic equation fit over the region $6 \le x \le 7.2$.

$$6 = a_0 + a_1(7.8-8) + a_2(7.8-8)^2 + a_3(7.8-8)^3 + a_4(7.8-8)^4 + a_5(7.8-8)^5$$

$$6 = a_0 + a_1(8.2-8) + a_2(8.2-8)^2 + a_3(8.2-8)^3 + a_4(8.2-8)^4 + a_5(8.2-8)^5$$

$$7 = a_0 + a_1(8.6-8) + a_2(8.6-8)^2 + a_3(8.6-8)^3 + a_4(8.6-8)^4 + a_5(8.6-8)^5$$

$$10 = a_0 + a_1(9-8) + a_2(9-8)^2 + a_3(9-8)^3 + a_4(9-8)^4 + a_5(9-8)^5$$

$$\begin{bmatrix} a_0 \\ a_1 \\ a_2 \\ a_3 \\ a_4 \\ a_5 \end{bmatrix} = \begin{bmatrix} 2 \\ 4 \\ 6 \\ 6 \\ 7 \\ 10 \end{bmatrix} * \begin{bmatrix} 1 & -0.8 & 0.64 & -0.512 & 0.4096 & -0.3277 \\ 1 & -0.6 & 0.36 & -0.216 & 0.1296 & -0.0778 \\ 1 & -0.2 & 0.04 & -0.008 & 0.0016 & -0.00032 \\ 1 & 0.2 & 0.04 & 0.008 & 0.0016 & 0.00032 \\ 1 & 0.6 & 0.36 & 0.216 & 0.1296 & 0.0778 \\ 1 & 1 & 1 & 1 & 1 & 1 \end{bmatrix}^{-1}$$

$$\begin{bmatrix} a_0 \\ a_1 \\ a_2 \\ a_3 \\ a_4 \\ a_5 \end{bmatrix} = \begin{bmatrix} 6.091 \\ -0.3398 \\ -2.3165 \\ 8.6259 \\ 1.1681 \\ -3.2285 \end{bmatrix}$$

The fifth-order polynomial equation obtained is

$$y = 6.091 - 0.3398(x - 8) - 2.3165(x - 8)^2 + 8.6259(x - 8)^3 + 1.1681(x - 8)^4 - 3.2285(x - 8)^5 \quad (1.14)$$

which when plotted over the original data points and for points in between from ($7.2 \leq x \leq 9$) provides a good curve fit as shown in Figure 1.10.

The two curves developed above cover the entire range of data providing curve fit domain over each associated region. However, there's a discontinuity at $x = 7.2$ where the sharp inflection occurs. No first-, second-, third-, and higher order derivative exists at that point. The data is not analytic at $x = 7.2$. If for some reason it were desired to transform this data into a paired domain by processing it through a 512 point DFT, the range ($6 \leq x \leq 9$) would need to be divided into 512 subregions each of width 0.0059 and a dependent variable value y would be calculated for each subregion. Over 6.6% of the 512 calculated values fall within ± 0.1 of the sharp inflection at $x = 7.2$. This discontinuity and

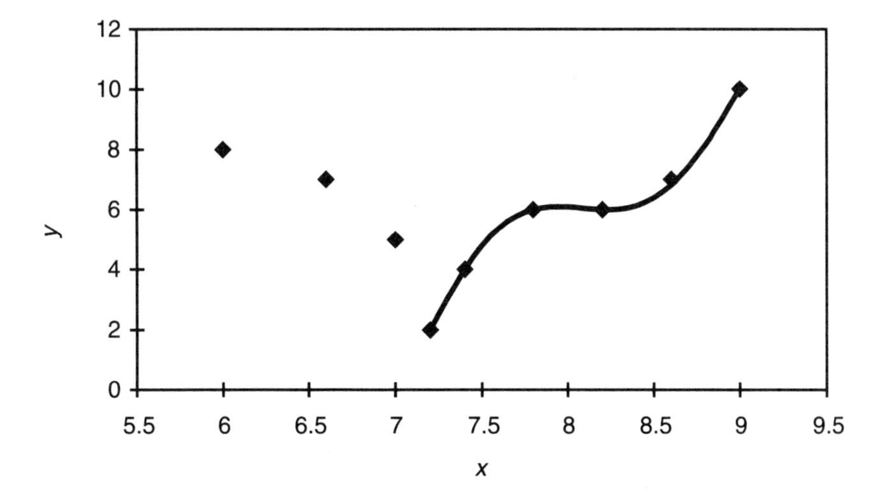

Figure 1.10 Fifth-order curve fit about $x = 8$ to region ($7.2 \leq x \leq 9$).

the valuation of interpolated data in its immediate vicinity can cause errors as high as ±4 dB (63%) in the data values transformed into the paired domain. Modeling a continuous domain immediately around the inflection is critical if transformation by DFT or convolution is to yield meaningful results.

Perform a fourth-order polynomial expansion around $x = 7.1$ to include the sharp inflection and two data points on either side of it. Again use matrix methods to solve the five simultaneous equations:

$$7 = a_0 + a_1(6.6 - 7.1) + a_2(6.6 - 7.1)^2 + a_3(6.6 - 7.1)^3 + a_4(6.6 - 7.1)^4$$

$$5 = a_0 + a_1(7 - 7.1) + a_2(7 - 7.1)^2 + a_3(7 - 7.1)^3 + a_4(7 - 7.1)^4$$

$$2 = a_0 + a_1(7.2 - 7.1) + a_2(7.2 - 7.1)^2 + a_3(7.2 - 7.1)^3 + a_4(7.2 - 7.1)^4$$

$$4 = a_0 + a_1(7.4 - 7.1) + a_2(7.4 - 7.1)^2 + a_3(7.4 - 7.1)^3 + a_4(7.4 - 7.1)^4$$

$$6 = a_0 + a_1(7.8 - 7.1) + a_2(7.8 - 7.1)^2 + a_3(7.8 - 7.1)^3 + a_4(7.8 - 7.1)^4$$

$$\begin{bmatrix} a_0 \\ a_1 \\ a_2 \\ a_3 \\ a_4 \end{bmatrix} = \begin{bmatrix} 7 \\ 5 \\ 2 \\ 4 \\ 6 \end{bmatrix} * \begin{bmatrix} 1 & -0.5 & 0.25 & -0.125 & 0.0625 \\ 1 & -0.1 & 0.01 & -0.001 & 0.0001 \\ 1 & 0.1 & 0.01 & 0.001 & 0.0001 \\ 1 & 0.3 & 0.09 & 0.027 & 0.0081 \\ 1 & 0.7 & 0.49 & 0.343 & 0.2401 \end{bmatrix}^{-1}$$

$$\begin{bmatrix} a_0 \\ a_1 \\ a_2 \\ a_3 \\ a_4 \end{bmatrix} = \begin{bmatrix} 2.937 \\ -15.677 \\ 57.812 \\ 67.7 \\ -156.25 \end{bmatrix}$$

The fourth-order polynomial equation

$$y = 2.937 - 15.677(x - 7.1) + 57.812(x - 7.1)^2 + 67.7(x - 7.1)^3 - 156.25(x - 7.1)^4 \quad (1.15)$$

is obtained from the solution and is plotted in Figure 1.11 over the applicable region of ($6.6 \leq x \leq 7.8$). Figure 1.11 illustrates the errors that rapidly develop when a polynomial solution is attempted for a radical curve fit.

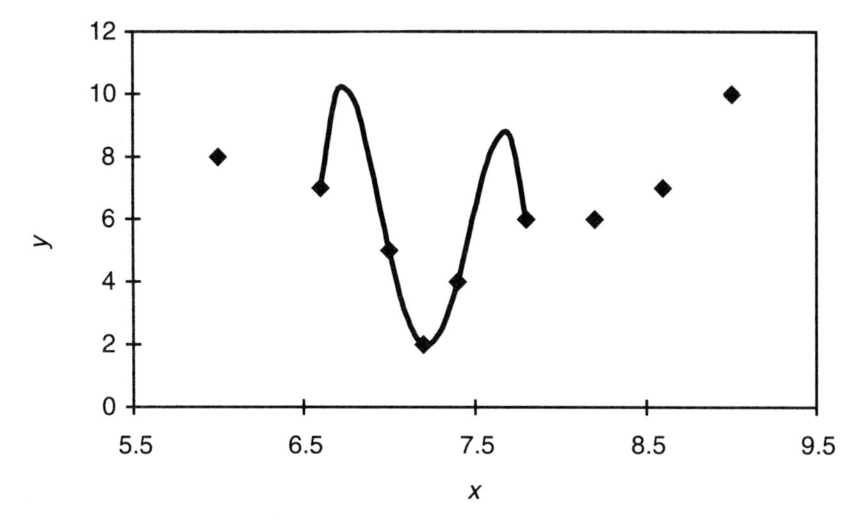

Figure 1.11 Fourth-order polynomial fit about $x = 7.1$ over the region $(6.6 \leq x \leq 7.8)$.

The errors become even more exaggerated when all of the data points in the range $(6 \leq x \leq 9)$ are used to generate an eighth-order polynomial expansion about $x = 7.1$ resulting in the equation

$$y = 2.74 - 15.2(x - 7.1) + 79.05(x - 7.1)^2 + 18.2(x - 7.1)^3 - 289.08(x - 7.1)^4 + 203.96(x - 7.1)^5 + 133.88(x - 7.1)^6 - 172.92(x - 7.1)^7 + 44.17(x - 7.1)^8 \quad (1.16)$$

which is plotted in Figure 1.12.

1.8 Spline Curve Fits

Various spline curve-fit techniques have been developed to minimize curve-fit errors that develop between measured data points where the data trace is a curve having rapid changes in first-, second-, third-, and higher order derivative values. These spline techniques are classified as B-spline, cubic-spline, fourth-order spline, and higher order splines. Each technique uses a second-order, third-order, fourth-order, or higher polynomial equation to accurately represent a small domain of data over a limited region of independent variable. The polynomials are forced to have the same values, and the same first, second, third, fourth, and higher derivatives at the transition points or junctions between

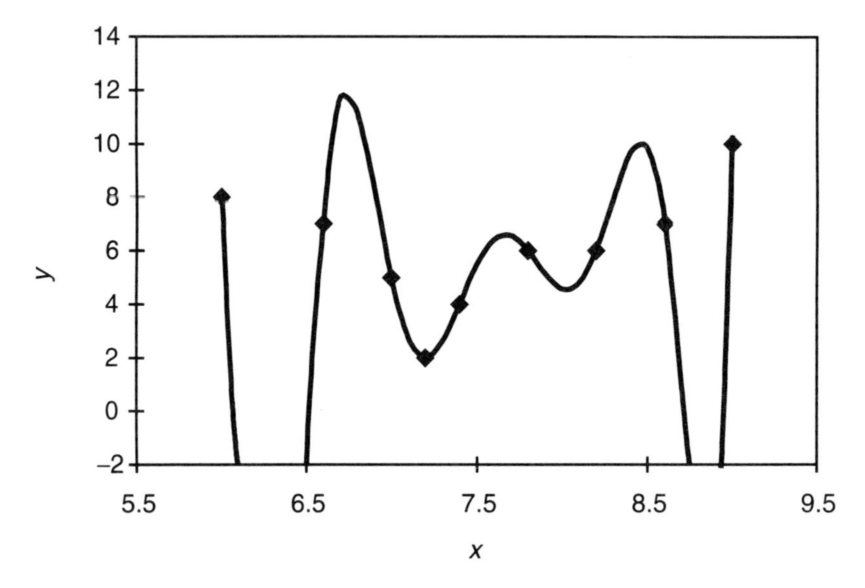

Figure 1.12 Curve fit using eighth-order polynomial expansion about $x = 7.1$.

independent variable regions. The resulting curve fits are acceptably accurate, but the number of equations needed to express the data over the full range can be unacceptably large and unwieldy to publish. No single closed-form equation results. This doesn't satisfy the behavioral modeling need for a single publishable equation that covers the entire range of independent variable. These spline techniques fall into the category of generalized additive models.

Utility chart wizards that graph computer spreadsheets often offer a choice between linear interpolation and spline curve-fitting techniques. The linear interpolation choice draws straight lines between data points, while the spline choice draws a smoothly flowing curve. Spline interpolation data is generated and used by the graphing utility to give continuity to the plotted function between entered data points. Figure 1.13 shows the curve fit that results when linear interpolation is selected as the plot technique in an EXCEL spreadsheet plot wizard utility.

The same data set plotted by selecting the EXCEL smoothed (spline) plotting technique has a smoother appearance as is shown in Figure 1.14. Note the curvature added by the graphing utility between the original data points. The spreadsheet does not give the user access to the data file generated by the spline curve fit. Obviously the data has been generated at data points between the original data set. That interpolated data is needed in order to perform a 512 point discrete Fourier transform on the original dataset.

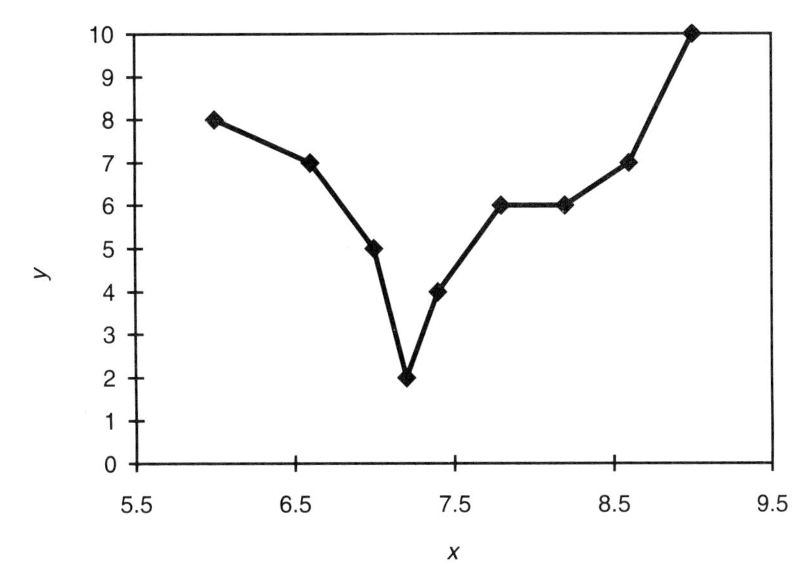

Figure 1.13 Linear interpolation curve fit.

1.9 Summary

The usefulness and limitations of linear regression and logarithmic regression curve-fitting techniques has been discussed. Application of these techniques will be demonstrated in chapters that follow. The power function has been discussed

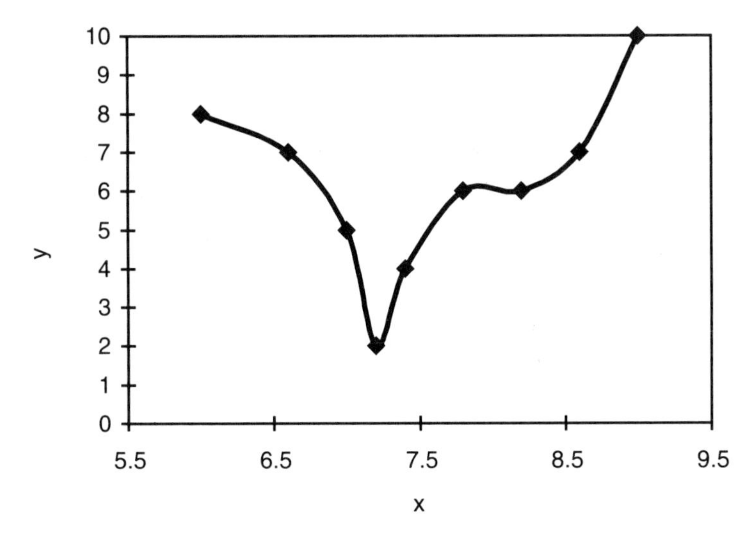

Figure 1.14 Smoothed spline curve fit to the same data.

and has been recognized as a valid modeling tool over limited signal ranges for some types of field effect transistors and for some vacuum tubes. Its usefulness is limited when large signal conditions cause saturation and cutoff. The exponential function has been recognized as a valid modeling tool for bipolar transistors and diodes over limited signal ranges. It, too, has limitations and cannot be used where large signal conditions cause saturation.

The polynomial function is recognized as a useful tool for modeling well behaved data traces over short regions of independent variable. An example illustrated the errors that develop when high-order polynomial equations are used to interpolate values between the measured data points.

Spline techniques were discussed and examples of spline-generated charts were shown. The lack of a single equation to represent the data over the entire range of independent variable makes spline techniques unacceptable for behavioral modeling. The desire is to find a method for generating a single equation that has acceptable accuracy in reproducing the existing data and interpolates between data points, giving curvature that represents true behavior of the measured object. A new curve-fitting approach is developed in the next chapter that satisfies that need.

1.10 Problem

Problem 1.1

An AT cut-quartz resonator's frequency change in parts per million as a function of temperature is measured relative to that at 25 degrees centigrade. Data

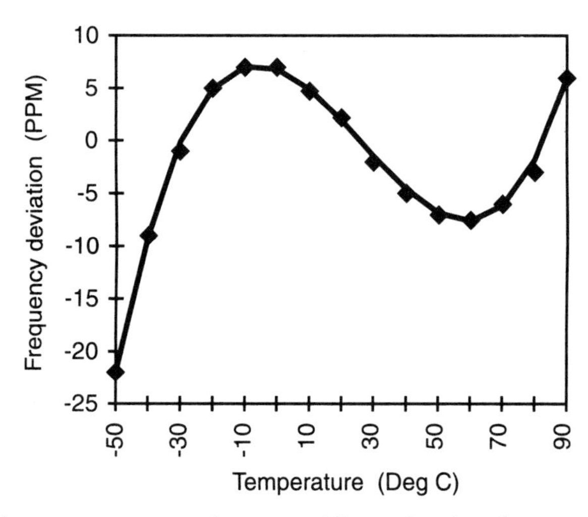

Figure 1.15 AT cut-quartz resonator frequency shift as a function of temperature.

Table 1.2
AT Cut-Quartz Resonator Data

T	Δf (PPM)
−50	−22
−40	−9
−30	−1
−20	5
−10	7
0	7
10	4.7
20	2.2
30	−2
40	−5
50	−7
60	−7.5
70	−6
80	−3
90	6

obtained is shown in Table 1.2 and plotted in Figure 1.15. Develop a third-order polynomial equation that models the data.

References

[1] Hastie, T. J. and R. J. Tibshirani, "Generalized Additive Models," *Monographs on Statistics and Applied Probability 43*, New York, NY: Chapman & Hall.

[2] Green, P. J. and B. W. Silverman, "Nonparametric Regression and Generalized Linear Models," *Monographs on Statistics and Applied Probability 58*, New York, NY: Chapman & Hall.

2

Traces for Numbers

2.1 A New Approach to Curve Fitting

Fundamental approaches to curve fitting discussed in Chapter 1 were shown to be inadequate for the purpose of modeling behavior that exhibits sharp discontinuity. A new approach is needed that results in a closed-form equation that exercises continuous domain over the entire independent variable range in spite of sharp discontinuities in the data trace. That new approach is developed in this chapter. Two new functions defined as right-hand and left-hand functions are used as elements in a new general additive approach resulting in an intuitive method for curve fitting. Once fully understood, the new general additive function approach can be used to design curves without the need of a dataset. Equations can be written representing any single valued function by sketching the desired curve, drawing asymptotes to various segments of the curve, noting the independent variable values where asymptote intersections occur, noting slope changes going from one asymptote to the next, and noting characteristics of curve-trace transition from one asymptote to the next. These few characteristic curve-trace parameters become coefficients in two newly defined right-hand and left-hand functions. A reference asymptote is selected, a linear equation is determined for it, and as many right-hand and left-hand functions as needed are added to the linear equation to complete the curve-fit equation. The new general additive approach curve-fit equation has the form

$$y(x) = A + Bx + \sum_{1}^{i} C_i * \log_{10}\left[\frac{10^{\frac{x-x_i}{c_i}}}{1 + 10^{\frac{x-x_i}{c_i}}}\right] + \sum_{1}^{k} D_k * \log_{10}\left[1 + 10^{\frac{x-x_k}{d_k}}\right] \quad (2.1)$$

The unique feature of this technique is that coefficients in the equation can be in units of commonly measured additive values such as current, voltage, power, and power ratio in decibels, gain in decibels, frequency, temperature, phase, and so forth. Equations generated by the new curve-fitting technique do not have discontinuities, do have first-, second-, third-, and higher order derivatives over the full range of independent variables, and do not generate excessive errors between data points. Interpolated data between measured data points is well behaved and smooth flowing, producing realistic results when processed through a DFT or convolved into a paired domain. Extension of the independent variable range beyond the limits of measured dataset is possible. The equations continue to trace a curve along the same slope established by the data points at each end of the independent variable range. The new curve-fit equations do not blow up outside of the measured data range as polynomial expansions do. As the equation is developed, the exponent value of terms like

$$u - 10^{\frac{z}{c}} \quad (2.2)$$

should not be allowed to exceed a computer's exponent value range, which is usually $(-499 \leq z/c \leq 499)$. Programmable hand-held calculators have less range with which to work.

An EXCEL spreadsheet utility included on a disk located inside the back cover of this book generates curve fit equations for curves having from two to twenty-one asymptotes. Instructions for using the curve fit utility are located on the disk.

2.1.1 Drawing Asymptotes

Begin the new curve-fitting technique by constructing asymptotes to the data trace that capture major changes in first derivative. Major change is roughly defined as a slope change greater than 2:1, and definitely a change in sign. The dataset from Table 1.1, in Chapter 1 is reproduced in Figure 2.1 and is used here as an example. Notice that five asymptotes are used to capture major changes in the first derivative.

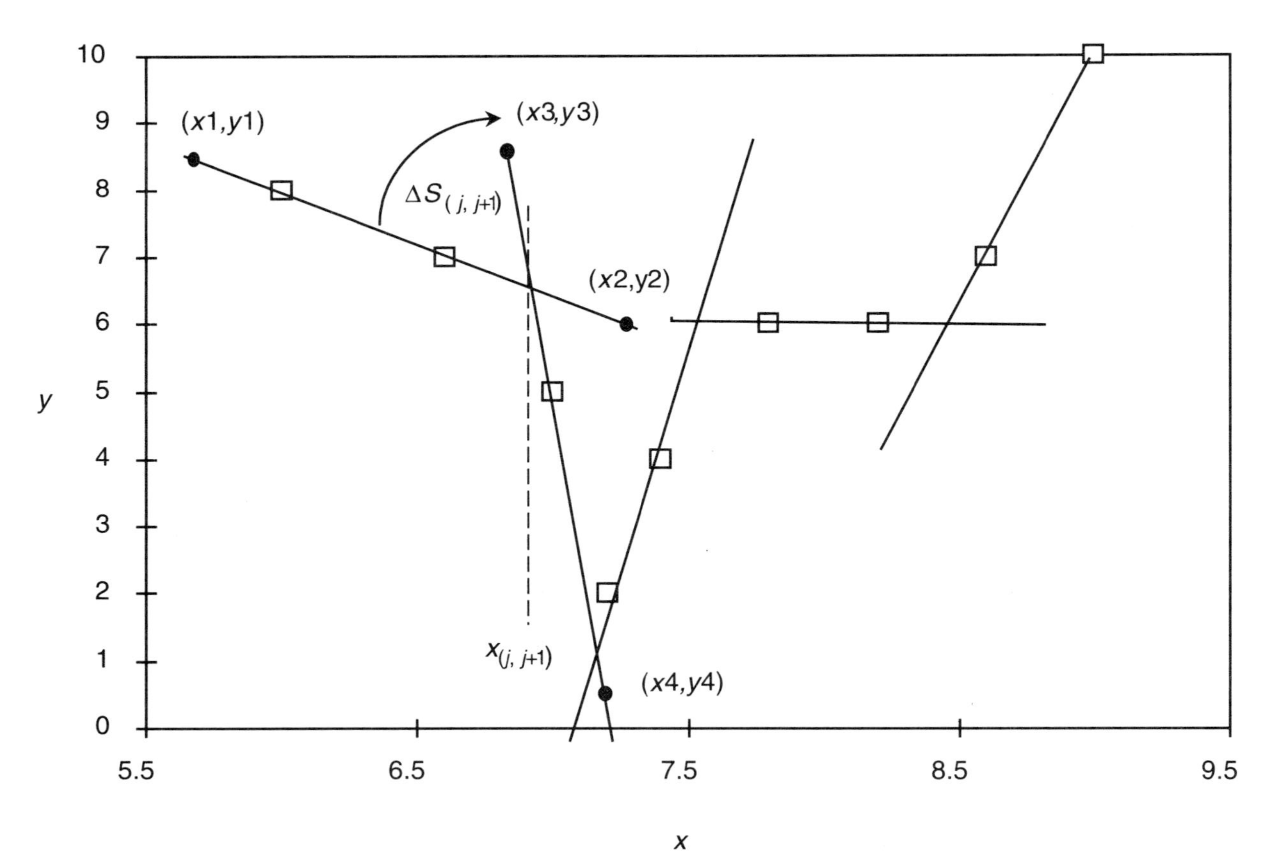

Figure 2.1 Construct asymptotes to capture first derivative changes.

Recall properties of lines. The equation

$$y = S_j x + m_j \tag{2.3}$$

defines line j passing through two points (x_j1, y_j1) and (x_j2, y_j2) in two dimensional space. The defined line has slope

$$S_j = \frac{(y_j2 - y_j1)}{(x_j2 - x_j1)} \tag{2.4}$$

and has value at $x = 0$ of

$$m_j = \frac{(x_j2)(y_j1) - (y_j2)(x_j1)}{(x_j2 - x_j1)} \tag{2.5}$$

Of interest is the independent variable value at the intersection of two asymptotes as shown in Figure 2.2.

Intersections occur where two line equations have the same solution as determined by

$$S_j x + m_j = S_{(j+1)} x + m_{(j+1)} \tag{2.6}$$

which when rearranged to solve for $x_{(j,j+1)}$, the intersection value, gives

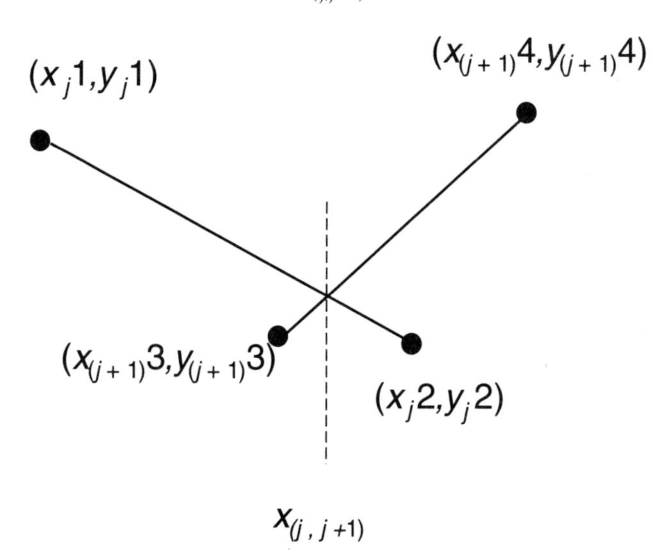

Figure 2.2 Determine independent variable values where asymptotes intersect.

$$x_{(j,j+1)} = \frac{(m_{(j+1)} - m_j)}{S_j - S_{(j+1)}} \tag{2.7}$$

The slope change

$$\Delta S_{(j,j+1)} = S_{(j+1)} - S_j \tag{2.8}$$

that occurs in transitioning from one asymptote to the next is also of interest, see Figure 2.3.

Take care to assign the proper sign value to the slope change $\Delta S_{(j,j+1)}$ at each asymptote intersection.

The data trace transitions from one asymptote to the next about each asymptote intersection. That transition is characterized by the extent of independent variable range $a_{(j,j+1)}$ over which half of the transition is accomplished as shown in Figure 2.4.

Precision is not necessary in assigning a value to $a_{(j,j+1)}$. The transition range value $a_{(j,j+1)}$, slope change $\Delta S_{(j,j+1)}$, and asymptote intersects $x_{(j,j+1)}$ are all independent coefficients that provide unique control of right-hand and left-hand functions.

2.1.2 The Right-Hand Function (RHF)

The right-hand function (RHF) is defined as

$$y = \Delta S_{(j,j+1)} a_{(j,j+1)} \log_{10}\left[1 + 10^{\frac{(x - x_{(j,j+1)})}{a_{(j,j+1)}}}\right] \tag{2.9}$$

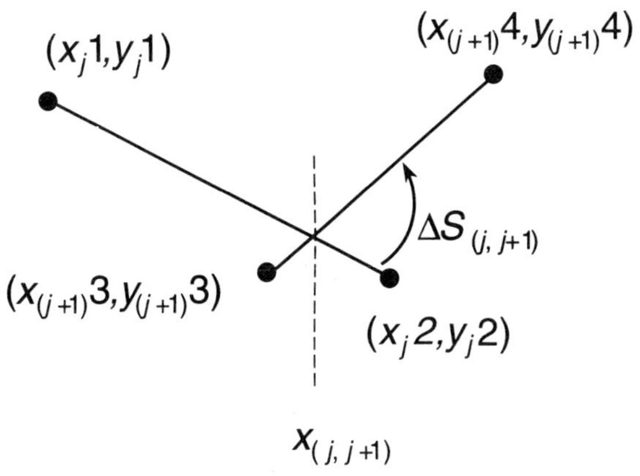

Figure 2.3 Determine slope change between intersecting asymptotes.

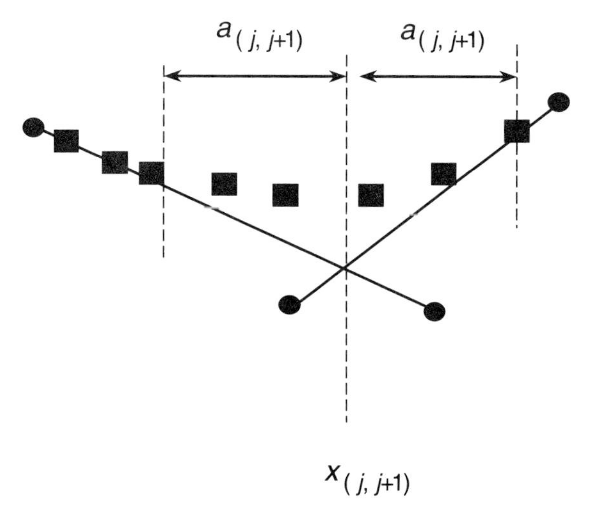

Figure 2.4 Transition range is characterized.

A natural logarithm equivalent to (2.9) exists and will be developed later. Examine the RHF and understand its characteristics. For independent variable values $x \ll x_{(j,j+1)}$, the value of the logarithm argument approaches unity, and the value of the logarithm itself approaches zero. At independent variable value $x = x_{(j,j+1)}$, the value of the logarithm argument is exactly 2. The value of the base 10 logarithm is then 0.30103. The value of the RHF at $x = x_{(j,j+1)}$ is then

$$y = 0.30103\Delta S_{(j,j+1)} a_{(j,j+1)} \tag{2.10}$$

For independent variable values $x \gg x_{(j,j+1)}$ the logarithm argument becomes greater than unity and approaches $10^{\frac{(x-x_{(j,j+1)})}{a_{(j,j+1)}}}$, the value of the base 10 logarithm approaches $\dfrac{(x - x_{(j,j+1)})}{a_{(j,j+1)}}$ and the value of the RHF approaches

$$y = \Delta S_{(j,j+1)}(x - x_{(j,j+1)}) \tag{2.11}$$

This is a straight line that is launched at the intersection of two asymptotes having a slope referenced to the slope of the asymptote on the left and continuing indefinitely as independent variable x increases.

To summarize, the value of the RHF is zero for independent variable values much less than the intersect point of two asymptotes; its value is $0.30103\Delta S_{(j,j+1)} a_{(j,j+1)}$ at the asymptote intersection, and its value grows or shrinks linearly for independent variable values much greater than the asymptote intersect point. These functions are used additively to model the transitions proceeding to the right along a data trace, hence the nomenclature, right-hand function (RHF).

Each of the three coefficients in the RHF act independently to control the way the function affects the curve trace. Figure 2.5 illustrates the independent control that slope change parameter $\Delta S_{(j,j+1)}$ gives to the RHF. Slope change values ranging from -5 to $+5$ are shown plotted in Figure 2.5, while transition range $a_{(j,j+1)} = 0.5$ and asymptote intersect $x_{(j,j+1)} = 7.2$ are held constant.

Note that if the RHF is added to an equation for a line, the slope of the line dominates for independent variable values less than the intersect point, but at independent variable values greater than the intersect point a new slope equal to $S_j + \Delta S_{(j,j+1)}$ is established. This is exactly how the RHF is used. For each transition onto a new asymptote for increasing values of x, another RHF is added to the equation. Each new RHF defines the change in slope going from one asymptote to the next at each successive asymptote intersection. Figure 2.6 illustrates the independent control that intersect values have on the RHF while slope change and transition range values are held constant.

The transition range parameter $a_{(j,j+1)}$ independently controls the way a transition from one asymptote to the next takes place. Small values of $a_{(j,j+1)}$ result in a very sharp transition, while large values provide a soft transition. Because the transition characteristic is independent of slope change and intersect point, it can be used as a fine-tuning tool to obtain the best fit to measured data. It will be shown in later chapters that this control over transition characteristic has a profound impact on the ability of behavioral modeling to account for measurable physical characteristics that conventional nonlinear modeling techniques fail to predict. Figure 2.7 illustrates the independent control the transition range parameter $a_{(j,j+1)}$ provides for curve fitting while slope change and intersect point values are held constant.

2.1.3 The Left-Hand Function (LHF)

The left-hand function (LHF) acts similar to the RHF, but for decreasing values of independent variables instead of increasing values. The LHF has zero value for independent variable values $x \gg x_{(i,i-1)}$ and develops a defined slope change for values of $x \ll x_{(i,i-1)}$ with a defined transition characteristic $a_{(i,i-1)}$. The left-hand function is defined as

$$y = \Delta S_{(i,i-1)} a_{(i,i-1)} \log_{10}\left[\frac{10^{\frac{(x-x_{(i,i-1)})}{a_{(i,i-1)}}}}{1 + 10^{\frac{(x-x_{(i,i-1)})}{a_{(i,i-1)}}}}\right] \tag{2.12}$$

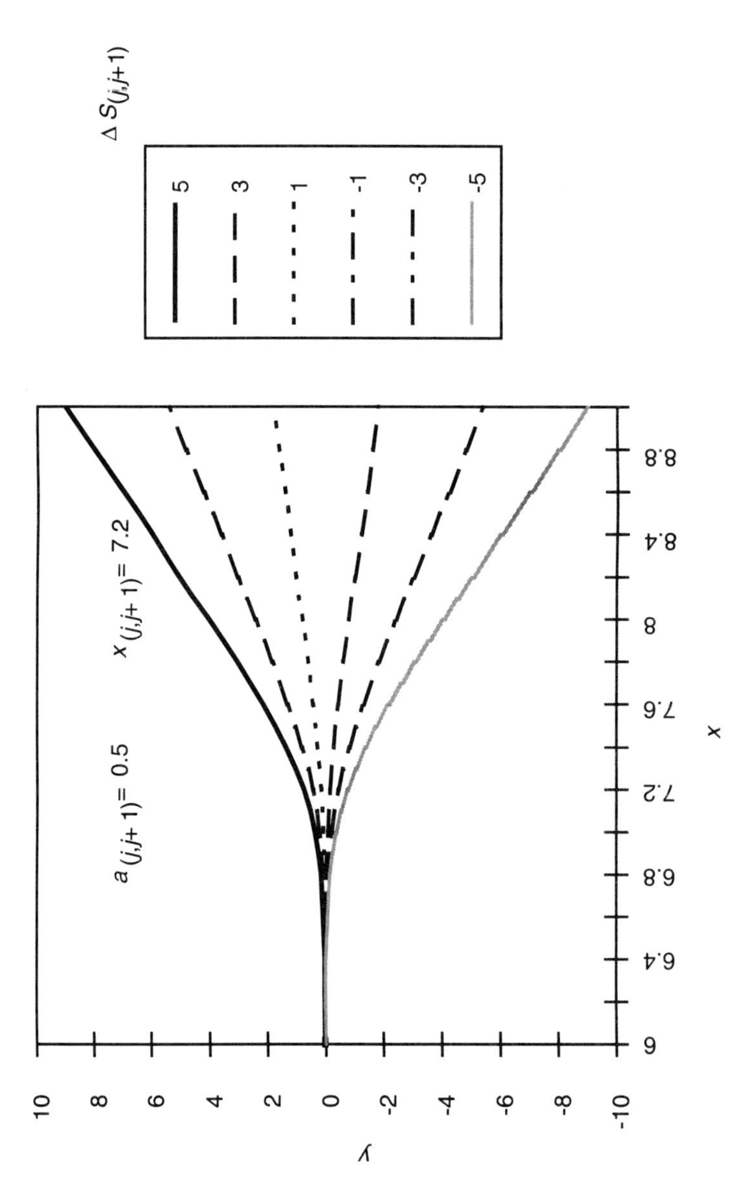

Figure 2.5 Slope change as an independent variable in the RHF.

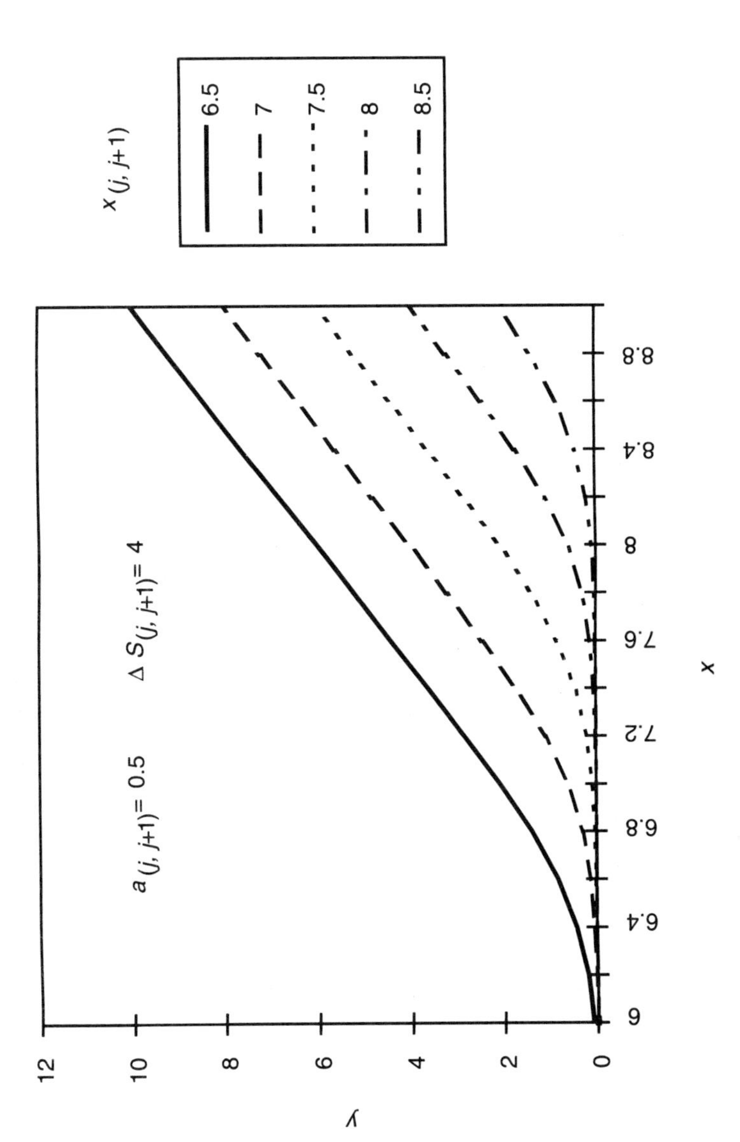

Figure 2.6 RHF characteristics as a function of intersect point values.

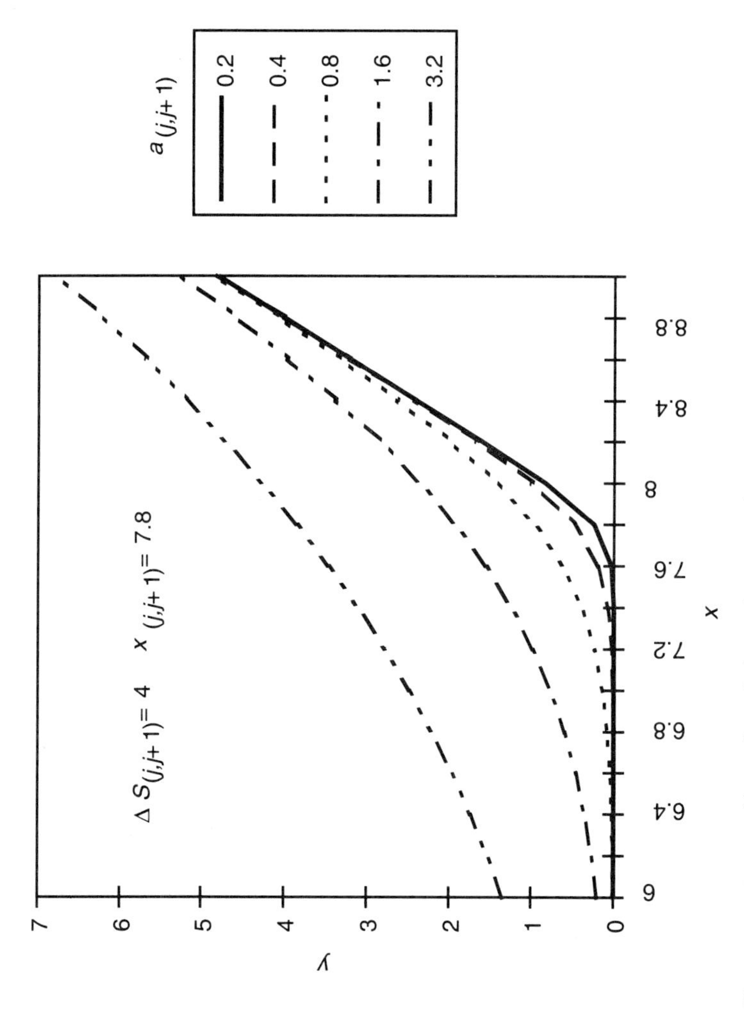

Figure 2.7 The transition parameter provides unique, independent control.

Coefficients in the LHF are defined the same as those in the RHF. Note that for values $x \gg x_{(i,i-1)}$, the denominator in the logarithm argument

$$1 + 10^{\frac{(x - x_{(i,i-1)})}{a_{(i,i-1)}}}$$

approaches the value of the numerator

$$10^{\frac{(x - x_{(i,i-1)})}{a_{(i,i-1)}}}$$

and the value of the logarithm argument approaches unity. The value of the logarithm approaches zero for $x \gg x_{(i,i-1)}$. All LHFs contribute nothing as the independent variable increases. At independent variable value $x = x_{(i,i-1)}$, the value of the logarithm argument becomes equal to 0.5, and the logarithm value is -0.30103. The value of the LHF at that point is

$$y = -0.30103\Delta S_{(i,i-1)}a_{(i,i-1)} \tag{2.13}$$

At independent variable values $x \ll x_{(i,i-1)}$, the logarithm argument's value approaches $10^{\frac{(x - x_{(i,i-1)})}{a_{(i,i-1)}}}$, the value of the logarithm approaches $\frac{(x - x_{(i,i-1)})}{a_{(i,i-1)}}$, and the value of the LHF approaches

$$y = \Delta S_{(i,i-1)}(x - x_{(i,i-1)}) \tag{2.14}$$

Be careful to determine correct slope change signs for LHFs. The three parameters, slope change, transition range, and intercept point value, provide control over the way the LHF traces data flow in exactly the same way that they work in the RHF. Figure 2.8 illustrates the LHF sensitivity to slope changes, while transition range and intersect point values are held constant. Notice that all curves go to zero at independent variable values $x \gg x_{(i,i-1)}$.

Asymptote intersect point is an independent variable in the LHF as it is in the RHF. A new LHF is added to the curve fit equation for each successive asymptote intersection encountered as independent variable x proceeds to the left. Figure 2.9 illustrates the independence that intersect point value has in the LHF while slope change and transition range values are held constant.

Transition range parameter $a_{(i,i-1)}$ works the same in the LHF as in the RHF. Figure 2.10 illustrates the independent control available in the LHF by

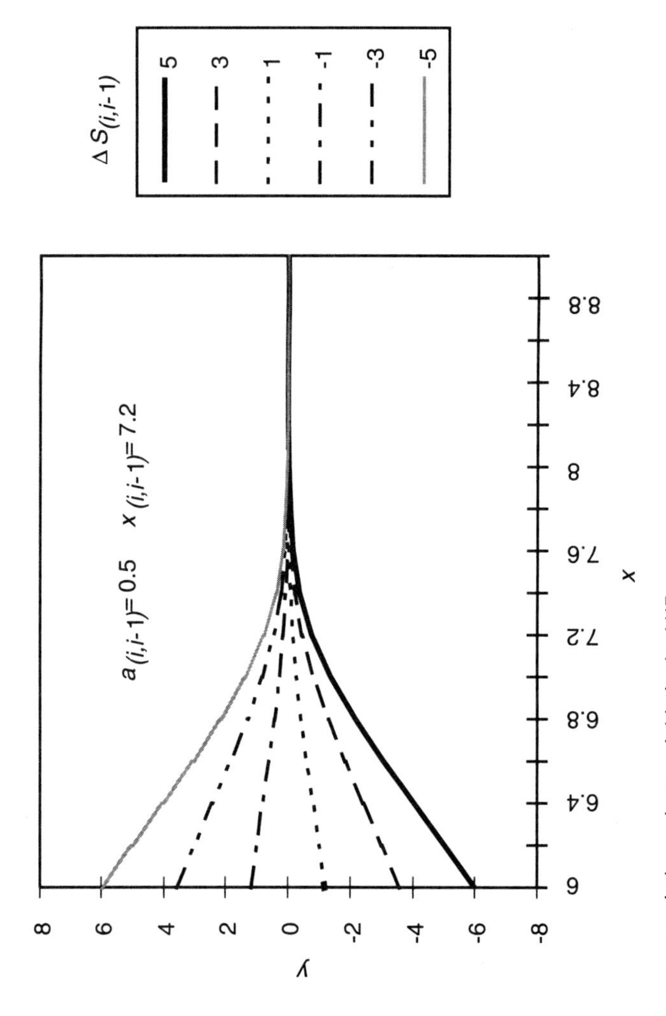

Figure 2.8 Slope change as an independent variable in the LHF.

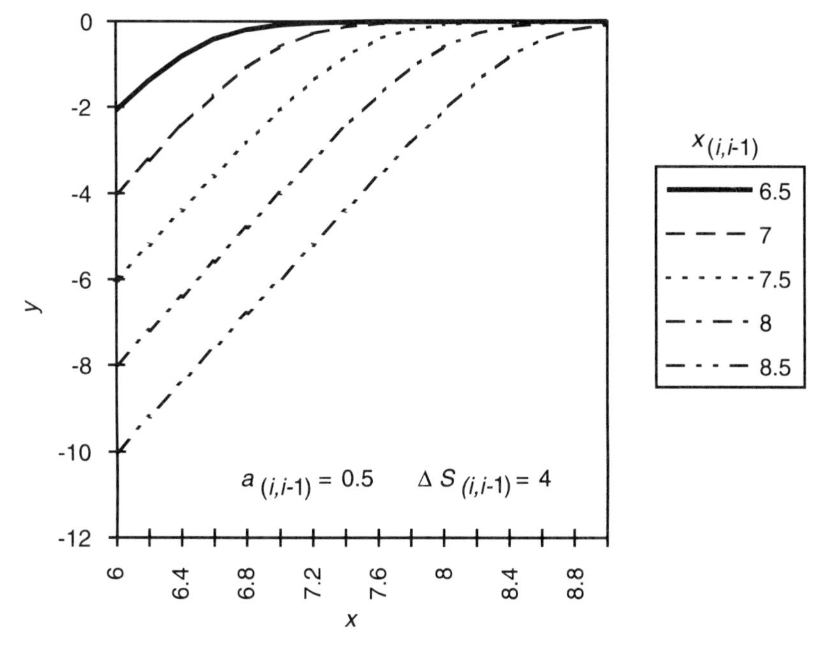

Figure 2.9 LHF characteristics as a function of intersect point values.

varying the transition range parameter value, while slope change and intersection point values are held constant. In practice, $a_{(i,i-1)}$ is fine tuned to obtain best curve fit to the data trace as it transitions between asymptotes.

2.1.4 Building the Curve-Fit Equation

Asymptotes are drawn over the data trace. Two sets of coordinates, (x_j1, y_j1) and (x_j2, y_j2), define each asymptote line, equations defining each asymptote are determined, asymptote intersections $x_{(j,j+1)}$ are calculated, slope changes $\Delta S_{(j,j+1)}$ are determined, and transition ranges are determined. A reference asymptote is selected, and the linear equation defining that asymptote

$$y = m_0 + S_0 x \tag{2.15}$$

becomes the first term of the curve fit equation. The choice of reference asymptote is arbitrary. Selecting specific asymptotes as the reference in certain modeling applications is advantageous resulting in fewer terms needed, or the significance that certain coefficients obtain. Coefficients are assigned to all right-hand functions representing all asymptote transitions to the right of the reference equation. These RHFs are added to the reference asymptote equation

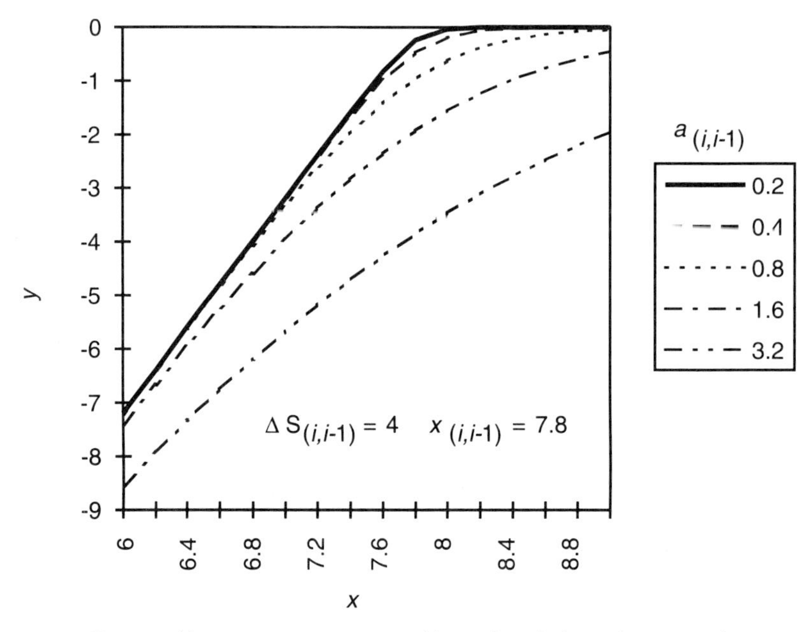

Figure 2.10 The transition parameter $a_{(i,i-1)}$ provides unique independent control.

$$y = m_0 + S_0 x + \sum \Delta S_{(j,j+1)} a_{(j,j+1)} \log_{10}\left[1 + 10^{\frac{(x - x_{(j,j+1)})}{a_{(j,j+1)}}}\right] \tag{2.16}$$

completing the curve fit to the right. Likewise, coefficients are assigned to all left-hand functions representing all asymptote transitions to the left of the reference equation. These LHFs are also added to the reference equation to complete the curve fit equation resulting in

$$y = m_0 + S_0 x + \sum \Delta S_{(j,j+1)} a_{(j,j+1)} \log_{10}\left[1 + 10^{\frac{(x - x_{(j,j+1)})}{a_{(j,j+1)}}}\right] +$$

$$\sum \Delta S_{(i,i-1)} a_{(i,i-1)} \log_{10}\left[\frac{10^{\frac{(x - x_{(i,i-1)})}{a_{(i,i-1)}}}}{1 + 10^{\frac{(x - x_{(i,i-1)})}{a_{(i,i-1)}}}}\right] \tag{2.17}$$

Since the choice of reference asymptote is purely arbitrary, it is possible to create a curve-fit equation with a linear equation and all RHFs or a linear equation and all LHFs. It is often useful to create a reference asymptote having slope

$S_0 = 0$. The reference asymptote equation then is a constant m_0. Reasons for making specific and unique reference asymptote choices become obvious as applications to behavioral modeling develop.

2.1.5 Software to Assist in Creating Curve-Fit Equations

A useful software application is included on the disk that is supplied with this book. The program is a Microsoft EXCEL Workbook that computes and displays the curve-fit equation that matches curves with 2 asymptotes to 21 asymptotes. "Curve Fit 1" is used for 2 to 13 asymptotes, and "Curve Fit 2" is used for 14 to 21 asymptotes. The program automatically selects an asymptote in the center of the independent variable range as the reference asymptote. Data entered are the (x,y) coordinates defining the asymptote lines, and the transition range values from asymptote to asymptote. The software then computes asymptote slopes, intersections, and RHF and LHF coefficients and presents the curve-fit equation and a plot of the equation.

2.1.6 Natural Logarithm Equivalent to the RHF

In this section and Section 2.1.7, the definitions of RHF and LHF are expanded into equivalent natural logarithm forms. The logarithm form used throughout this book is base 10. These sections are added for completeness, and for the interest of the purist, and are not necessary reading in order to understand the remaining chapters.

Consider a parameter u that can be expressed as a power of 10 and also as a power of e by the equality

$$u = 10^v = e^{\beta v} \tag{2.18}$$

Take the natural logarithm of each term in the equality and obtain

$$\ln(u) = v \ln(10) = \beta v$$

which gives the identity

$$\beta = \ln(10) \tag{2.19}$$

Let $v = \dfrac{(x - x_0)}{a}$. The equality expressed in equation (2.18) then becomes

$$e^{\frac{(x-x_0)\beta}{a}} = 10^{\frac{(x-x_0)}{a}}$$

This identity permits the substitution of $e^{\frac{(x-x_0)\beta}{a}}$ for $10^{\frac{(x-x_0)}{a}}$ in the logarithm argument in both the RHF and the LHF. Examine the RHF in terms of natural logarithm and logarithm base 10. Multiply the natural logarithm by the constant $\frac{1}{B}$ and declare that the two RHF forms are equal

$$\frac{\Delta Sa}{B} \ln\left[1 + e^{\frac{(x-x_0)\beta}{a}}\right] = \Delta Sa \log_{10}\left[1 + 10^{\frac{(x-x_0)}{a}}\right] \quad (2.20)$$

Now verify that the two forms are in fact equal. At values of $x \ll x_0$, the values of $e^{\frac{(x-x_0)\beta}{a}}$ and $10^{\frac{(x-x_0)}{a}}$ both approach zero. The arguments of the natural logarithm and the log base 10 both approach unity, and the logarithm values approach zero. Where $x = x_0$, the values of $e^{\frac{(x-x_0)\beta}{a}}$ and $10^{\frac{(x-x_0)}{a}}$ both become unity, and the arguments of the two logarithms become 2 giving the equation

$$\frac{1}{B} \ln[2] = \log_{10}[2] \quad (2.21)$$

which when rearranged to solve for B gives

$$B = \frac{\ln[2]}{\log_{10}[2]} = \ln[10] = \beta \quad (2.22)$$

Proof that $B = \ln[10]$ is left as an exercise for the student. For values of $x \gg x_0$, the values of $e^{\frac{(x-x_0)\beta}{a}}$ and $10^{\frac{(x-x_0)}{a}}$ both become much greater than unity, and the arguments of the respective logarithms approach $e^{\frac{(x-x_0)\beta}{a}}$ and $10^{\frac{(x-x_0)}{a}}$ in the limit. The postulated equality still holds

$$\frac{\Delta Sa(x - x_0)\beta}{\beta a} = \frac{\Delta Sa(x - x_0)}{a} \quad (2.23)$$

The natural logarithm equivalent to the RHF is

$$y = \frac{\Delta Sa}{\ln[10]} \ln\left[1 + e^{\frac{(x-x_0)\ln(10)}{a}}\right] \tag{2.24}$$

2.1.7 Natural Logarithm Equivalent to the LHF

Using the same approach used in developing the natural logarithm equivalent of the RHF, it can be shown that the natural logarithm equivalent of the LHF is

$$y = \frac{\Delta Sa}{\ln(10)} \ln\left[\frac{e^{\frac{(x-x_0)\ln(10)}{a}}}{1 + e^{\frac{(x-x_0)\ln(10)}{a}}}\right] \tag{2.25}$$

2.1.8 Applying the New Curve-Fit Technique

Experience in applying the new curve-fit technique is best gained by studying an example, then working problems.

Example 2.1

Use the data presented in Table 1.1 in Chapter 1 and develop a curve fit equation that provides domain over a trace through the data points. Figure 2.1 of this chapter shows a set of asymptotes drawn through the data points taken from Table 1.1. Coordinate pairs that describe the asymptote lines shown in Figure 2.1 are listed in Table 2.1. The asymptotes are numbered 1 through 5. In this example asymptotes 1, 4, and 5 are drawn directly through sequential data points. Phantom data points are created for asymptotes 2 and 3 at $(x2, y2) = (7.15, 2)$ and $(x3, y3) = (7.3, 2)$ respectively to permit the curve trace transition to pass through the inflection data point at (7.2, 2) with minimal error. Coordinate pairs listed in Table 2.1 are used to calculate asymptote slopes and constants that complete the information needed to obtain asymptote linear equations. Arbitrarily select asymptote 3 as the reference.

Estimate transition range $a_{(j,j+1)}$ values for transitions between slopes 3 to 4 and 4 to 5, and values for and $a_{(i,i-1)}$ at transitions 3 to 2 and 2 to 1. Calculate and tabulate asymptote intersections, changes in slope, transition

Table 2.1
Asymptote Coordinate Values, Slopes and Constants

Asymptote	(x_i1, y_i1)	(x_i2, y_i2)	S_i	m_i
1	(6, 8)	(6.6, 7)	−1.667	18
2	(7, 5)	(7.15, 2)	−20	145
3	(7.3, 2)	(7.4, 4)	20	−144
4	(7.8, 6)	(8.2, 6)	0	6
5	(8.6, 7)	(9, 10)	7.5	−57.5

ranges, and slope change-transition range products for all asymptotes. Table 2.2 shows the estimated and calculated values for the five asymptotes.

Reference asymptote 3 is represented by the linear equation

$$y = -144 + 20x$$

Proceeding to the right, transition from asymptote 3 to 4 is described by the RHF

$$-4\log_{10}\left[1 + 10^{\frac{(x-7.5)}{0.2}}\right]$$

Transition from asymptote 4 to 5 is described by the RHF

$$+1.5\log_{10}\left[1 + 10^{\frac{(x-8.467)}{0.2}}\right]$$

Table 2.2
Asymptote to Asymptote Slope Change, Intersect Point, and Transition Range

Asymptote Transition	ΔSlope_{ij}	a_{ij}	x_{ij}	$\Delta S_{ij}a_{ij}$
2 to 1	18.33	0.2	6.927	3.667
3 to 2	−40	0.1	7.225	−4
3 to 4	−20	0.2	7.5	−4
4 to 5	7.5	0.2	8.467	1.5

Moving from reference asymptote 3 toward the left, transition from asymptote 3 to 2 is described by the LHF

$$-8 \log_{10}\left[\frac{10^{\frac{(x-7.225)}{0.2}}}{1+10^{\frac{(x-7.225)}{0.2}}}\right]$$

and the transition from asymptote 2 to 1 is described by the LHF

$$+3.667 \log_{10}\left[\frac{10^{\frac{(x-6.927)}{0.2}}}{1+10^{\frac{(x-6.927)}{0.2}}}\right]$$

The sum of the linear equation, two RHFs and two LHFs is the curve-fit equation that provides domain over the trace for the given dataset. A plot of this curve-fit equation is shown in Figure 2.11.

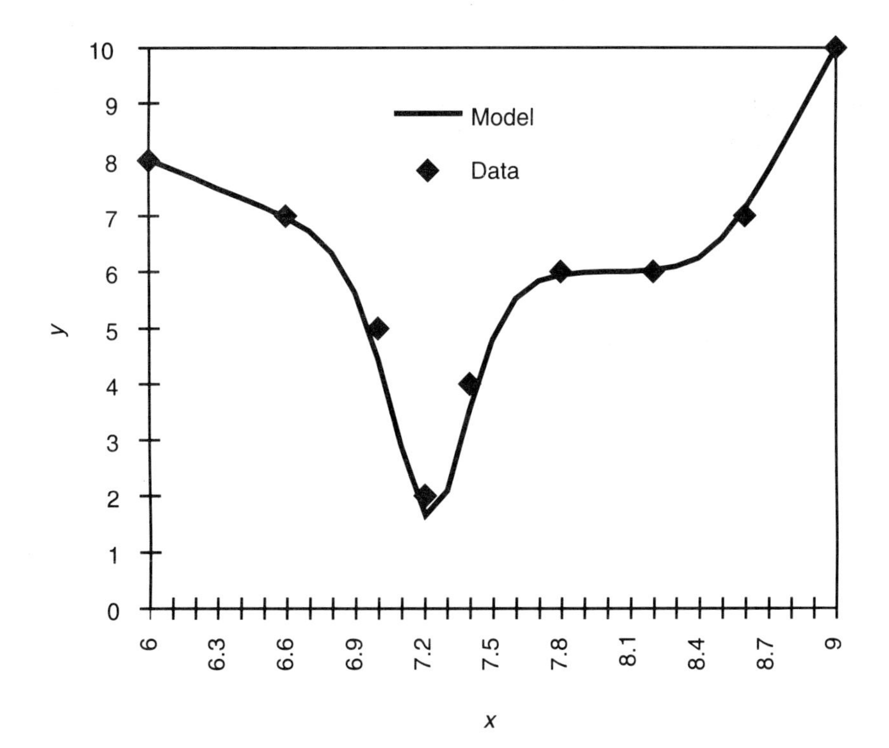

Figure 2.11 The model curve-fit equation fits data.

2.2 Adding Dimension to the Curve-Fit Equation

The curve fit (2.16) is two-dimensional. Additional dimensions are easily added by determining how coefficients vary as a function of the added dimension. Coefficient variability functions can be linear, logarithmic, exponential, power exponent, polynomial expansions, and even more RHF and LHFs. Data is often given or taken for an object describing performance over frequency at several temperatures, as a function of frequency at several bias voltage or bias current levels, or at a series of input power levels. Figure 2.12 shows a set of curves all related to the same object, which has sensitivity to variable *U*. The goal of behavioral modeling is to create one equation that describes the object's behavior over independent variable x for any simultaneous value U. The resulting equation is useful for generating a surface of three dimensions.

A procedure for developing a single equation having multiple independent variables is to first determine the maximum number of asymptotes needed for any one trace over x for any one given variable U value. The number of asymptotes will determine the number of additive functions required in the equation. Select a reference asymptote which minimizes the number of coefficients and the number of terms in the final equation; in this case the left most asymptote is selected as a reference, and only RHFs are used. Asymptote equations are developed, slope changes and intercept points are calculated, and transition ranges are determined for each curve trace over variable x at each value of U. Multiple equations are developed for each curve trace.

$$
\begin{aligned}
y(U1) = {} & m_0(U1) + S_0(U1) + \Delta S_1(U1)a_1(U1)\log_{10}\left[1 + 10^{\frac{(x - x_1(u1))}{a_1(U1)}}\right] \\
& + \Delta S_2(U1)a_2(U1)\log_{10}\left[1 + 10^{\frac{(x - x_2(u1))}{a_2(U1)}}\right] + \circ\circ\circ \\
& + \Delta S_3(U1)a_3(U1)\log_{10}\left[1 + 10^{\frac{(x - x_3(u1))}{a_3(U1)}}\right] + \Delta S_4(U1)a_4(U1) \\
& \log_{10}\left[1 + 10^{\frac{(x - x_4(u1))}{a_4(U1)}}\right] + \circ\circ\circ
\end{aligned}
\tag{2.26}
$$

$$
\begin{aligned}
y(U2) = {} & m_0(U2) + S_0(U2) + \Delta S_1(U2)a_1(U2)\log_{10}\left[1 + 10^{\frac{(x - x_1(u2))}{a_1(U2)}}\right] \\
& + \Delta S_2(U2)a_2(U2)\log_{10}\left[1 + 10^{\frac{(x - x_2(u2))}{a_2(U2)}}\right] + \circ\circ\circ
\end{aligned}
$$

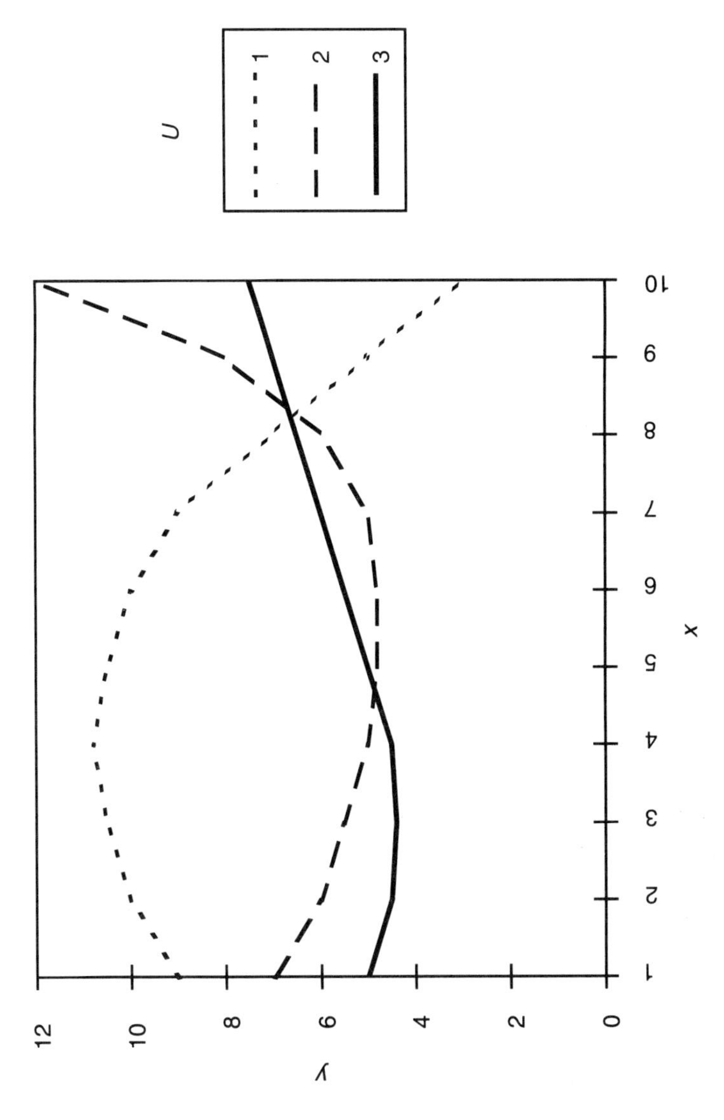

Figure 2.12 Some objects have sensitivity to multiple independent variables.

$$+ \Delta S_3(U2)a_3(U2)\log_{10}\left[1 + 10^{\frac{(x-x_3(u2))}{a_3(U2)}}\right] + \Delta S_4(U2)a_4(U2)$$

$$\log_{10}\left[1 + 10^{\frac{(x-x_4(u2))}{a_4(U2)}}\right] + \text{ooo} \tag{2.27}$$

$$y(U3) = m_0(U3) + S_0(U3) + \Delta S_1(U3)a_1(U3)\log_{10}\left[1 + 10^{\frac{(x-x_1(u3))}{a_1(U3)}}\right]$$

$$+ \Delta S_2(U3)a_2(U3)\log_{10}\left[1 + 10^{\frac{(x-x_2(U3))}{a_2(U3)}}\right] + \text{ooo}$$

$$+ \Delta S_3(U3)a_3(U3)\log_{10}\left[1 + 10^{\frac{(x-x_3(u3))}{a_3(U3)}}\right] + \Delta S_4(U3)a_4(U3)$$

$$\log_{10}\left[1 + 10^{\frac{(x-x_4(U3))}{a_4(U3)}}\right] + \text{ooo} \tag{2.28}$$

and if there are more curves representing $U4$ through Un, develop the equations

$$y(U4) = \bullet\bullet\bullet\bullet\bullet$$

$$\bullet$$

$$\bullet$$

$$\bullet$$

$$y(Un) = \bullet\bullet\bullet\bullet$$

The values for each coefficient as a function $f(U)$ are plotted, and equations are found that define curve fits to those coefficients as a function $f(U)$. If the coefficient behavior as a function $f(U)$ is not linear but is well behaved, a low-order polynomial expansion is the first choice of the coefficient model; it is the easiest equation to generate. The result is a set of equations for coefficients

$$m_0(U) = c_0 + c_1U + c_2U^2 + c_3U^3$$

$$S_0(U) = d_0 + d_1U + d_2U^2 + d_3U^3$$

$$\Delta S_1(U) = g_0 + g_1U + g_2U^2 + g_3U^3$$

$$a_1(U) = h_0 + h_1U + h_2U^2 + h_3U^3$$

$$x_1(U) = k_0 + k_1 U + k_2 U^2 + k_3 U^3$$

$$\Delta S_2(U) = l_0 + l_1 U + l_2 U^2 + l_3 U^3$$

$$a_2(U) = m_0 + m_1 U + m_2 U^2 + m_3 U^3$$

$$x_2(U) = p_0 + p_1 U + p_2 U^2 + p_3 U^3$$

$$\bullet$$
$$\bullet \qquad (2.29)$$
$$\bullet$$

$$\Delta S_n(U) = r_0 + r_1 U + r_2 U^2 + r_3 U^3$$

$$a_n(U) = s_0 + s_1 U + s_2 U^2 + s_3 U^3$$

$$x_n(U) = z_0 + z_1 U + z_2 U^2 + z_3 U^3$$

These coefficients are substituted back into a single equation

$$y(U) = m_0(U) + S_0(U) + \Delta S_1(U) a_1(U) \log_{10}\left[1 + 10^{\frac{(x - x_1(u))}{a_1(U)}}\right] + \Delta S_2(U) a_2(U) \log_{10}\left[1 + 10^{\frac{(x - x_2(U))}{a_2(U)}}\right] + \Delta S_3(U) a_3(U) \log_{10}\left[1 + 10^{\frac{(x - x_3(u))}{a_3(U)}}\right] + \Delta S_4(U3) a_4(U) \log_{10}\left[1 + 10^{\frac{(x - x_4(U))}{a_4(U)}}\right] + \circ\circ\circ \qquad (2.30)$$

to give a three-dimensional curve-fit capability that results in a surface. The surface representation of the data plotted in Figure 2.12 is shown in Figure 2.13.

2.3 The Step Function—A Useful Combination of Right-Hand Functions

The step function is easily modeled using two right-hand functions. Two models are needed, one for the step-up case having the value zero for all independent

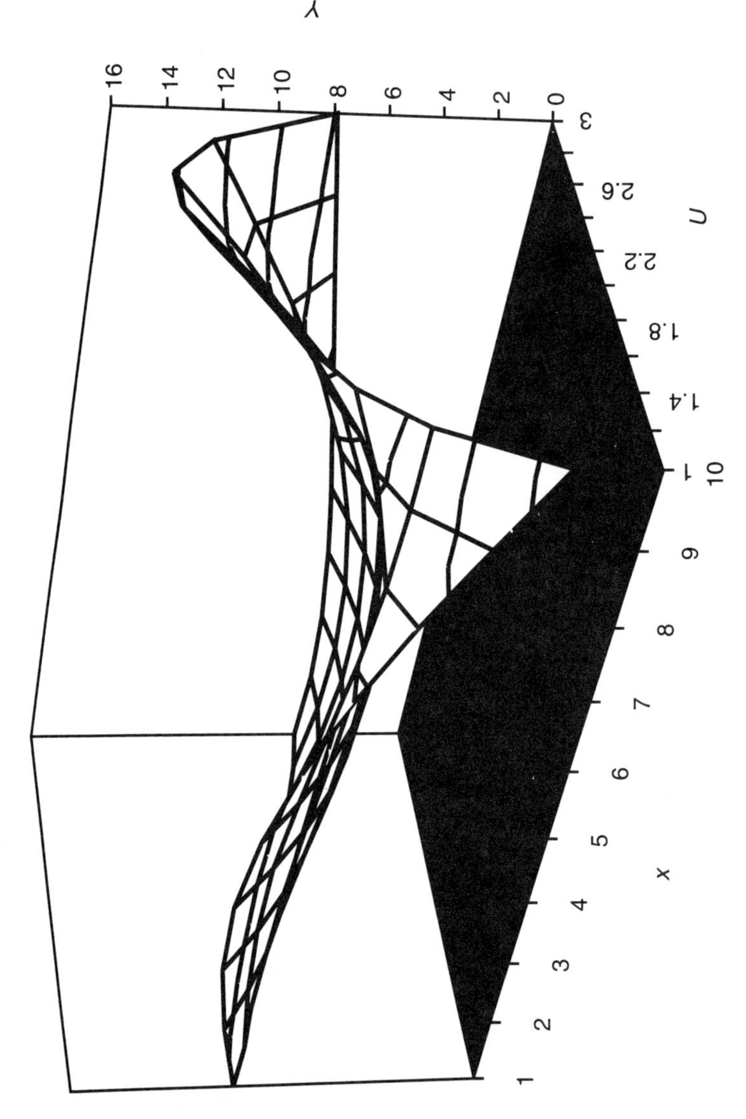

Figure 2.13 Three-dimensional surface plot of Figure 2.12 data.

variable values up to the point of step, then having the value A for all independent variable values after the step, and the second, the step-down, having the value A for all independent variable values before the step and zero for all independent variable values after the step. Both models are derived from the equation

$$y = A(0) + \Delta S_0 a_0 \log_{10}\left[1 + 10^{\frac{(x-x_0)}{a_0}}\right] - \Delta S_0 a_0 \log_{10}\left[1 + 10^{\frac{(x-x_1)}{a_0}}\right] \quad (2.31)$$

only the coefficient values vary. Consider the step-up case where the value is zero before the step, and the step transitions from zero to A over the short independent variable range Δx. The change in slope ΔS is the slope of the transition

$$\Delta S = \frac{A}{\Delta x} \quad (2.32)$$

The transition range a_0 can be no softer than $0.5\Delta x$ and can be as sharp as needed, keeping in mind the limits imposed by the computer's ability to handle extremely small exponents. See (2.2) for an example.

The first intersect point is at $x = x_0$, the second is at $x = x_0 + \Delta x$. The value $A(0) = 0$ for all $x \leq 0$. Substitute the coefficient values into (2.31); realize that the coefficients multiplying the logarithms are the same except for sign. Therefore, logarithms can be combined to obtain

$$y = \frac{Aa_0}{\Delta x} \log_{10}\left[\frac{1 + 10^{\frac{(x-x_0)}{a_0}}}{1 + 10^{\frac{(x-x_0-\Delta x)}{a_0}}}\right] \quad (2.33)$$

for the step-up function equation which is illustrated in Figure 2.14.

The step-down function begins with $y = A(0) = A$, which holds for all independent variable values $x < x_0$. At $x = x_0$ the transition to $x = 0$ begins, and the function obtains the value of zero for all $x > x_0 + \Delta x$. The transition slope

$$\Delta S = \frac{-A}{\Delta x} \quad (2.34)$$

is negative. Transition range a_0 has the same limitations imposed on the step-up function transition range. Substituting step-down coefficients into (2.31), combining logarithms, and adding the value A to the RHFs, obtain

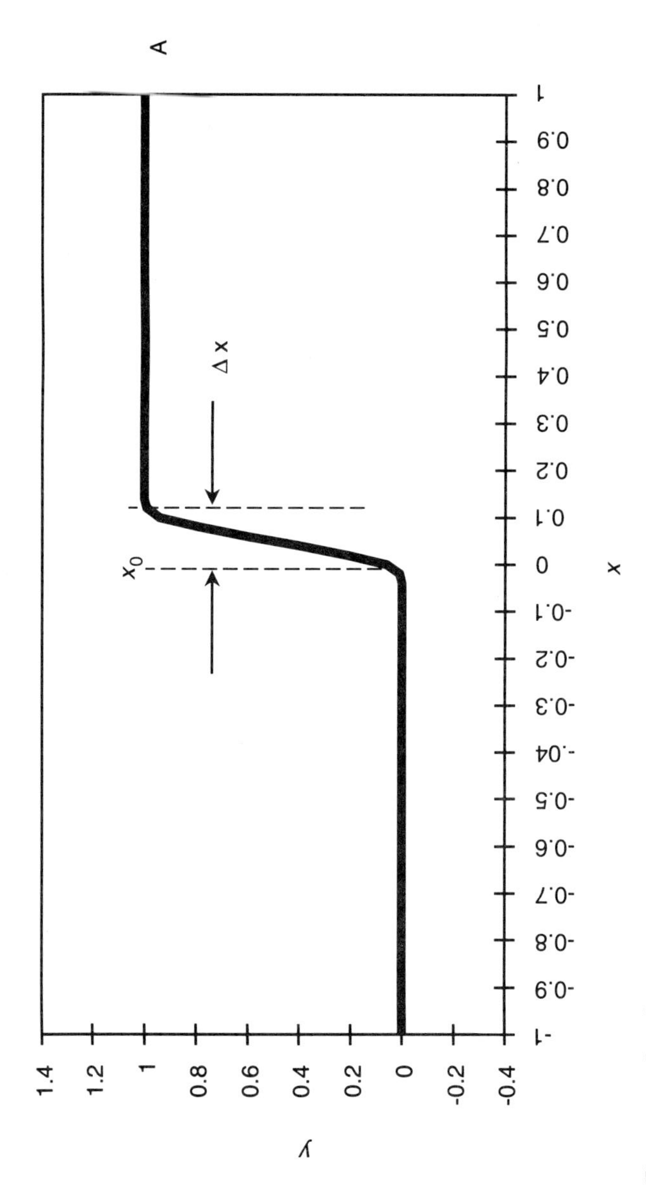

Figure 2.14 The step-up function.

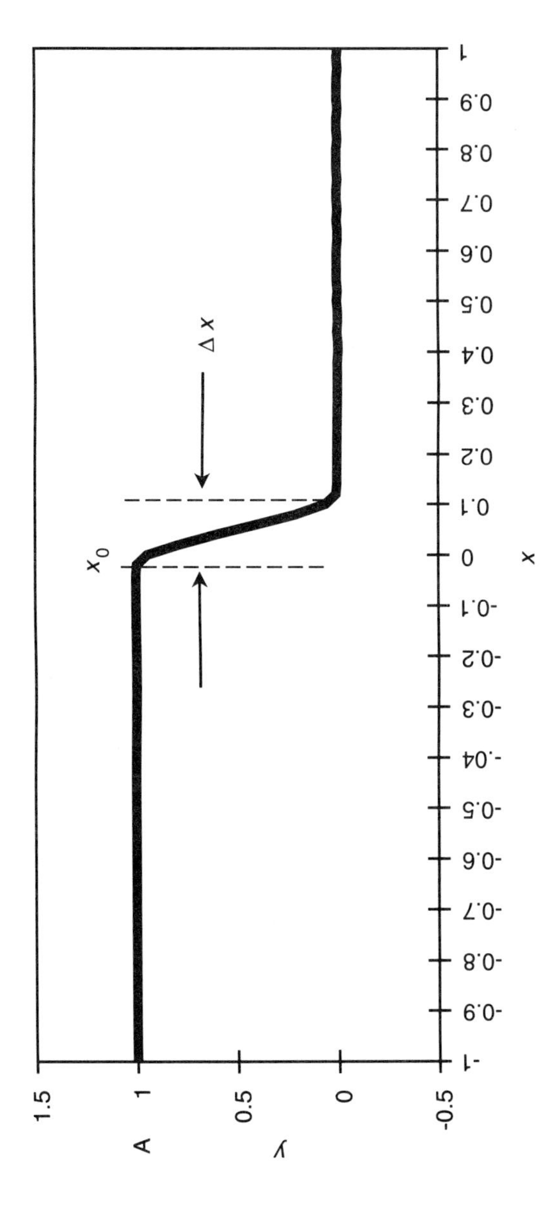

Figure 2.15 The step-down function.

$$y = A\left\{1 - \frac{a_0}{\Delta x}\log_{10}\left[\frac{1 + 10^{\frac{(x-x_0)}{a_0}}}{1 + 10^{\frac{(x-x_0-\Delta x)}{a_0}}}\right]\right\} \quad (2.35)$$

as the equation for a step-down function, which is illustrated in Figure 2.15.

2.4 Summary

A new curve fit technique has been described which can be used to develop a single equation that provides domain over a trace that can have sharp inflections and can be a function of multiple orthogonal variables. Right-hand functions (RHFs) and left-hand functions (LHFs) have been defined as the general additive elements of the new curve-fitting technique. Natural and base 10 logarithm versions of RHFs and LHFs were defined and proven to be interchangeable.

2.5 Problems

Problem 2.1

Create a single equation that approximates one sine wave cycle in time from $-\infty \leq t \leq +\infty$. The sine wave cycle begins at $\omega t = 0$, ends at $\omega t = 2\pi$ and has peak amplitude A. Hint: The sine wave has slope A at $\omega t = 0$, slope $-A$ at $\omega t = \pi$, and slope A at $\omega t = 2\pi$. Use 4 RHFs.

Problem 2.2

Using the data in Figure 2.12, follow the instructions in Section 2.2 to develop a single equation that generates the surface shown in Figure 2.13.

3

Current Source Behavior

3.1 Modeling Transistor Device Current Sources

The new curve-fit method developed in Chapter 2 is used in this chapter to model the nonlinearity of bipolar and field-effect transistor device current sources. Bipolar transistors obey an exponential characteristic when small signals are applied, and field-effect transistors tend to follow a square law characteristic. Walter Curtice [1] and H. Statz et. al. [2] have proposed MESFET models that work reasonably well for small signal levels, but fail to follow device behavior satisfactorily where saturation and cutoff occur. The work of Curtice is used as a starting point, then is modified to obtain a large signal MESFET behavioral model. The Ebers-Moll [3] model is then modified to develop a large signal behavioral model for the bipolar transistor current source.

The new behavioral models are tested at different bias conditions and for different input signal levels at each bias condition. DC bias conditions range from 10 percent of saturation current to 90 percent of saturation current. Modeled output current waveform of the transistor is computed as a function of time as input signal levels range from small signal linear to large signal where hard saturation and cutoff occur. The modeled time-based waveform is processed through a discrete Fourier transform (DFT) generating values for average DC current, power gain, fundamental signal output level, compression of fundamental power output, and harmonic component levels as a function of bias and input power. Results obtained from the tests are studied and become the basis for amplifier behavioral modeling in Chapter 4.

3.2 The Curtice Square Law MESFET Model

Walter Curtice proposed a square law model

$$i_{ds} = \beta(v_{gs} - v_{po})^2(1 + \lambda v_{ds}) \tanh(\alpha v_{ds}) \tag{3.1}$$

which gives drain current i_{ds} as a function of gate to source voltage v_{gs}, drain to source voltage v_{ds}, pinch-off voltage v_{po}, and several constants α, β, and λ. When gate-to-source voltage is limited to the range $v_{po} \leq v_{gs} \leq 0$, and appropriate values for α, β, and λ are used, the model generates a family of curves like those shown in Figure 3.1.

Coefficient β is proportional to the transistor's saturation current i_{dss}. The value of β ranges from 0.01 for small transistors to 10 for really large transistors. Coefficient λ controls the slope of drain current i_{ds} as a function of drain voltage v_{ds} for any value of gate-to-source voltage v_{gs} and has a nominal value of 0.03. Curtice's α coefficient controls the knee voltage v_k, which is the drain-to-source voltage above which drain current i_{ds} becomes more a function of gate voltage v_{gs}, and less a function of drain-to-source voltage v_{ds}. A typical value α = 1.7, gives knee voltage in the range 0.5 V $\leq v_k \leq$ 1.0 V.

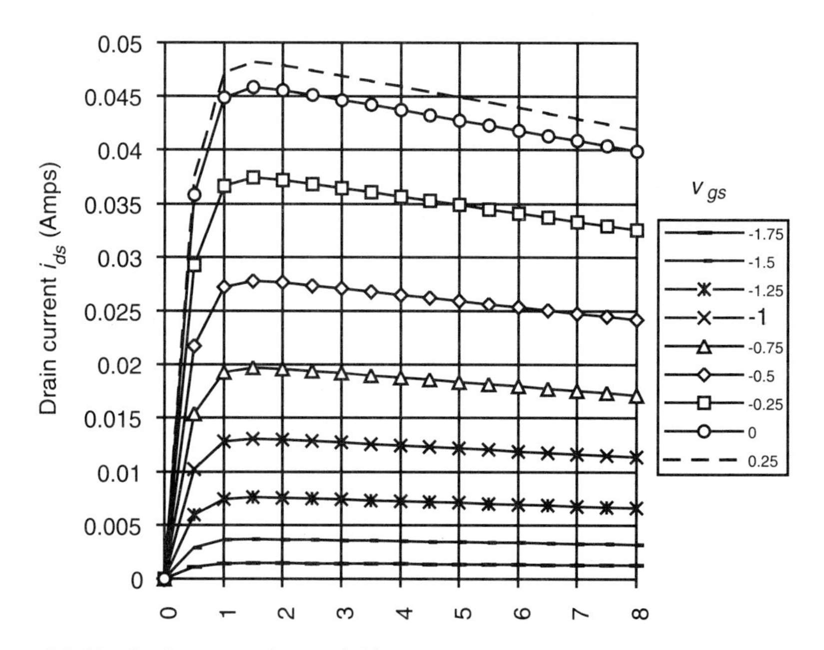

Figure 3.1 The Curtice square law model is valid over a limited range.

A transistor becomes a power gain device when delivering current into a properly selected load impedance. Load impedance value might be high if a high-power gain ratio is desired, or load impedance might be low if the design goal is to obtain maximum power output. In either case, think of the transistor as a current source that generates maximum deliverable current at zero gate voltage and zero output current where gate voltage is less than pinch-off v_{po}. Curtice's square law model, (3.1), fails to satisfy either of these extremes. Curtice has also proposed a cubic model [1] to more accurately model MESFET behavior where $v_{gs} < v_{po}$. The cubic model has limited valid domain into the pinch-off region and, like the square law model, fails to adequately reproduce saturation characteristics. The cubic nature of his second model gives a third-order term that is useful in computing third-order intermodulation and cross modulation sidebands where two signals are simultaneously introduced into the device. Unfortunately, a simple third-order model does not model the device's square law behavior accurately, and overtones greater than the third harmonic cannot be generated.

Unique doping profiles are often designed into MESFET devices to produce very linear characteristic behavior over gate voltages ranging from just above pinch-off to just below saturation. Curtice's models do not fit behavior of specially doped devices. Figure 3.2 shows a comparison between Curtice's square law model and the desired behavioral model in the regions of pinch-off and saturation.

The transistor behavioral model equations developed in this chapter address the nonlinear behavior of pinch-off and saturation and can be substituted into the Curtice model (3.1) in place of the term $\beta(v_{gs} - v_{po})^2$. Curtice's terms that determine knee voltage and slope of drain current versus drain-to-source voltage are retained in the new behavioral model.

3.3 Developing a MESFET Current Source Behavioral Model

Assume that a MESFET device load impedance is selected such that saturation current is i_{dss} at gate voltage $v_{gs} = 0$ volts. The value of load impedance is then

$$R_l = \frac{2(v_{ds} - v_k)}{I_{dss}} \tag{3.2}$$

The square law behavior of a MESFET's drain current as a function of gate-to-source voltage between pinch-off and saturation is adequately modeled by a variation of Curtice's square law model where

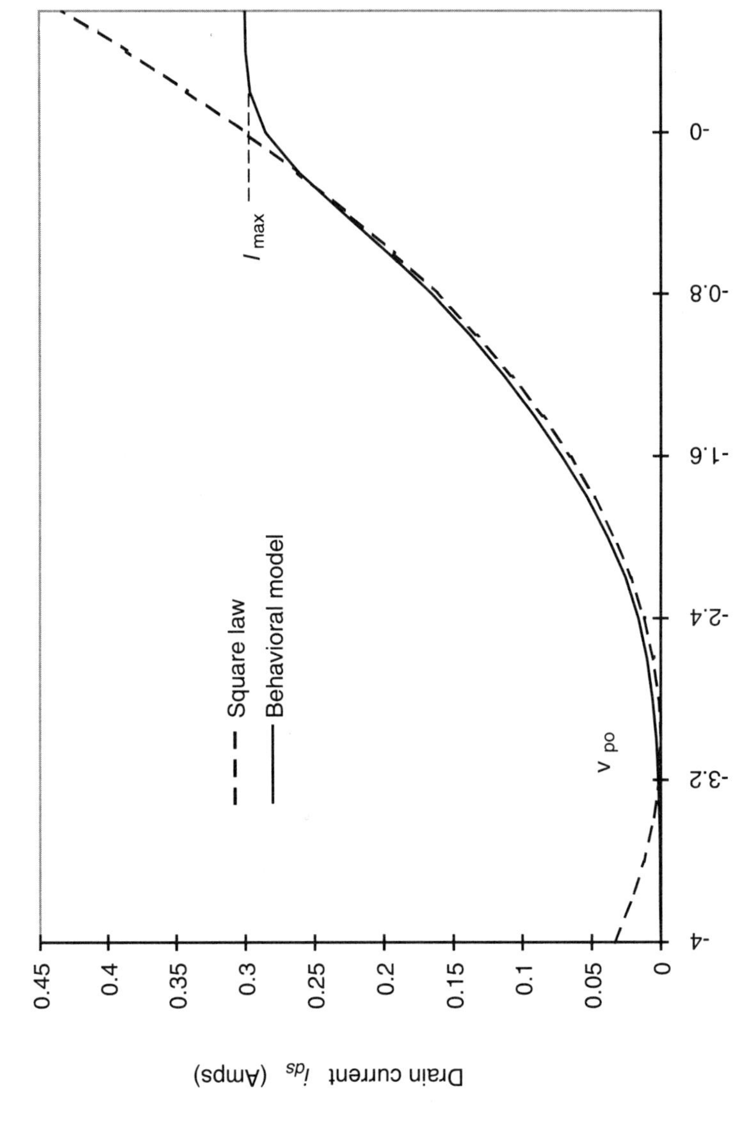

Figure 3.2 Comparison of Curtice's square law model to real behavior.

$$i_{ds} = i_{dss}\left[1 - \frac{2v_{gs}}{v_{po}} + \left(\frac{v_{gs}}{v_{po}}\right)^2\right] \quad (3.3)$$

It will be useful to note that the first derivative of this square law function

$$g_m = \frac{\partial i_{ds}}{\partial v_{gs}} = \frac{2i_{dss}}{v_{po}}\left[\frac{v_{gs}}{v_{po}} - 1\right] \quad (3.4)$$

is defined to be the transistor's transconductance g_m. Transconductance g_m values at gate bias voltages where $v_{gs} = 0$ and where $v_{gs} = 0.5v_{po}$ are of particular interest.

Where $v_{gs} = 0$, the first derivative has the value

$$g_m(0) = -\frac{2i_{dss}}{v_{po}} \quad (3.5)$$

Where $v_{gs} = 0.5v_{po}$ the first derivative has the value

$$g_m\left(\frac{v_{po}}{2}\right) = -\frac{i_{dss}}{v_{po}} \quad (3.6)$$

Note that the value of g_m varies with bias setting. The values represented by (3.5) and (3.6) are slopes of tangents to the square law function at the two points cited. These two tangents to the square law function are used as two of the asymptotes (**A2** and **A3**) needed to construct a behavioral model. Two more asymptotes required are one having zero slope and running through the pinch-off point where drain current goes to zero (asymptote **A1**), and the other also having zero slope representing saturation current i_{dss} (asymptote **A4**). The four asymptotes are illustrated in Figure 3.3.

Choose **A1** as the reference asymptote. This reduces the number of terms in the model since the equation for **A1** is **A1** = 0. This choice also has the advantage of forcing asymptotes **A2**, **A3**, and **A4** to be RHFs, again reducing the number of terms in the model equation. The equation for asymptote **A2**, tangent to the trace at $v_{gs} = 0.5v_{po}$, is

$$y = i_{dss}\left[\frac{3}{4} - \frac{v_{gs}}{v_{po}}\right] \quad (3.7)$$

and the intersect point of **A1** and **A2** is determined to be

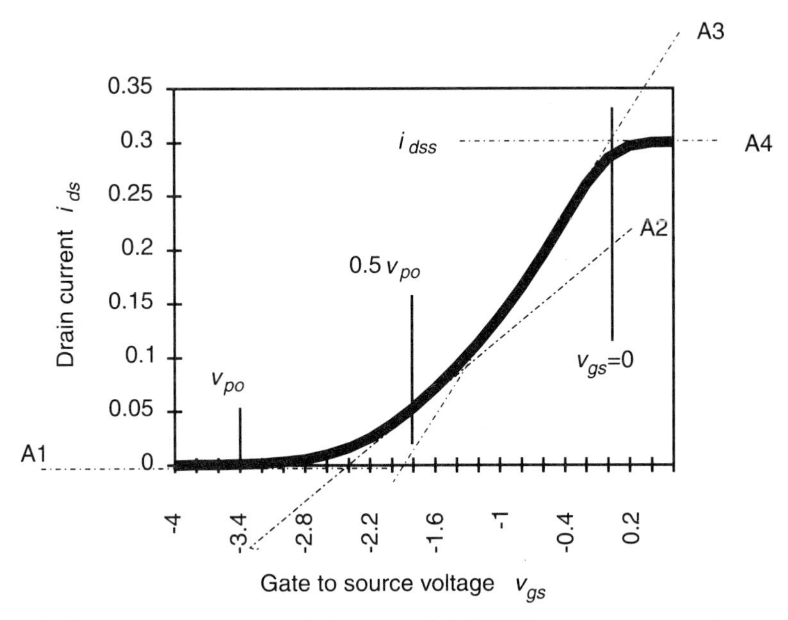

Figure 3.3 Four asymptotes are needed to model full MESFET current source behavior.

$$v_{gs_{1,2}} = \frac{3v_{po}}{4} \tag{3.8}$$

Slope change from **A1** to **A2** is determined by subtracting the slope of **A1** (which is = 0) from the value of the first derivative of **A2** at the point $v_{gs} = 0.5v_{po}$ which is (3.6). The value of pinch-off voltage v_{po} is negative; therefore, slope change is positive.

Equation (3.6) is the slope change from **A1** to **A2**. Arbitrarily select the transition range from asymptote **A1** to **A2** to be

$$a_{1,2} = \frac{|v_{po}|}{4} \tag{3.9}$$

This gives continuous curvature to the entire trace from pinch-off to the point of tangency of **A1**.

The model's first RHF is now defined to be

$$y = -\left(\frac{i_{dss}}{v_{po}}\right)\left(\frac{|v_{po}|}{4}\right)\log_{10}\left[1 + 10^{\frac{4v_{gs}-3v_{po}}{|v_{po}|}}\right] \tag{3.10}$$

Asymptote **A3**, tangent to the trace at $v_{gs} = 0$, is identified by the linear equation

$$y = i_{dss}\left[1 - \frac{2v_{gs}}{v_{po}}\right] \tag{3.11}$$

having a slope at $v_{gs} = 0$ described by (3.5). The intersection of **A2** and **A3** is determined to be at

$$v_{gs_{2,3}} = \frac{v_{po}}{4} \tag{3.12}$$

Use the same transition range selected for the transition from asymptote **A1** to **A2**, which is defined by (3.9) to give continuous curvature from tangency point of **A2** to tangency point of **A3**, which is at $v_{gs} = 0$. Slope change from asymptote **A2** to **A3** is determined by subtracting (3.6) from (3.5) obtaining slope change

$$\Delta A_{2,3} = -\frac{2i_{dss}}{v_{po}} - \left(-\frac{i_{dss}}{v_{po}}\right) = -\frac{i_{dss}}{v_{po}} \tag{3.13}$$

The value of pinch-off voltage v_{po} is negative; therefore, slope change is positive. Sufficient information now exists to formulate the model's second RHF which is

$$y = -\left(\frac{i_{dss}}{v_{po}}\right)\left(\frac{|v_{po}|}{4}\right)\log_{10}\left[1 + 10^{\left(\frac{4v_{gs} - v_{po}}{|v_{po}|}\right)}\right] \tag{3.14}$$

The third and final RHF describes the change from the slope of asymptote **A3** to a slope of zero, a constant i_{dss}. The value of the slope change is the negative value of (3.5). The asymptote intersect point is at $v_{gs} = 0$ volts. Arbitrarily assign a sharp transition range from **A3** to **A4** to be

$$a_{3,4} = \frac{|v_{po}|}{10} \tag{3.15}$$

thus giving the final RHF the value

$$y = \left(\frac{2i_{dss}}{v_{po}}\right)\left(\frac{|v_{po}|}{10}\right)\log_{10}\left[1 + 10^{\left(\frac{10v_{gs}}{|v_{po}|}\right)}\right] \tag{3.16}$$

Combine the three RHFs realizing that v_{po} is a negative number to obtain a MESFET current source behavioral model equation

$$i_{ds} = \frac{i_{dss}}{4}\left\{\log_{10}\left[1 + 10^{\left(\frac{4v_{gs} - 3v_{po}}{|v_{po}|}\right)}\right] + \log_{10}\left[1 + 10^{\left(\frac{4v_{gs} - v_{po}}{|v_{po}|}\right)}\right] - \frac{4}{5}\log_{10}\left[1 + 10^{\left(\frac{10v_{gs}}{|v_{po}|}\right)}\right]\right\} \quad (3.17)$$

which describes sharp cutoff and solid pinch-off behavior. Equation (3.17) is used in Curtice's model in place of the term $\beta(v_{gs} - v_{po})^2$ for the case where load impedance is equal to that described by Equation (3.2). Figure 3.4 compares the square law model and the new behavioral model.

The difference between square law and behavioral model drain current, within the square law region, as a function of gate voltage between pinch-off and saturation, is uniform and has no high-order ripple. The difference expressed in terms of percent of i_{dss} is less than -4.7% of i_{dss} for $v_{po} < v_{gs} < 0$ as shown in Figure 3.5.

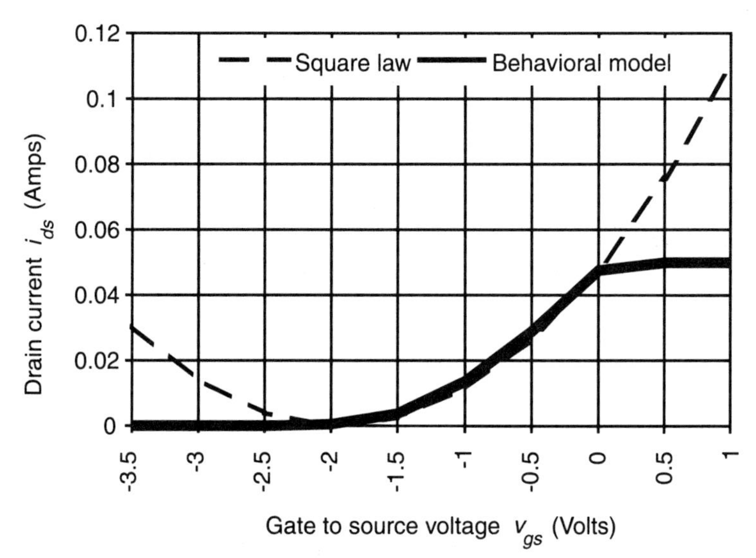

Figure 3.4 A comparison of MESFET square law and behavioral models.

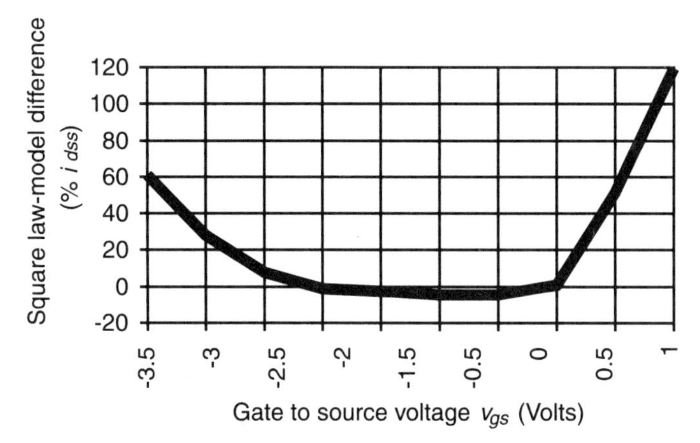

Figure 3.5 The difference between square law and behavioral model.

3.4 An Alternate MESFET Current Source Behavioral Model

An alternate technique used in developing behavioral models involves nulling the primary function over ranges where extreme nonlinearity occurs, and inserting a function that models the nonlinearity. This is accomplished by multiplying the primary function by a unit step function that goes to zero at the appropriate independent variable values and adding in the nonlinear characteristic, which is also multiplied by a unit step function that becomes equal to unity causing it to become active at the same appropriate independent variable values. Step-up and step-down functions described in Chapter 2, (2.32) and (2.34), are used to switch in and out the nulling or zeroing functions.

The primary function describing MESFET drain current as a function of gate-to-source voltage and pinch-off voltage, (3.3), is made to have zero value at $v_{gs} < v_{po}$ by multiplying it by a unit step-up function that has zero value for $v_{gs} < v_{po}$. The step up from zero to unity begins at $v_{gs} = v_{po}$ with transition range $a = 0.01|v_{po}|$, and reaches unity at $v_{gs} = v_{po} + 0.02|v_{po}|$. Step function change in slope at $v_{gs} = v_{po}$ is equal to transition slope

$$\frac{A}{\Delta x} = \frac{1}{(v_{po} + 0.02|v_{po}|) - v_{po}} = \frac{1}{0.02|v_{po}|} \tag{3.18}$$

Remember that v_{po} is a negative value. The complete step-up function equation has the form

$$y = \frac{0.01|v_{po}|}{0.02|v_{po}|}\log_{10}\left[\frac{1 + 10^{\left(\frac{v_{gs}-v_{po}}{0.01|v_{po}|}\right)}}{1 + 10^{\left(\frac{v_{gs}-v_{po}-0.02|v_{po}|}{0.01|v_{po}|}\right)}}\right] \tag{3.19}$$

which when multiplied by the square law function, (3.3), gives

$$i_{ds} = \frac{i_{dss}}{2}\log_{10}\left[\frac{1 + 10^{\left(\frac{v_{gs}-v_{po}}{0.01|v_{po}|}\right)}}{1 + 10^{\left(\frac{v_{gs}-v_{po}-0.02|v_{po}|}{0.01|v_{po}|}\right)}}\right]\left[1 - \frac{2v_{gs}}{v_{po}} + \left(\frac{v_{gs}}{v_{po}}\right)^2\right] \tag{3.20}$$

Equation (3.20) adequately describes behavior below pinch-off as shown in Figure 3.6.

The MESFET's drain current limits at i_{dss} when driven into saturation working into the load described by (3.2). Equation (3.20) needs further modification to model the MESFET's saturation characteristic. In its present form, drain current increases quadricly without bound when gate-to-source voltage increases and exceeds zero volts. Saturation nonlinearity is modeled by nulling the primary function, (3.20),with a unity step-down function, and adding in a value for i_{dss} for $v_{gs} > 0$ volts with a unity step-up function. A unity step-down function is modeled to step down at $v_{gs} = 0$, obtain zero value at $v_{gs} \geq 0.1$ volts, have transition slope $\Delta S = -10$, and transition range $a = 0.05$,

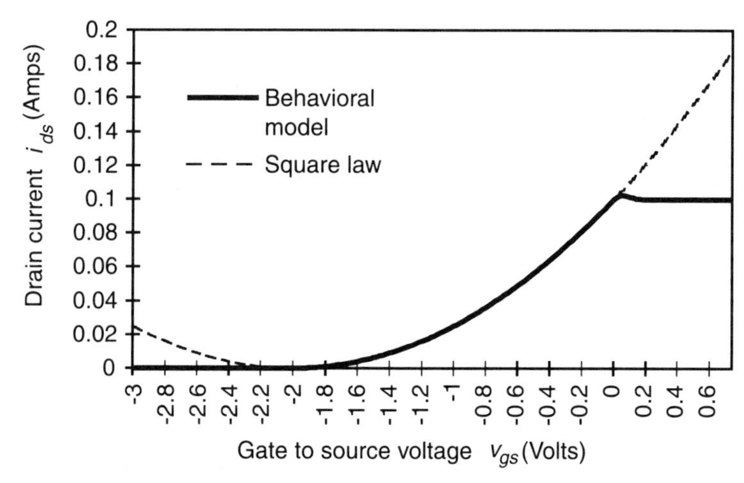

Figure 3.6 Alternate MESFET current source behavioral model using nulling functions.

$$y = 1 + \frac{1}{2}\log_{10}\left[\frac{1 + 10^{\left(\frac{v_{gs}-0.1}{0.05}\right)}}{1 + 10^{\left(\frac{v_{gs}}{0.05}\right)}}\right] \tag{3.21}$$

A unity step-up function that begins to step up at $v_{gs} = 0$, obtains unity at $v_{gs} > 0.1$, having transition slope $\Delta S = 10$ and transition range $a = 0.05$, has the form

$$y = \frac{1}{2}\log_{10}\left[\frac{1 + 10^{\left(\frac{v_{gs}}{0.05}\right)}}{1 + 10^{\left(\frac{v_{gs}-0.1}{0.05}\right)}}\right] \tag{3.22}$$

Multiply (3.20) by (3.21) to null the primary function at $v_{gs} > 0$ volts, and multiply (3.22) by i_{dss} and add it to the product of (3.20) and (3.21) to complete a behavioral model that saturates at i_{dss} for $v_{gs} > 0$ volts and pinches off at $v_{gs} < v_{po}$

$$i_{ds} = \frac{i_{dss}}{2}\log_{10}\left[\frac{1 + 10^{\left(\frac{v_{gs}-v_{po}}{0.01|v_{po}|}\right)}}{1 + 10^{\left(\frac{v_{gs}-v_{po}-0.02|v_{po}|}{0.01|v_{po}|}\right)}}\right]\left[1 - \frac{2v_{gs}}{v_{po}} + \left(\frac{v_{gs}}{v_{po}}\right)^2\right] *$$

$$\left\{1 + \frac{1}{2}\log_{10}\left[\frac{1 + 10^{\left(\frac{v_{gs}-0.1}{0.05}\right)}}{1 + 10^{\left(\frac{v_{gs}}{0.05}\right)}}\right]\right\} + \frac{i_{dss}}{2}\log_{10}\left[\frac{1 + 10^{\left(\frac{v_{gs}}{0.05}\right)}}{1 + 10^{\left(\frac{v_{gs}-0.1}{0.05}\right)}}\right] \tag{3.23}$$

Saturation is abrupt due to the choice of transition range $a = 0.05$ and slope $= 10$ of the step-up and step-down functions (3.21) and (3.22) as is illustrated in Figure 3.6. If a softer transition range is used, a sizeable overshoot develops as the modeled function transitions into saturation.

Difference between the square law function and the new behavioral model equation over the range $v_{po} < v_{gs} < 0$ is zero except for a small overshoot near $v_{gs} = 0$. The behavioral model drain current overshoot of 0.14 percent of i_{dss} at $v_{gs} = 0$ volts illustrated in Figure 3.7 generates a negligible high-order error.

Equation (3.23) is substituted back into Curtice's model in place of the term $\beta(v_{gs} - v_{po})^2$ for devices terminated with loads described by (3.2).

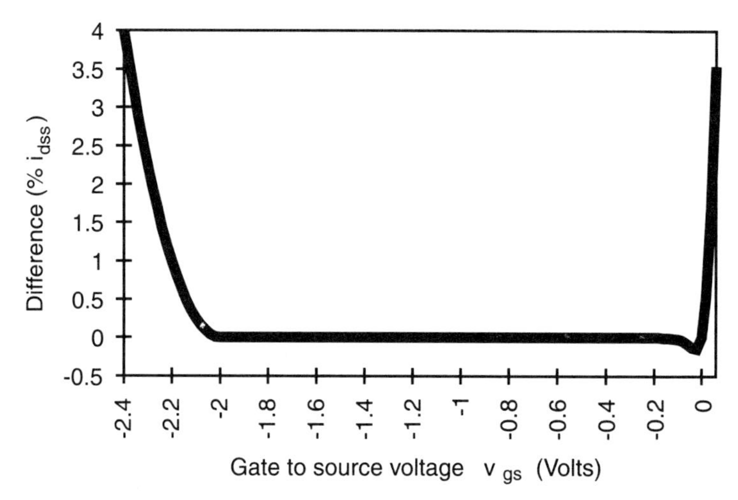

Figure 3.7 Difference between nulling behavioral model and the square law function.

3.5 A General MESFET Current Source Behavioral Model

Unique doping profiles are often implemented in the design of MESFET transistors to improve linearity, to increase power output, and to improve power-added efficiency. The resulting MESFET drain current characteristics exhibit square law behavior for gate-to-source voltages ranging from pinch-off to a few tenths of a volt above pinch-off. The response is then linear from a few tenths of a volt above pinch-off to saturation at $v_{gs} = 0$ volts. A general MESFET behavioral model is now developed that has the flexibility to adapt to the characteristics of various doping profiles for different size transistors. Figure 3.8 illustrates the behavior that needs to be characterized using the newly developed curve-fit technique.

Nulling functions cannot be used efficiently to model the step-doped MESFET because two separate equations, one quadratic and one linear with a RHF for saturation, would have to be switched in and out. The square law approximation approach is used. Using this approach, a new general MESFET behavioral model equation is developed that collapses to the previously developed square law behavioral model, (3.17), in the limit where the linear region shrinks to zero.

The previously developed square law MESFET behavioral model, (3.17), is derived from four asymptotes, one equal to zero representing pinch-off current, the second, tangent to the square law function at a point halfway between pinch-off v_{po} and saturation at $v_{gs} = 0$ volts, the third tangent to the square law function at $v_{gs} = 0$ volts, and the fourth equal to I_{dss} for $v_{gs} > 0$ volts. The first asymptote, equal to pinch-off current, is the reference. Coefficients for RHFs

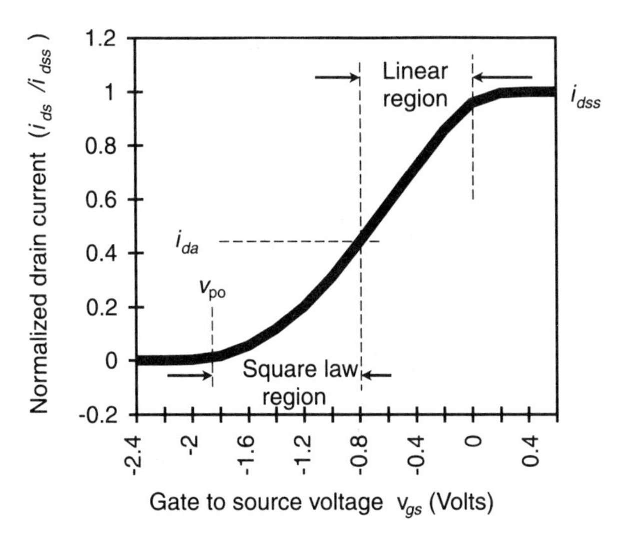

Figure 3.8 Step-doped MESFETs exhibit a linear region that requires a new model.

are determined transitioning to the right through the simulated square law region and into saturation.

The same approach is taken here using asymptote **A1** to represent pinch-off current, and two asymptotes, **A2** and **A3**, to simulate the square law region. In the general MESFET model, the second square law region asymptote **A3** is extended to become the new model's linear range. The final asymptote, **A4**, is I_{dss}, is the saturation current. Figure 3.9 illustrates the asymptotes used in developing the general MEFET model.

Transition from square law to linear occurs at arbitrary gate voltage v_{ga}. Asymptote **A3** representing the linear range has slope equal to that of the square law curve at v_{ga}. The value of that slope needs to be determined. The generalized square law curve

$$i_{ds} = H(v_{gs} - v_{po})^2 \tag{3.24}$$

has first derivative

$$\frac{\partial i_{ds}}{\partial v_{gs}} = 2H(v_{gs} - v_{po}) \tag{3.25}$$

Asymptote **A3**, which is a tangent to the square law function at gate voltage v_{ga} (see Figure 3.9), has slope

$$S(A3) = 2H(v_{ga} - v_{po}) \tag{3.26}$$

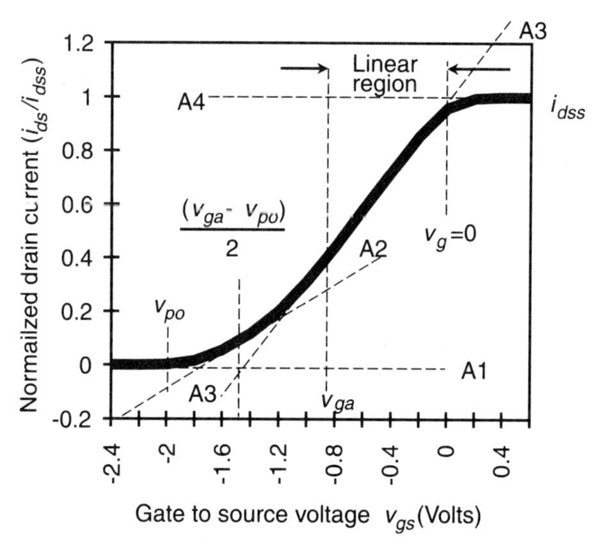

Figure 3.9 Asymptote constructions used in the general MESFET model.

Asymptote **A3**, the tangent to the square law function at gate voltage v_{ga}, also passes through i_{dss} at gate voltage $v_{gs} = 0$ volts. These two points define a line

$$y = 2H(v_{ga} - v_{po})v_{gs} + i_{dss} \tag{3.27}$$

Solve for H by setting (3.24) and (3.27) to be equal at gate voltage v_{ga}.

$$H(v_{ga} - v_{po})^2 = 2H(v_{ga} - v_{po})v_{ga} + i_{dss}$$

$$H = \frac{-i_{dss}}{(v_{ga}{}^2 - v_{po}{}^2)} \tag{3.28}$$

The linear equation describing asymptote **A3** is then

$$A3 = i_{dss}\left(1 - \frac{2v_{gs}}{(v_{ga} + v_{po})}\right) \tag{3.29}$$

The slope of **A3**

$$S(A3) = \frac{-2i_{dss}}{(v_{ga} + v_{po})} \tag{3.30}$$

Asymptote **A3** has value equal to zero at gate voltage

$$v_{gb} = \frac{(v_{ga} + v_{po})}{2} \tag{3.31}$$

which is a point halfway between v_{po} and v_{ga} (see Figure 3.9). This is the gate voltage at which asymptote **A2** is tangent to the square law function. The slope of **A2** is the first derivative of the square law function, (3.25), at gate voltage v_{gb} from (3.31)

$$S(A2) = 2\left(\frac{-i_{dss}}{(v_{ga}^{\ 2} - v_{po}^{\ 2})}\right)(v_{gb} - v_{po}) = 2\left(\frac{-i_{dss}}{(v_{ga}^{\ 2} - v_{po}^{\ 2})}\right)\left(\frac{v_{ga} + v_{po}}{2} - v_{po}\right)$$

$$S(A2) = \frac{-i_{dss}}{(v_{ga} + v_{po})} \tag{3.32}$$

The line equation for asymptote **A2** is

$$y = \frac{-i_{dss}}{(v_{ga} + v_{po})} v_{gs} + b \tag{3.33}$$

Solve for the constant b by setting (3.33) equal to the square law function, (3.24), at gate-to-source voltage $v_{gs} = v_{gb}$ obtaining

$$\frac{-i_{dss}}{(v_{ga}^{\ 2} - v_{po}^{\ 2})}(v_{gb} - v_{po})^2 = \frac{-i_{dss}}{(v_{ga} + v_{po})} v_{gb} + b$$

and substituting (3.31) for value v_{gb}

$$b = \frac{i_{dss}}{4}\left[\frac{v_{ga} + 3v_{po}}{v_{ga} + v_{po}}\right] \tag{3.34}$$

The equation for asymptote **A2** (3.33) is rewritten using (3.34) to obtain a new line equation

$$A2 = \frac{i_{dss}}{(v_{ga} + v_{po})}\left(\frac{v_{ga} + 3v_{po}}{4} - v_{gs}\right) \tag{3.35}$$

Asymptote **A2** has value equal to zero at

$$v_{gs} = \frac{v_{ga} + 3v_{po}}{4} \tag{3.36}$$

Equation (3.36) is the intersection of asymptote **A1** and asymptote **A2**. Asymptote **A1** is the constant zero representing pinch-off current.

The intersection of asymptotes **A2** and **A3** is determined by setting (3.29) equal to (3.35)

$$i_{dss}\left(1 - \frac{2v_{gs}}{(v_{ga} + v_{po})}\right) = \frac{i_{dss}}{(v_{ga} + v_{po})}\left(\frac{v_{ga} + 3v_{po}}{4} - v_{gs}\right)$$

and solving for

$$v_{gs}(2, 3) = \frac{3v_{ga} + v_{po}}{4} \tag{3.37}$$

Transition range

$$a = \left|\frac{v_{ga} - v_{po}}{4}\right| \tag{3.38}$$

between **A1** and **A2** and between **A2** and **A3** is arbitrarily set to be half the distance between the points where asymptote **A2** and asymptote **A3** are tangent to the square law function. Transition range into saturation is arbitrarily set to be the same as that which is used for the square law MESFET model, see (3.15).

All of the coefficients for the general MESFET behavioral model are now defined. The first term is zero. An RHF transitions out of pinch-off onto asymptote **A2** with a slope change defined by (3.32), a transition range from (3.38), at an intersection defined by (3.36). The first RHF

$$\frac{-i_{dss}}{(v_{ga} + v_{po})}\left|\frac{(v_{ga} - v_{po})}{4}\right| \log_{10}\left[1 + 10^{\frac{4v_{gs} - v_{ga} - 3v_{po}}{|v_{ga} - v_{po}|}}\right]$$

is defined. The model transitions onto asymptote **A3** at intersection $v_{gs}(2,3)$ defined by (3.37), with transition range defined by (3.38), and slope change equal to the difference between (3.30) and (3.32). The second RHF

$$\frac{-i_{dss}}{(v_{ga} + v_{po})}\left|\frac{(v_{ga} - v_{po})}{4}\right| \log_{10}\left[1 + 10^{\frac{4v_{gs} - 3v_{ga} - v_{po}}{|v_{ga} - v_{po}|}}\right]$$

is defined. Saturation occurs at $v_{gs} = 0$ volts with a transition range defined by (3.15), and a slope change equal to the negative value of (3.30). The third RHF

$$\frac{2i_{dss}}{(v_{ga} + v_{po})}\left|\frac{v_{po}}{10}\right|\log_{10}\left[1 + 10^{\frac{10v_{gs}}{|v_{po}|}}\right]$$

is defined.

The general MESFET behavioral model is the sum of the three RHFs. Adding the RHFs and collecting terms to simplify, obtain

$$i_{ds} = \frac{i_{dss}}{(v_{ga} + v_{po})}\left|\frac{v_{po}}{5}\right|\log_{10}\left[1 + 10^{\frac{10v_{gs}}{|v_{po}|}}\right] - \frac{i_{dss}}{(v_{ga} + v_{po})}\left|\frac{v_{ga} - v_{po}}{4}\right| *$$
$$\left\{\log_{10}\left[1 + 10^{\frac{4v_{gs} - v_{ga} - 3v_{po}}{|v_{ga} - v_{po}|}}\right] + \log_{10}\left[1 + 10^{\frac{4v_{gs} - 3v_{ga} - v_{po}}{|v_{ga} - v_{po}|}}\right]\right\} \quad (3.39)$$

Notice that when $v_{ga} = 0$ volts, the general MESFET behavioral model (3.39) reduces to the square law MESFET behavioral model (3.17). Equation (3.39) can be substituted back into Curtice's model in place of the term $\beta(v_{gs} - v_{po})^2$ for the case where load impedance is equal to that described by (3.2). Figure 3.10 illustrates the modeling of a step-doped MESFET that has $v_{po} = -2$ volts, $v_{ga} = -0.7$ volts, and $i_{dss} = 0.1$ amperes.

3.6 Modeling the Bipolar Transistor Current Source

Bipolar transistor collector current is related to base current as described by the Ebers-Moll [3] model. Use a modified version of the Ebers-Moll model

$$I_c = I_s\beta\left(e^{\frac{qv_{be}}{kT}} - 1\right) \quad (3.40)$$

where I_c is collector current, I_s is reverse or saturation current, β is base-current to collector-current amplification ratio, q = electron charge (1.6×10^{-19} coulomb), v_{be} is applied base to emitter voltage, k = Boltzman's constant (1.38×10^{-23} watt sec/degrees K), and T is absolute temperature in degrees Kelvin. At room temperature $T = 300$ degrees, and the exponent $q/kT = 38.6$ volts. This bipolar transistor current source model characterizes pinch-off nicely, but collector current rises exponentially without bound as base to emitter voltage

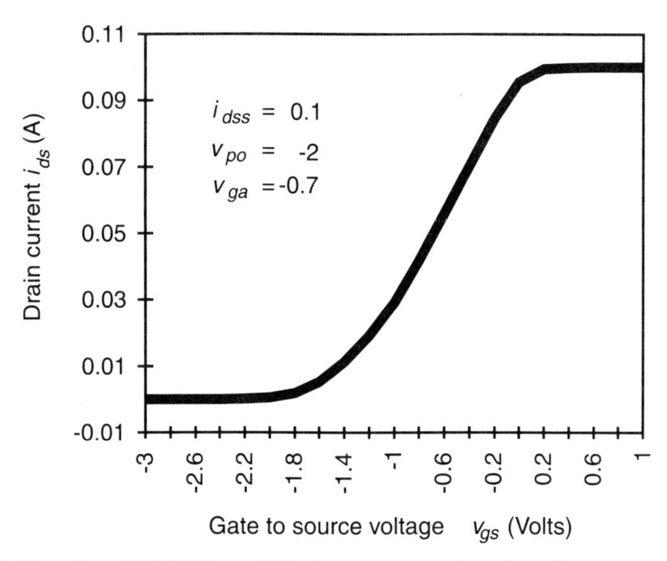

Figure 3.10 Step-doped MESFET model drain current as a function of v_{gs}.

increases. Saturation fails to be modeled adequately. When the collector is loaded with impedance Z_l, a maximum collector current

$$I_{max} = \frac{2(v_{ce} - v_k)}{Z_l} \tag{3.41}$$

flows at saturation, where v_{ce} is collector to emitter voltage, and v_k is knee voltage. Typical bipolar transistor behavior of collector current as a function of collector voltage and base current with a load described by (3.41) is illustrated in Figure 3.11.

One approach to modeling the bipolar transistor's behavior is realized by nulling the primary function, (3.40), by multiplying it with a unity step down to zero as the device goes into saturation, and adding in saturation current I_{max} at that point by multiplying it with a unity step-up function. An equally acceptable approach is to cancel out the primary function with a switched-in mirror function and adding in a switched I_{max}. In either approach the base-to-emitter voltage that generates saturation current I_{max} needs to be determined. Rearrange (3.40) and solve for the base voltage

$$v_{bm} = \frac{kT}{q} \ln\left(\frac{I_{max} - \beta I_s}{\beta I_s}\right) \tag{3.42}$$

that generates I_{max}.

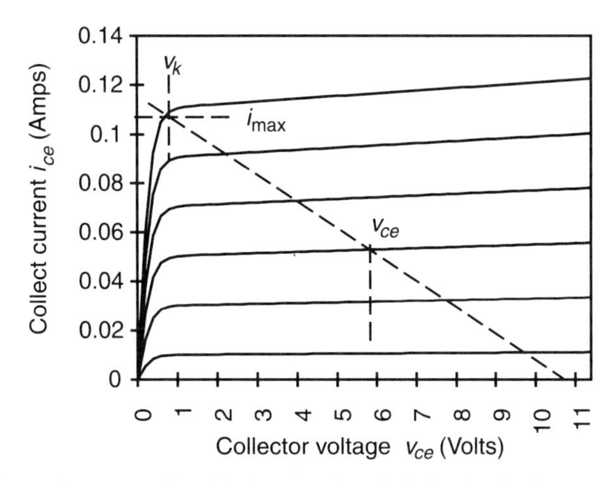

Figure 3.11 Saturation current in a bipolar transistor is defined to be I_{max}.

The mirror function approach requires only one unity step up and has fewer terms than the primary function nulling approach. The mirror function, the negative of the bipolar exponential function (3.40), is switched on using a unity step-up function that begins to rise at base voltage v_{bm}. The sum of the exponential function and the mirrored exponential function is zero for all $v_{be} > v_{bm}$. Maximum drain current I_{max} is also switched on by the same unity step-up function when it is added to the mirrored exponential function. The primary function's rapid exponential rate of increase requires a very short switch interval to transition into saturation without significant overshoot. Arbitrarily let step-up function transition range $a = 0.0008$ volts to keep overshoot less than 1 percent of I_{max}. The total transition from zero to unity takes place over an interval $\Delta v_b = 2a$. Slope change through the transition

$$\Delta S = \frac{1}{2a} \tag{3.43}$$

The unity step-up function that transitions into saturation consists of two RHFs

$$\Delta S a \log_{10}\left[1 + 10^{\frac{v_{be} - v_{bm}}{a}}\right] - \Delta S a \log_{10}\left[1 + 10^{\left(\frac{v_{be} - v_{bm} - 2a}{a}\right)}\right]$$

$$= 0.5 \log_{10}\left[\frac{1 + 10^{\left(\frac{v_{be} - v_{bm}}{a}\right)}}{1 + 10^{\left(\frac{v_{be} - v_{bm} - 2a}{a}\right)}}\right] \tag{3.44}$$

The bipolar transistor behavioral model describing collector current as a function of base voltage is

$$I_c = \beta I_s\left(e^{\frac{qv_{be}}{kT}} - 1\right)$$
$$+\left[I_{max} - \beta I_s\left(e^{\frac{qv_{be}}{kT}} - 1\right)\right] + 0.5\log_{10}\left[\frac{1 + 10^{\left(\frac{v_{be}-v_{bm}}{a}\right)}}{1 + 10^{\left(\frac{v_{be}-v_{bm}-2a}{a}\right)}}\right] \quad (3.45)$$

Figure 3.12 shows modeled collector current as a function of base emitter voltage, (3.45), for a device loaded with Z_l that produces I_{max} as defined by (3.41) where base emitter voltage swings well into pinch-off and saturation.

3.7 Behavioral Model Examples

Example 3.1 A Square Law MESFET

Given: A square law MESFET having $v_{po} = -2V$, $v_{ga} = 0V$, $i_{dss} = 0.1A$, $v_{ds} = 5.0V$, and $v_k = 1V$, input impedance $R_{in} = 100\Omega$, load impedance $R_l = 80\Omega$, is connected in a common source configuration.

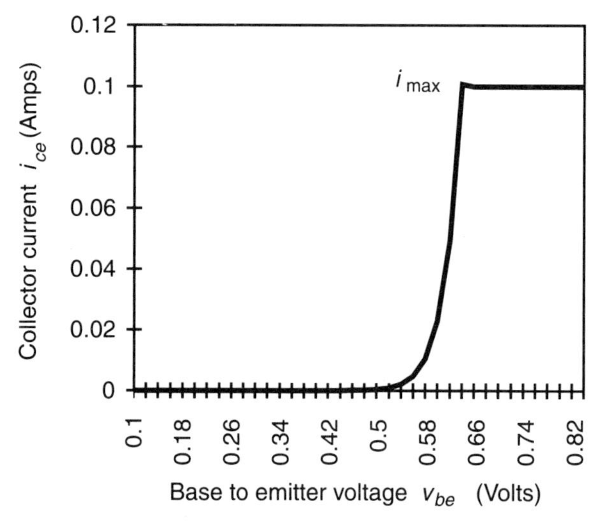

Figure 3.12 Bipolar transistor behavioral model showing saturation and pinch-off.

Required: Bias the transistor at nine different gate voltages such that quiescent drain currents I_{ds} at the nine bias voltages are 0.01A, 0.02A, 0.03A, 0.04A, 0.05A, 0.06A, 0.07A, 0.08A, and 0.09A. Vary power input to the common emitter device from -20 dBm to $+16$ dBm in 2 dB steps. Determine average drain current as a function of power input. Determine fundamental power output as a function of power input. Determine second harmonic power output as a function of power input. Determine third harmonic power output as a function of power input.

Procedure: Using a spreadsheet that has a Fourier transform data analysis tool (for instance, Microsoft's Excel) establish reference cells for entry of each of the given parameters and assign labels to them. Establish reference cells for power input P_{in} in dBm and peak amplitude of the input signal sine wave (A) where peak sine wave amplitude

$$A = \sqrt{2R_{in}10^{\left(\frac{P_{in}-30}{10}\right)}}$$

is calculated from input power in units of dBm and input impedance in units of ohms. Establish a single cell into which an arbitrary gate bias voltage can be entered and enter a gate voltage value of -1.0 volts. Next to this cell, establish a cell containing the general MESFET behavioral model (3.39) where gate voltage entered into the equation is taken from the arbitrary gate bias voltage cell, and all other variable values are taken from reference cells containing given data. Use the solver tool utility in the spreadsheet, entering values for quiescent drain current i_{ds} of 0.01, 0.02, 0.03, 0.04, 0.05, 0.06, 0.07, 0.08, and 0.09 sequentially and solving for gate voltage values that give the corresponding required quiescent drain currents. These gate voltage values will be used to establish the required test drain currents. Establish a vertical column labeled RADIANS having 512 cells. The first cell contains zero. Each radian cell increases in value in increments by

$$\Delta r = \frac{16*2\pi}{512}$$

to complete 16 cycles of input waveform in the vertical column. Establish another column labeled v_{in} to the right of the radian column and enter the sum of gate bias voltage v_{gs} and the value of the input signal sine wave corresponding to the radian value

$$v_{in} = v_{gs} + A\sin(r)$$

Enter the general MESFET model (3.39) that calculates i_{ds} in the next column to the right and copy down 512 cells referencing data cells previously established for given data input. Create a graph of at least two cycles of the output drain current waveform from these cells. Use the Fourier transform utility to generate a 512-point discrete Fourier transform (DFT) of the drain current waveform. The first output cell of the DFT is proportional to average magnitude of drain current. The second term, eight cells down, is proportional to the peak value of the fundamental component of the drain current waveform, the third term, eight more cells down, is proportional to the peak value of the second harmonic component of the drain current waveform, and the fourth term, another eight cells down, is proportional to the third harmonic component of the drain current waveform. Divide all DFT output cells by 512, determine the absolute value of complex numbers and obtain RMS current values. Calculate and save average drain current and power delivered to the load impedance $R_l = 80\Omega$ in the fundamental and each harmonic in terms of dBm for each power input from -20 dBm $\leq P_{in} \leq +16$ dBm. Repeat the above process for each of the nine required quiescent drain currents.

Results: Plot drain current as a function of power input for each gate bias and obtain data shown in Figure 3.13.

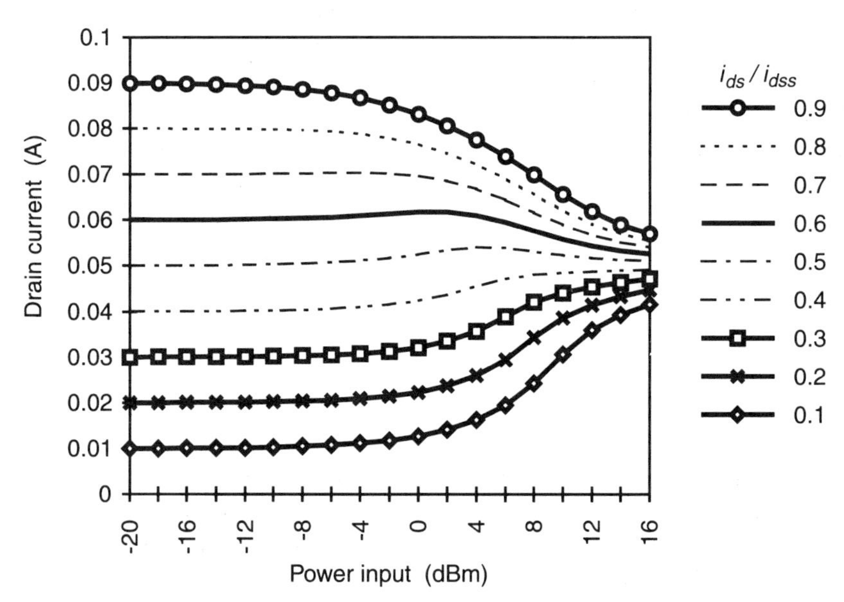

Figure 3.13 Square law MESFET drain current as a function of power input at different gate bias voltages.

Drain current always tends toward 50 percent i_{dss} as the device is driven hard into saturation, regardless of gate-bias voltage, because the output current waveform approaches a square wave with its base at zero current and its maximum at i_{dss} as shown in the drain current waveform Figure 3.14.

Power contained in the fundamental component of the drain current waveform is usually of greatest interest unless the device is used as a harmonic generator. Plot the fundamental component power as a function of power input for each of the gate-bias voltages to obtain data shown in Figure 3.15. Notice the difference in compression characteristic for each gate bias as the device is driven into saturation. Compression occurs very softly in the square law MESFET where gate-bias voltage produces quiescent drain currents greater than 50 percent i_{dss}. As gate-bias voltage is moved toward pinch-off and the device is biased for class AB operation, compression characteristic becomes very sharp with some evidence of gain expansion. Compression characteristic is quantified in Chapter 4, where amplifier modeling is developed.

Power contained in the second harmonic of the drain current waveform varies significantly as a function of power input for different gate-bias voltages and as a function of quiescent drain current. Lowest second harmonic content occurs when quiescent drain current is 70 percent i_{dss}. Highest second harmonic content occurs when quiescent drain current is 90 percent i_{dss}. Small signal level second harmonic power increases 2 dB for every dB increase in power input.

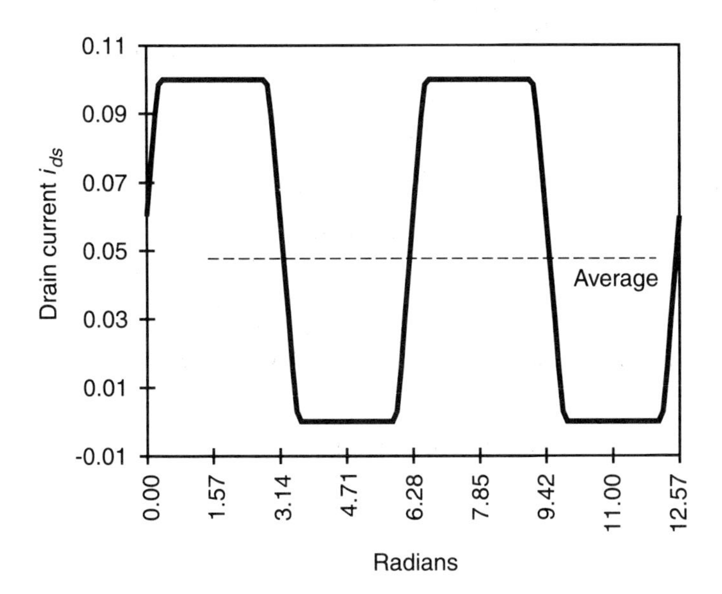

Figure 3.14 Square law MESFET drain current waveform at hard saturation.

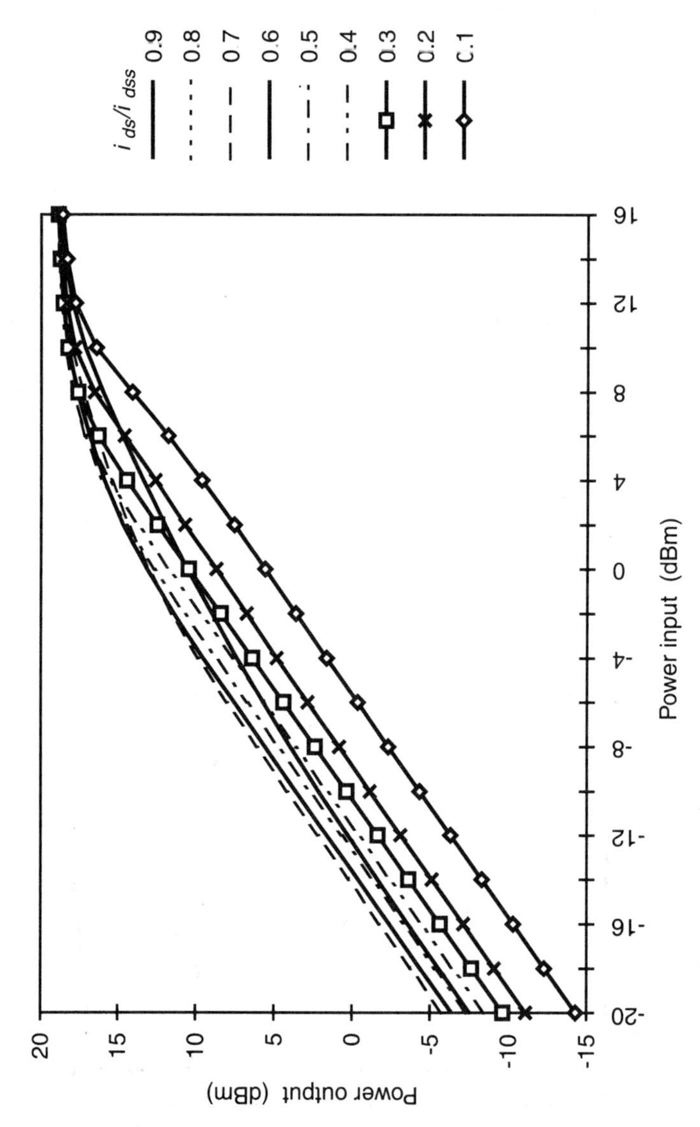

Figure 3.15 Square law MESFET drain current waveform fundamental power output component as a function of power input for different gate-bias voltages.

However, where gate bias establishes a drain current quiescent value in the range of 50 percent to 70 percent i_{dss}, drain current waveform second harmonic content behaves abnormally. A sharp null in second harmonic power output occurs as the device goes into compression. At quiescent drain current less than 40 percent i_{dss} a soft peaking of second harmonic power output occurs as the device goes into compression as is indicated by the data plotted in Figure 3.16.

The null in second harmonic power output that occurs with quiescent drain current 50 percent to 70 percent i_{dss} is useful if an amplifier design specification calls for exceptionally low second harmonic component in the output.

Power contained in the third harmonic component of the drain current waveform (see Figure 3.17) increases at a rate of 3 dB per each 1 dB of power input increase at small signal levels. Third harmonic amplitude in the drain current waveform for small signals is lowest at quiescent drain current 30 percent i_{dss}. Low small signal third-order content is maintained for quiescent drain bias up to 60 percent i_{dss}. However, the third harmonic component acts abnormally where quiescent drain current settings are equal to and less than 60 percent i_{dss} as power output approaches compression. The third harmonic softly peaks as the device goes into compression where quiescent drain current is 60 percent i_{dss}, and third harmonic component increases at a rate greater than 3 dB per 1 dB of input power. Where quiescent drain current is less than 60 percent i_{dss},

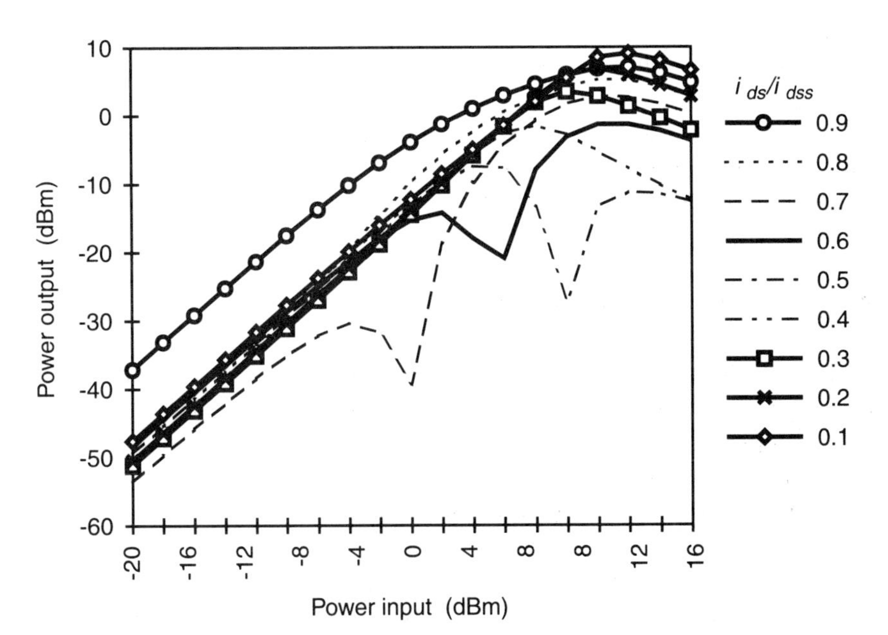

Figure 3.16 Square law MESFET drain current second harmonic content as a function of power input and gate-bias voltage.

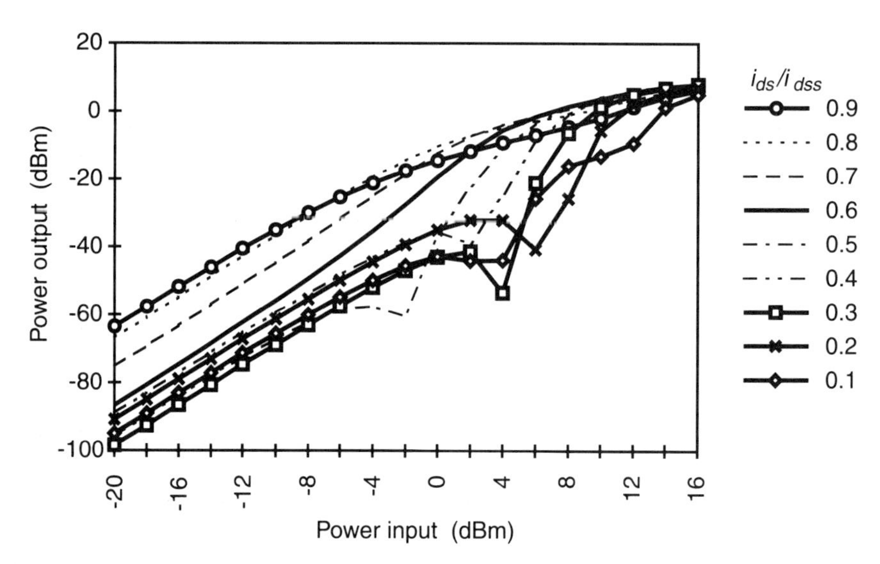

Figure 3.17 Square law MESFET third harmonic power output as a function of power input at different gate-bias voltages.

the third harmonic power output dips sharply as the device goes into compression then rises faster than the expected 3 dB per 1 dB of power input as shown in Figure 3.17.

Behavior of fundamental, second harmonic, and third harmonic power outputs as a function of input power level is also related to the specific doping designed into the transistor device. This example focused on a square law device. The next example focuses on a step-doped MESFET having a linear range of operation where $-0.7\text{V} \leq v_{gs} \leq 0\text{V}$.

Example 3.2 A Step-Doped MESFET

Given: Use all of the transistor parameters of Example 3.1 above except input a value $v_{ga} = -0.7$ volts to define a linear range of operation $-0.7\text{V} \leq v_{gs} \leq 0\text{V}$.

Required: Solve for all of the transistor behavioral characteristics required of the square law device in Example 3.1 above. Do this for each of the nine quiescent drain current values at input power levels from -20 dBm to $+16$ dBm in 2 dB steps.

Procedure: Use the spreadsheet developed for the previous example. Change the value of v_{ga} from 0 volts to -0.7 volts. Solve for new gate voltage values that give quiescent drain currents of 0.01, 0.02, 0.03, 0.04, 0.05, 0.06, 0.07, 0.08, and 0.09 amperes. They will be different than those used for the square law device. Recalculate drain current waveforms as a function of power input for each of the quiescent drain current values. Perform a DFT on the waveform for each combination of power input and quiescent drain current setting calculating average drain current, as well as RMS current contained in the fundamental, second, and third harmonics. Build tables of fundamental and harmonic power level data as a function of power input at each quiescent drain current setting for plotting.

Results: Plot average drain current as a function of power input for each of the nine quiescent drain current settings obtaining the data set shown in Figure 3.18. There is not much difference between drain current behavior of the square law MESFET and the step-doped MESFET. Both approach 50 percent i_{dss} at hard saturation. Compare Figures 3.18 and 3.13.

The step-doped MESFET's drain current waveform fundamental component amplitude as a function of power input for different gate-bias voltages is well behaved (see Figure 3.19). Variation in step-doped MESFET small signal gain is about 2 dB less than square law MESFET small signal gain as quiescent drain current is varied. Gain expansion occurs as power input increases in the step-doped device where quiescent drain current is 10 percent i_{dss}. Compression is sharp for quiescent drain currents less than 50 percent i_{dss} and soft for quiescent drain currents greater than 50 percent i_{dss}. Compression characteristic is quantified in Chapter 4 where behavioral models of nonlinear amplifiers are developed. Saturated power output $P_{sat} = +18.9$ dBm is essentially the same as that of the square law device biased under the same conditions.

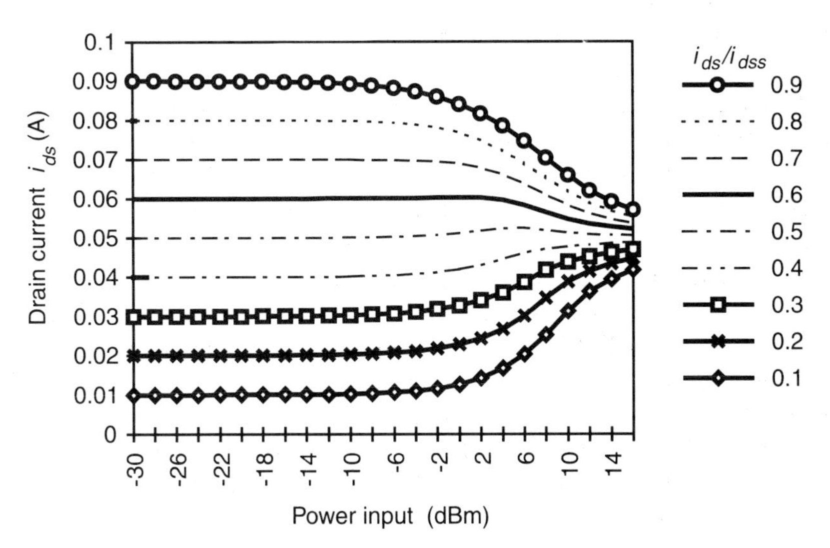

Figure 3.18 Step-doped MESFET drain current as a function of power input for different gate-bias voltages.

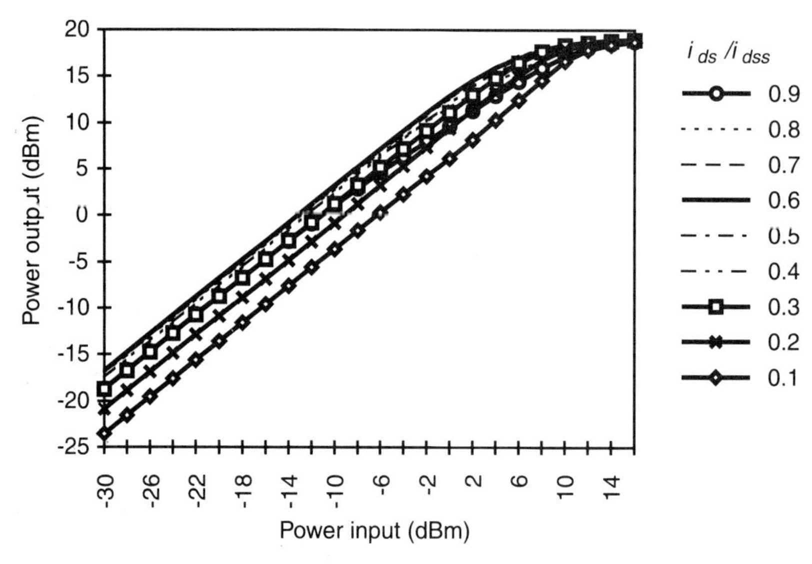

Figure 3.19 Step-doped device fundamental power output as a function of power input for different gate-bias voltages.

Step-doped devices exhibit unique drain current waveform harmonic content as a function of input power. The second harmonic component increases 2 dB for each 1 dB increase in power input at small signal levels. Where drain current quiescent bias is 50 percent i_{dss} to 60 percent i_{dss}, the drain current waveform's second harmonic component dips in amplitude as power input is increased toward device compression. Second harmonic component amplitude is lowest at small power input signal levels when quiescent drain current is 70 percent i_{dss}. However, as the device goes into compression when biased between 70 percent i_{dss} and 80 percent i_{dss}, the second harmonic component of the drain current waveform increases at a rate faster than 3 dB per 1 dB increase in power input. Second harmonic component is greatest at small signal input levels when quiescent drain current is 90 percent i_{dss}. Figure 3.20 illustrates drain current waveform second harmonic amplitude as a function of power input at various quiescent drain current bias levels for the step-doped MESFET.

Step-doped MESFET drain current waveform third harmonic component increases 3 dB for every 1 dB increase of input power at small signal levels (see Figure 3.21). Lowest third-order component amplitude is realized at quiescent drain current bias of 60 percent i_{dss}, and at quiescent drain current bias less than 20 percent i_{dss}. Third harmonic component amplitude in the drain current waveform is highest when quiescent drain current bias is 90 percent i_{dss}. The third harmonic component amplitude dips at a single input power as the device is driven into compression when biased for 20 percent i_{dss} quies-

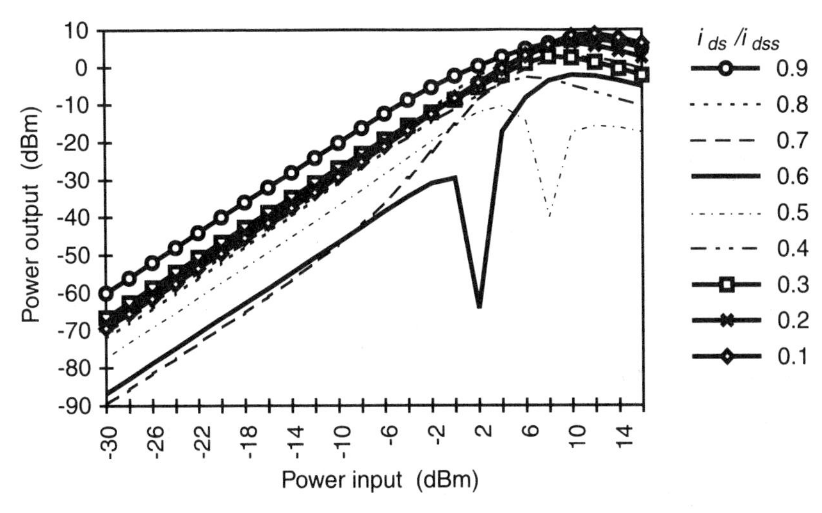

Figure 3.20 Step-doped MESFET drain current second harmonic amplitude as a function of power input at different quiescent drain currents.

cent drain current. The third harmonic amplitude dips at two input power levels as the device is driven into compression when biased at 10 percent i_{dss} quiescent drain current as shown in Figure 3.21. Third-order harmonic amplitude increases at a rate greater than 3 dB per 1dB increase in input power as the

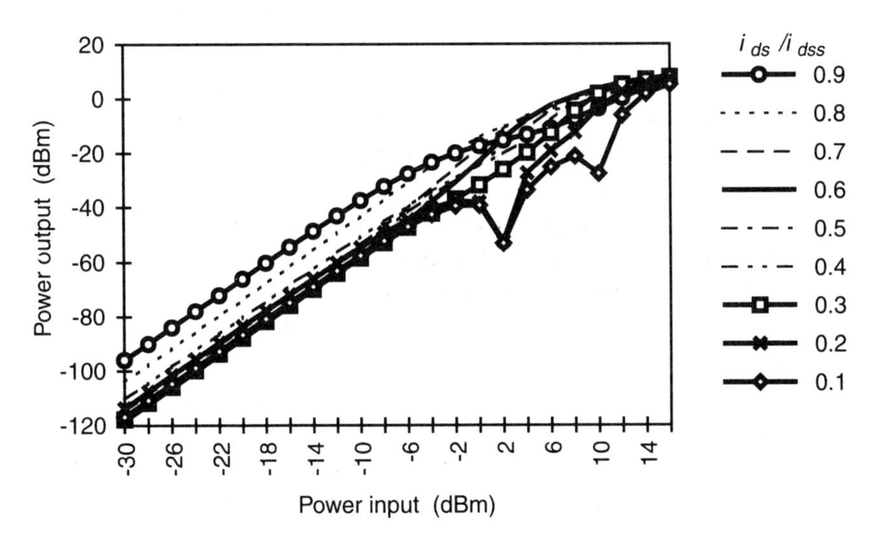

Figure 3.21 Step-doped MESFET drain current third-order harmonic component as a function of power input at different drain current quiescent bias levels.

device is driven into compression at quiescent drain bias levels ranging from 30 percent i_{dss} to 70 percent i_{dss}.

Example 3.3 A Bipolar Transistor

Given: A bipolar transistor characterized by parameters $\beta = 20$, $I_s = 10^{-13}$, operating in the common emitter configuration with $v_{ce} = 5$ volts, $v_k = 1$ volt, $R_l = 80\Omega$, $I_{max} = 0.1$ A, $T = 300$ degrees Kelvin. Common emitter base to emitter resistance to signal input $R_{be} = \left(\frac{kT}{I_s q}\right) e^{\frac{-qv_{be}}{kT}}$.

Required: Bias the transistor at nine different base voltages such that collector currents i_{ce} at the nine bias voltages are 0.01A, 0.02A, 0.03A, 0.04A, 0.05A, 0.06A, 0.07A, 0.08A, and 0.09A. Vary power input to the common emitter device from -20 dBm to $+16$ dBm in 2 dB steps. Determine average collector current as a function of power input. Determine fundamental power output as a function of power input. Determine second harmonic power output as a function of power input. Determine third harmonic power output as a function of power input.

Procedure: Design a spreadsheet similar to the one explained in Example 3.1 using the given parameters above as constants to be placed in reference cells. Follow the procedure of Example 3.1 in calculating the base voltages needed to establish the required collector currents. The bipolar behavioral model (3.45) is used in place of the general MESFET behavioral model. Follow the procedure used in Example 3.1 to establish data over 16 cycles of signal input. Follow the procedure used in Example 3.1 to establish a DFT of the collector current waveform. Generate the required data files as a function of power input at the required collector current levels and plot data.

Results: The bipolar device average collector current as a function of power input for different quiescent collector currents appears different from MESFET data due to the exponential increase in collector current as base voltage rises. Average collector current at low collector quiescent currents (10 percent i_{max}), increases rapidly as power input increases due to the exponential function. Average collector current at high collector quiescent currents (90 percent i_{max}) decreases very slowly as power input increases because the exponential function forces the output waveform to dwell near i_{max} most of the time as Figure 3.22 illustrates. In the limit, average collector current obtains 50 percent i_{max} as the collector current waveform approaches a square wave.

Bipolar transistor collector current waveform at 3 dB gain compression is plotted in Figure 3.23(b) where quiescent collector current is 50 percent i_{max}. It

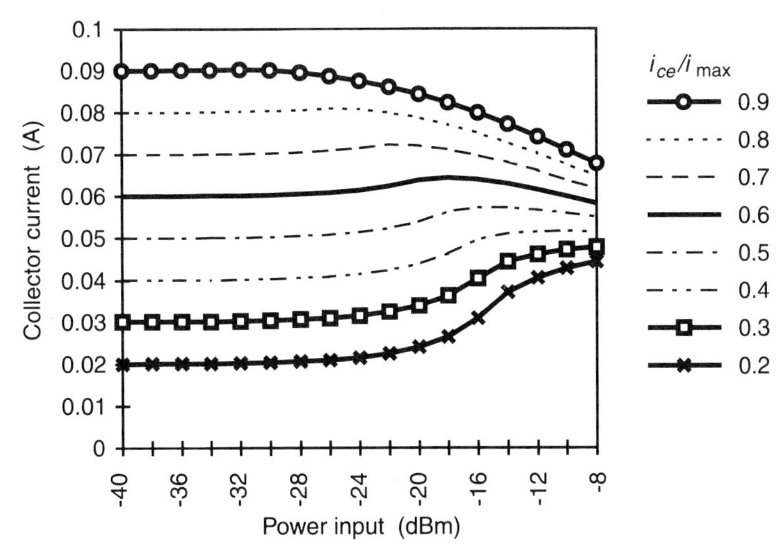

Figure 3.22 Bipolar transistor collector current as a function of power input for different quiescent collector current levels.

is compared to square law MESFET drain current waveform at 3 dB gain compression where quiescent drain current is 50 percent i_{dss} (see Figure 3.23(a)). It is clear that bipolar transistor harmonic content is quite different than that of the MESFET. The bipolar device waveform has more energy in its harmonic content and less in its fundamental. The result is approximately 3.0 dB lower fundamental saturated power output for the same peak-to-peak current waveform excursion when compared to the MESFET. The three examples, (3.1), (3.2), and (3.3) all use the same load resistance, the same knee voltage, and the same prime power voltage. Saturated power output for the two MESFET examples is +18.8 dBm, and saturated power output for the bipolar model is only +15.8 dBm

Bipolar transistor collector current waveform fundamental component experiences gain expansion when increasing power input drives the device into compression and quiescent collector current bias is 10 percent i_{max}. Compression occurs sharply with 10 percent i_{max} bias. Power gain is lowest at low quiescent bias currents. Transistor power gain increases with increasing quiescent collector current bias and compression characteristic becomes more soft. Quantification of compression characteristic is covered in Chapter 4 in the development of behavioral models for nonlinear amplifier characteristics. Soft compression simply means device power gain is reduced significantly by gain compression well before saturated power output is obtained. Sharp compression occurs when modest gain compression is experienced and power output is only

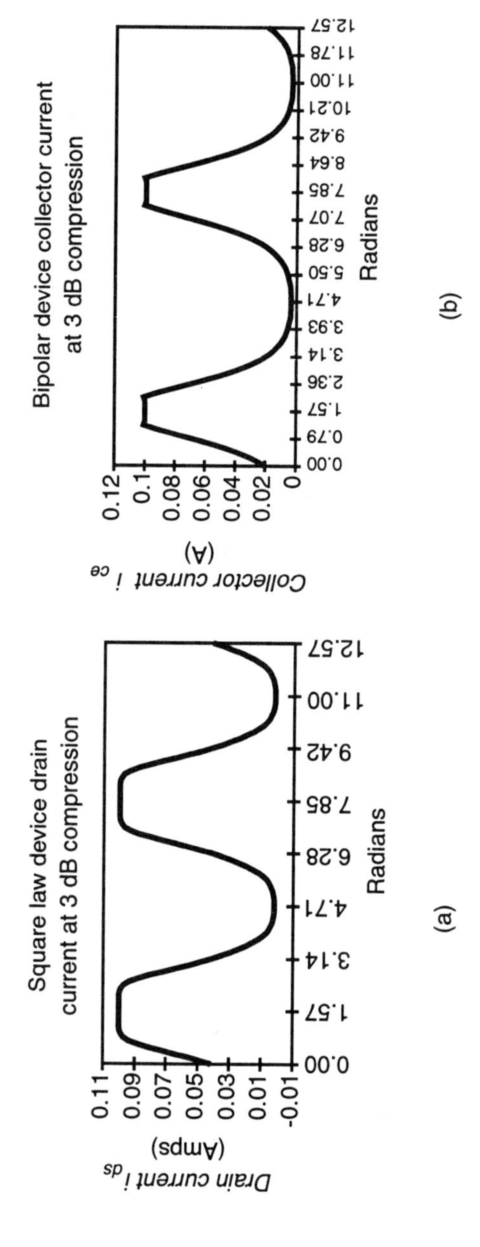

Figure 3.23 Comparison of MESFET and bipolar current waveforms at 3 dB gain compression and 50 percent i_{max} bias.

a few tenths of a dB below P_{sat}. Figure 3.24 shows collector current waveform fundamental component amplitude as a function of power input for various quiescent collector current bias levels. Notice the differences in compression characteristic for different quiescent collector currents.

The bipolar transistor's collector current waveform second harmonic content increases in power 2 dB per 1dB of input power increase at small signal levels (see Figure 3.25). This behavior is maintained where quiescent collector current is less than 40 percent i_{max} until the device goes into gain compression. Where quiescent collector current is greater than 40 percent i_{max} second harmonic component experiences a dip in power output as power input is increased. The location of the dip depends on quiescent collector current value. The higher the quiescent collector current, the earlier the dip occurs as power input is increased. This behavior of the bipolar transistor is illustrated in Figure 3.25. It is interesting to note that there is no one quiescent collector current that generates a lower second harmonic content at small signal levels.

Bipolar transistor collector current waveform third harmonic content exhibits unique behavior. At small signal levels, there is the expected 3 dB increase per 1 dB power input increase (see Figure 3.26). Lowest third harmonic content at small signal levels is experienced at quiescent collector current values in the range of 60 percent i_{max} to 80 percent i_{max}. As power input increases, the third harmonic amplitude rises sharply where quiescent collector current is greater than 50 percent i_{max}. The sharpest rise occurs at 90 percent i_{max} where at relatively low power input the slope changes from 3 dB per 1 dB power input

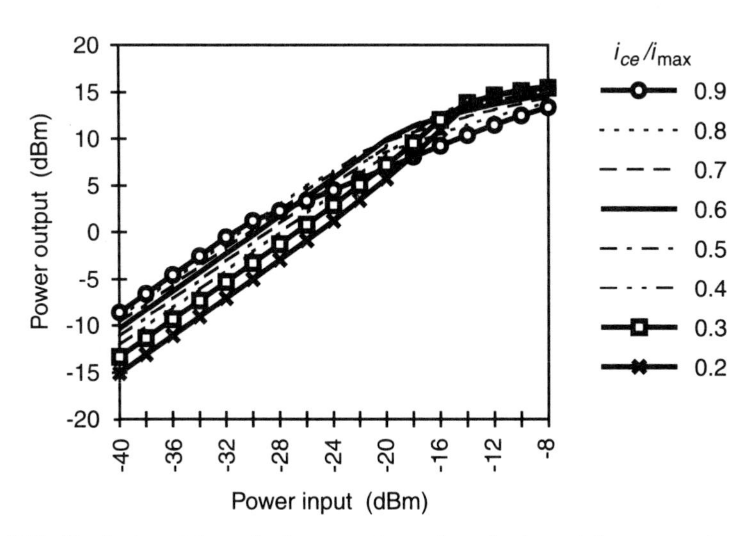

Figure 3.24 Bipolar transistor collector current waveform fundamental component as a function of power input for different quiescent collector currents.

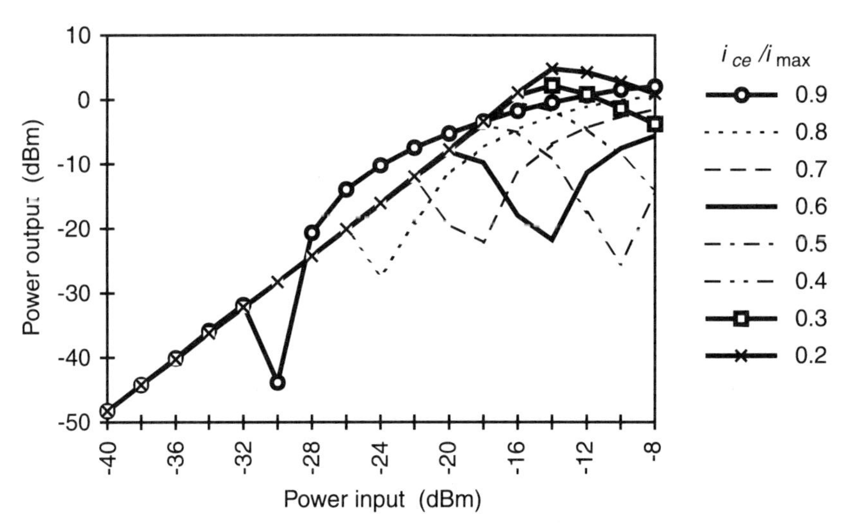

Figure 3.25 Bipolar transistor collector current waveform second harmonic component as a function of power input for different quiescent collector currents.

to 15 dB per 1 dB power input. Third harmonic component amplitude dips as the device goes into compression where quiescent collector current is less than 40 percent i_{max}. The power input at which the dip occurs is dependent on quiescent collector current value. It is clear that 50 percent i_{max} quiescent collector current offers the best compromise for low third harmonic content over the full range of power input from small signal to saturation as shown in Figure 3.26.

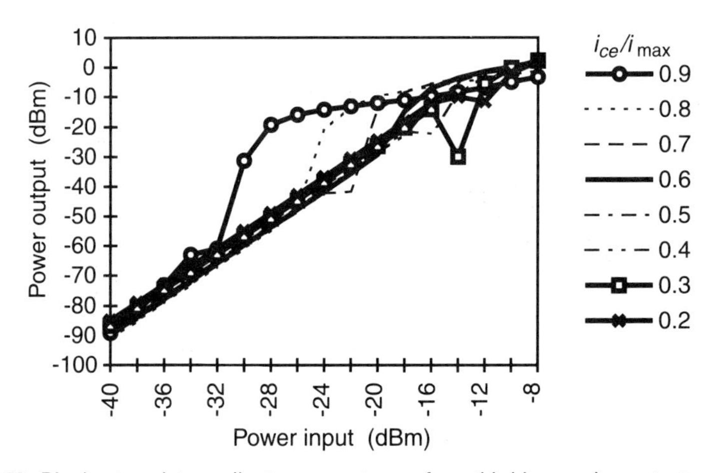

Figure 3.26 Bipolar transistor collector current waveform third harmonic content as a function of power input for different quiescent collector currents.

3.8 Summary

A technique for modeling transistor current source behavior has been developed using new curve-fitting methods described in Chapter 2. A general behavioral model has been developed for MESFET devices that can be adjusted to fit linearized step-doped as well as square law behavior. A behavioral model for bipolar transistors has also been developed. Time waveforms representing current delivered to a load from each type of transistor were calculated using the newly developed models. Transistor quiescent bias levels were adjusted for values ranging from 10 percent i_{max} to 90 percent i_{max} and input signal levels were varied from small signal linear to large signal saturated. Output current waveforms were calculated for each combination of quiescent bias and input signal level. Output waveforms were analyzed for frequency spectrum content by performing discrete Fourier transforms. Average DC current, fundamental output power, and harmonic output power were tabulated and plotted to gain understanding of relationships between bias conditions and nonlinear operations. Changes in average DC current as a function of compression depth were found to occur. Differences in compression characteristic of fundamental frequency power output were noted. Unusual harmonic behavioral at certain bias levels was discovered. All of these characteristics are found in actual device behavior. The device characteristics modeled in this chapter are used as a basis for the development of class A and class AB amplifier behavioral models in Chapters 4 and 5.

3.9 Problem

Problem 3.1

Create a spreadsheet for the general MESFET behavioral model. Use device values given in Example 3.1.

a) Plot at least two cycles of drain current waveform for power input levels from small signal linear into compression for quiescent drain current values from 10 percent i_{dss} to 90 percent i_{dss}.

b) Develop the ability to perform a 512-point DFT on 16 cycles of drain current waveform. Using the DFT output determine device fundamental component power gain as a function of power input for various quiescent drain currents.

c) Plot a three-dimensional surface of fundamental power gain as a function of power input and quiescent drain current values.

d) Using the DFT output reproduce data files for second and third harmonic component as functions of power input and quiescent drain current values. Create three-dimensional plots of second and third harmonic power as a function of power input and quiescent drain current values. The three-dimensional plot gives insight into the topology of dips in third harmonic.

References

[1] Curtice, W. R. "GaAs MESFET Modeling and Nonlinear CAD," *IEEE Transactions on Microwave Theory and Techniques*, Vol. 36, No. 2, February 1988, p. 220.

[2] Statz, H., P. Newman, I. Smith, R. Pucel, and H. Haus, "GaAs FET Device and Circuit Simulation in SPICE," *IEEE Transactions on Electronic Devices*, Vol. 34, February 1987, pp. 822–852.

[3] Ebers, J. J. and J. L. Moll, "Large Signal Behavior of Junction Transistors," *Proceedings of the IRE*, Vol. 45, December 1954, pp. 1761–1772.

4

Amplifier Behavior

4.1 Modeling the Nonlinear Class A Amplifier

Transistor models developed in Chapter 3 reveal output current waveform distortions that occur when devices are biased at different quiescent currents and driven into saturation and cutoff. Analysis of waveform fundamental and harmonic frequency content at various power input levels and different quiescent current levels is obtained by discrete Fourier transform. Data from discrete Fourier transforms (DFTs) are organized into tables from which behavior of the device as a function of power input and quiescent bias current is obtained. Waveform analysis results show fundamental power output as a function of power input consists of a linear region and a compression region. The rate at which compression occurs varies from soft to sharp depending on the quiescent bias conditions and power input. In some instances gain expansion is observed. In all cases, there is a saturated power output beyond which transistor power output cannot be driven unless power supply voltage and transistor maximum current or i_{dss} is increased. There are indications that second harmonic and third harmonic amplitudes as a function of power input behave in an unexpected manner under certain quiescent bias current and power input conditions.

When transistors are used in amplifiers, they become current sources that work into complex networks which shape amplitude response as a function of frequency. The current waveform generated by the amplifier transistor follows the behavior described in Chapter 3 and has absolute harmonic amplitudes as described, but amplitudes of its harmonic content are modified by complex matching and load impedance networks before arriving at the load. Relative amplitudes of the waveform's harmonic content at the load as a function of power input and quiescent bias current still follow the behavior described in Chapter 3.

Knowledge of specific transistor details is not always available when measured amplifier data is taken, particularly where an amplifier is packaged and sealed and has to be treated as a black box. Amplifier behavioral models have to be developed independent of transistor models developed in Chapter 3, but they still have to agree with results indicated by the transistor models developed. Fundamental power output as a function of power input has to exhibit a linear range where output increases 1 dB for every 1 dB of input power increase. As power input increases, gain compression begins to occur and power output begins to saturate. In some instances, gain expansion occurs before saturation. Phase shift is experienced as compression occurs. Compression coefficient might be soft as indicated by a very slow approach to saturation as input power increases, or compression coefficient might be sharp as indicated by a very rapid transition from linear into saturation. The only way to know what black box characteristics exist is to measure behavior as a function of power input, frequency, temperature, and any other variable such as gain control and phase control.

The nonlinear class A amplifier behavioral model needed to consolidate measured parameters is developed in this chapter. An equation giving power output as a function of power input is the basis of most of what follows in succeeding chapters. This power-out versus power-in equation includes a coefficient K that describes compression coefficient. Compression coefficient is defined and quantified. Phase-out as a function of power input is also characterized with a newly developed nonlinear behavioral model equation. These equations are unique in that their coefficients are measurable parameters such as power input, small signal gain, saturated power output, drain-to-source voltage, knee voltage, and drain current. Equations for power-added efficiency as a function of RF power input are developed where DC power varies as drain current changes with power input. Third-order intermodulation (IM) side bands are discussed, equations are developed that compute IM side band level as a function of power input and compression coefficient. IM side band growth rates as a function of power input are discussed. Dependence of IM side band growth rate on compression coefficient is shown. Noise figure rule of thumb models are proposed. The third-order intercept point is defined, and relationships between 3rdIOP, P_{1dB}, and P_{sat} are developed. Figure 4.1 illustrates all of the amplifier parameters that enter into the model, and the characteristics that are modeled. Most important is the interrelationship of all of the parameters. It is this interrelationship that facilitates the development of unique trade spaces which lead to optimum cascaded amplifier stage design in later chapters.

4.1.1 Power-Out Versus Power-In

The class A amplifier is typically biased at 50 percent i_{dss} or 50 percent i_{max} where drain or collector current does not vary significantly as a function of

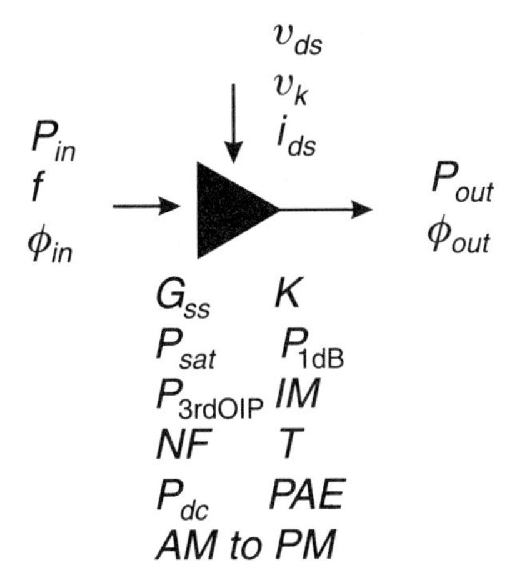

Figure 4.1 Measurable amplifier parameters and characteristics used in behavioral modeling.

power input. Small signal power output P_{out} is directly related to power input P_{in} by amplifier small signal gain G_{ss} expressed in decibel terms

$$P_{out} = P_{in} + G_{ss} \tag{4.1}$$

Equation (4.1) is a linear equation defining a line having unity slope. In practice, power output increases linearly until compression occurs, and finally saturation is obtained.

Saturated power output is easily estimated by constructing a load line over transistor I-V curves and assigning values of peak-to-peak current and peak-to-peak voltage that the output waveform could possibly obtain. Figure 4.2 shows the excursion of current and voltage on a load line biased for class A operation. Saturated power output in units of watts is easily estimated

$$p_{sat} = \frac{(v_{ds} - v_k)i_{dss}}{4} \tag{4.2}$$

The convention used here assigns lower case letter parameters to units of watts, volts, amperes, and so forth while capital letters are assigned to parameters of dBm, dB, temperature, constants, and so forth. For example, saturated power output in terms of dBm

$$P_{sat} = 10 * \log_{10}\left[\frac{(v_{ds} - v_k)i_{dss}}{4}\right] + 30 \tag{4.3}$$

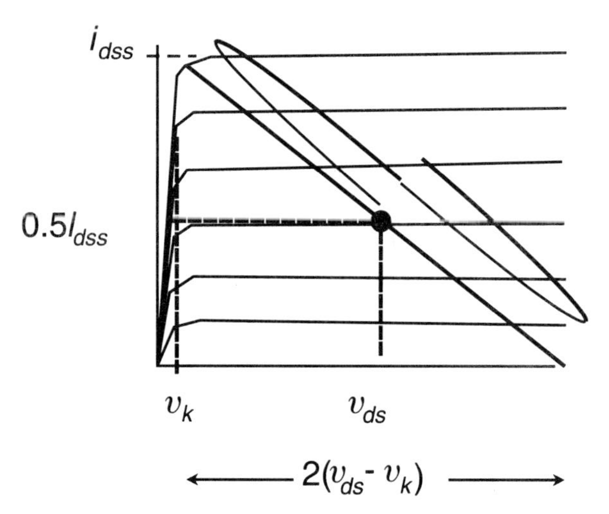

Figure 4.2 Saturated power output is easily estimated.

Saturated power output of bipolar transistors was seen to be about half that of MESFETs for the same peak-to-peak current swing due to the extremely rapid exponential rise in collector current as a function of drain voltage. The bipolar transistor collector current waveform was observed to have less power in the fundamental and more in the harmonic content (see Example 3.3).

An amplifier's power output versus power input behavioral model is obtained by constructing a graph having two asymptotes, the linear equation (4.1), and saturated power output, (4.3). Amplifier fundamental power output as a function of power input traces a curve along the linear for small signals, then transitions onto saturated power output as power input increases. Transition range *a* can be either sharp, soft, or anything between. Figure 4.3 illustrates the model concept thus far developed.

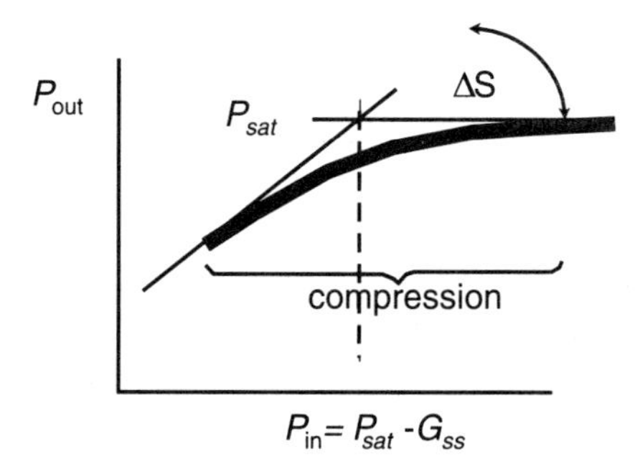

Figure 4.3 Amplifier power-out versus power input traces a curve between two asymptotes as signal level varies.

4.1.1.1 Compression Coefficient K is Defined

The asymptotes shown in Figure 4.3 are connected by a RHF having change in slope $\Delta S = -1$, transition range $a = K$, and asymptote intercept at $P_{in} = (P_{sat} - G_{ss})$. Transition range K is defined to be compression coefficient. The value of K determines how softly or sharply compression occurs. Determination of transition range K value will be quantified later. Saturated power output and small signal gain are in terms of dBm and dB respectively. The elements of an amplifier model are now defined. An equation can now be written describing power output as a function of power input.

$$P_{out} = P_{in} + G_{ss} - K\log_{10}\left[1 + 10^{\left(\frac{P_{in} + G_{ss} - P_{sat}}{K}\right)}\right] \tag{4.4}$$

The first two terms of (4.4) describe the amplifier's linear range, and the third term describes the amplifier's compression depth

$$\text{Compression} = K\log_{10}\left[1 + 10^{\left(\frac{P_{in} + G_{ss} - P_{sat}}{K}\right)}\right] \tag{4.5}$$

Compression can be either gain compression or power output compression.

4.1.1.2 Quantifying Compression Coefficient K

Compression coefficient K value is determined by measuring amplifier power output $P_{out}(K)$ when power input is set at the value $P_{in} = (P_{sat} - G_{ss})$. The logarithm argument in (4.4) then has the value of 2, and the logarithm is equal to 0.3010. Compression coefficient

$$K = \frac{P_{sat} - P_{out}(K)}{0.3010} \tag{4.6}$$

The square law MESFET used in Example 3.1, Chapter 3, exhibits differences in compression coefficient ranging from soft to sharp compression as a function of quiescent drain current. Power output $P_{out}(K)$ at $P_{in} = (P_{sat} - G_{ss})$ is interpolated from spreadsheet data files generated by waveform DFT analysis as functions of quiescent current and power input. Interpolated data is used in (4.5) to generate values for compression coefficient K. Compression coefficient K values as a function of quiescent bias point for the square law MESFET are plotted here in Figure 4.4. Notice that as quiescent bias is set for smaller percentage of i_{dss}, the value of compression coefficient K becomes smaller and compression becomes sharper.

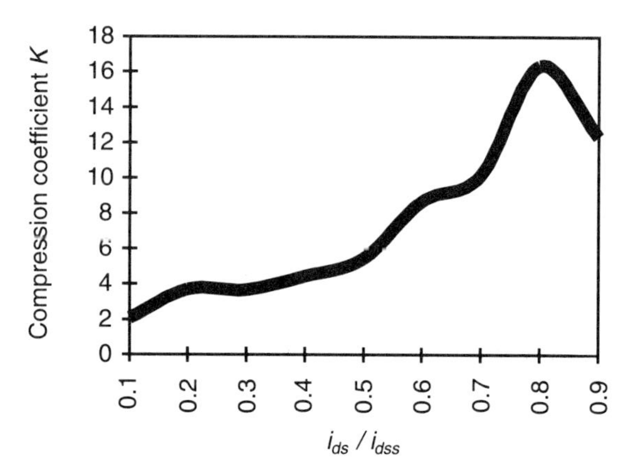

Figure 4.4 Compression coefficient K as a function of quiescent drain current for the square law MESFET of Example 3.1.

Compression coefficient K as a function of quiescent drain current for the step-doped MESFET of Example 3.2 has likewise been computed from spreadsheet data and is shown here in Figure 4.5.

The bipolar transistor used in Example 3.3 has similar compression coefficient K as a function of quiescent collector current. Compression coefficient K is largest at 90 percent i_{max}, indicating very soft compression and smallest at 20 percent i_{max}, indicating very sharp compression as seen in Figure 3.24 of Chapter 3. Compression coefficient K as a function of quiescent collector current of the bipolar transistor used in Example 3.3 is plotted in Figure 4.6.

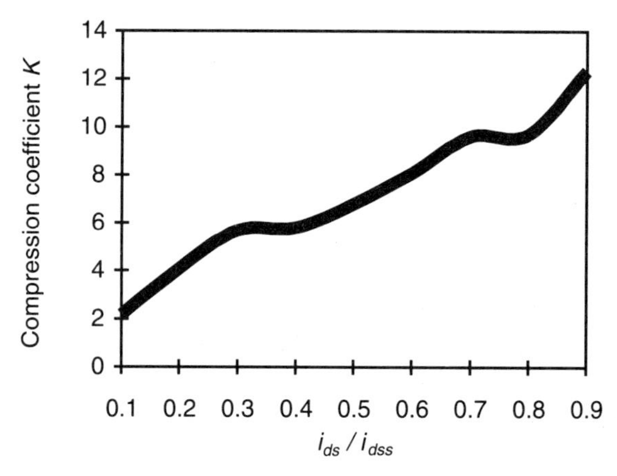

Figure 4.5 Compression coefficient K as a function of quiescent drain current for the step-doped MESFET of Example 3.2.

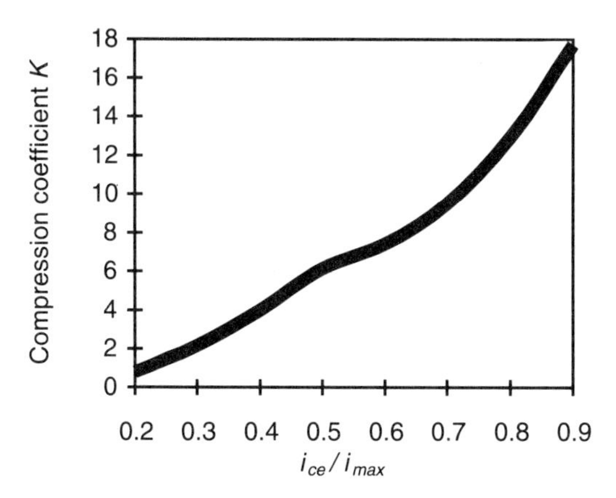

Figure 4.6 Compression coefficient K as a function of quiescent collector current of the bipolar transistor used in Example 3.3.

4.1.2 The Relationship Between P_{1dB} and P_{sat}

The recognition of differences in amplifier compression coefficients, and the definition of a compression coefficient K is particularly powerful for behavioral modeling of amplifiers. Compression coefficient is not quantified by any presently existing computer-aided design and analysis software tool. Recognition of the compression term, (4.5), gives the ability to determine a relationship between the one dB compression point P_{1dB} power output and saturated power output P_{sat}. The definition of P_{1dB} is illustrated by Figure 4.7.

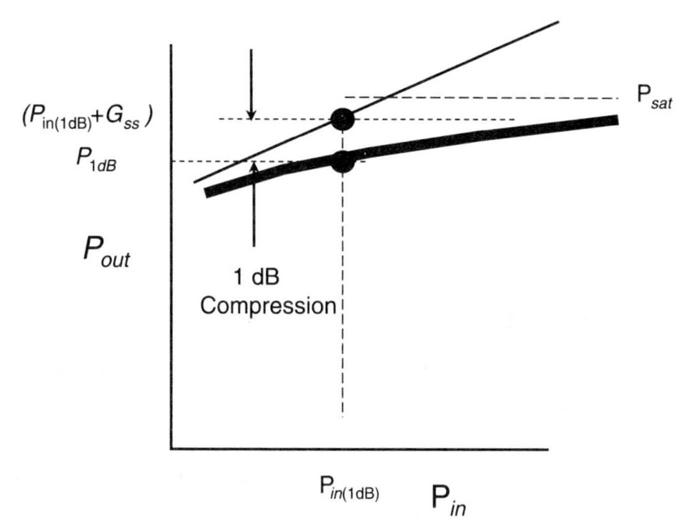

Figure 4.7 The P_{1dB} point is defined.

Determine the relationship between P_{1dB} and P_{sat} by first setting the compression term, (4.5), equal to 1, which occurs at power input $P_{in}(1dB)$

$$1 = K^* \log_{10}\left[1 + 10^{\left(\frac{P_{in}(1dB)+G_{ss}-P_{sat}}{K}\right)}\right]$$

divide both sides of the equation by K and take the antilog

$$10^{\frac{1}{K}} = 1 + 10^{\left(\frac{P_{in}(1dB)+G_{ss}-P_{sat}}{K}\right)}$$

subtract 1 from both sides, take the logarithm, then multiply both sides by K

$$K^* \log_{10}\left[10^{\frac{1}{K}} - 1\right] = P_{in}(1dB) + G_{ss} - P_{sat}$$

finally add P_{sat} to both sides to obtain

$$P_{sat} + K^* \log_{10}\left[10^{\frac{1}{K}} - 1\right] = P_{in}(1dB) + G_{ss}$$

Where power input is P_{in}(1 dB), and compression is equal to 1 dB, (4.4) gives $P_{1dB} = P_{in}(1dB) + G_{ss} - 1$. Substitute $P_{in}(1\ dB) + G_{ss}$ from the equation developed above into this equality and obtain the relationship between 1 dB compressed output power P_{1dB} and P_{sat}.

$$P_{1dB} = P_{sat} - 1 + K^* \log_{10}\left[10^{\frac{1}{K}} - 1\right] \tag{4.7}$$

This relationship is extremely useful in determining a similar relationship between P_{sat}, P_{1dB}, and the amplifier's output third-order intercept point (P_{3rdOIP}).

4.1.3 Defining Amplifier Third-Order Intercept Point

Amplifier output third-order intercept (3rd OIP) is usually determined by introducing two equal amplitude tones into an amplifier at frequencies $f1$ and $f2$ and increasing their amplitudes equally, driving the amplifier into compression. Frequency difference $f2$-$f1$ must be less than the amplifier's active band width such that both tones are equally amplified. When the two tones are increased in power input through the linear range and into compression, the amplifier's nonlinearity generates intermodulation (IM) side bands at frequencies $2f1\text{-}f2$ and $2f2\text{-}f1$. These intermodulation side bands $2f1\text{-}f2$ and $2f2\text{-}f1$ must also fall within the amplifier's active bandwidth. Amplitude of the two tones $f1$ and $f2$ in the amplifier's output increases 1 dB for every dB increase in power input through the linear range. In the ideal amplifier, amplitude of the intermodulation side bands at frequencies $2f1\text{-}f2$ and $2f2\text{-}f1$ increases 3 dB for every 1 dB increase in power input.

Amplifier output third-order intercept point is a value constructed by projecting third-order intermodulation side band amplitude growth at either $2f1\text{-}f2$ or $2f2\text{-}f1$ at a 3 dB per 1 dB slope and projecting linear signal $f1$ and $f2$ output amplitude growth at a 1 dB per dB slope to a point of intersection. That point of intersection is defined to be the amplifier's output 3rd OIP and is valued at whatever output power level the intersection occurs.

Consider what is happening as a result of amplifier nonlinearity. The two signals at frequencies $f1$ and $f2$ represented by vectors $\underset{e_{c1}}{\longrightarrow}$ and $\underset{e_{c2}}{\longrightarrow}$ are compressed by coherent intermodulation vectors also at frequencies $f1$ and $f2$ $\underset{e_{im}}{\longrightarrow}$, but having smaller amplitudes and having opposite sign. Figure 4.8 illustrates the two tones $f1$ and $f2$ and the intermodulation (IM) side bands attributed to each carrier due to the nonlinearity. Each $f1$ and $f2$ vector is compressed to a new amplitude

$$\underset{e_{r1}}{\longrightarrow} = \underset{e_{c1}}{\longrightarrow} + \underset{e_{m}}{\longleftarrow}$$

and the reduction in power output at $f1$ and $f2$ due to compression is

$$\Delta dB = 20 * \log_{10}\left[\frac{e_{c1} - e_{im}}{e_{c1}}\right] \tag{4.8}$$

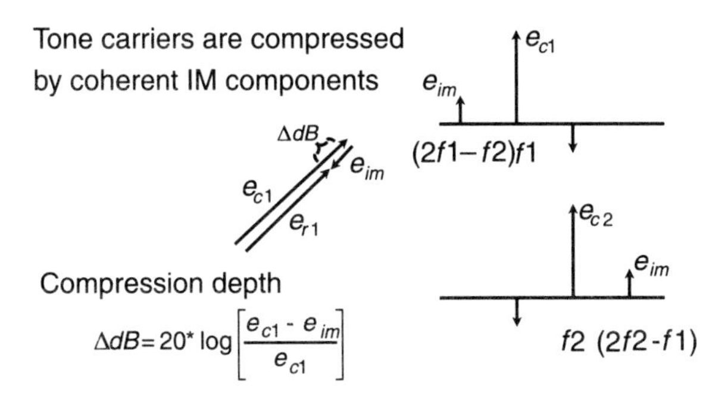

Figure 4.8 Tone carriers are compressed by coherent IM side bands.

Amplifier compression, the negative value of compression depth from (4.5), and reduction in tone output power due to coherent IM side band component compression, (4.8), can be set equal.

$$20 * \log_{10}\left[\frac{e_c - e_{im}}{e_c}\right] = -K * \log_{10}\left[1 + 10^{\left(\frac{P_{in} + G_{ss} - P_{sat}}{K}\right)}\right]$$

where P_{in} is total power in both tones in dBm. Solve this equality for the decibel value of the ratio (e_{im}/e_c). Divide both sides of the equation by 20 and take the anti-logarithm.

$$1 - \frac{e_m}{e_c} = \left[1 + 10^{\left(\frac{P_{in} + G_{ss} - P_{sat}}{K}\right)}\right]^{\frac{-K}{20}}$$

Subtract 1 from both sides, multiply both sides by minus 1, and take the logarithm, finally multiply both sides by 20 to obtain the ratio (e_{im}/e_c) in decibels

$$20 * \log_{10}\left[\frac{e_{im}}{e_c}\right] = 20 * \log_{10}\left\{1 - \left[1 + 10^{\left(\frac{P_{in} + G_{ss} - P_{sat}}{K}\right)}\right]^{\frac{-K}{20}}\right\} \tag{4.9}$$

Consider the difference in amplitude $IM\Delta dB$ between the output tones at $f1$ and $f2$ and the IM side bands at $2f2$-$f1$ and $2f1$-$f2$ where amplitudes of $f1$ and $f2$ are small signal, well below compression, and resulting IM side bands are so small that compression of tones $f1$ and $f2$ is negligible. Figure 4.9 illustrates such a condition by showing the summation of the two spectra illustrated in Figure 4.8. The difference in power level between the tones and the IM side bands expressed in decibels is

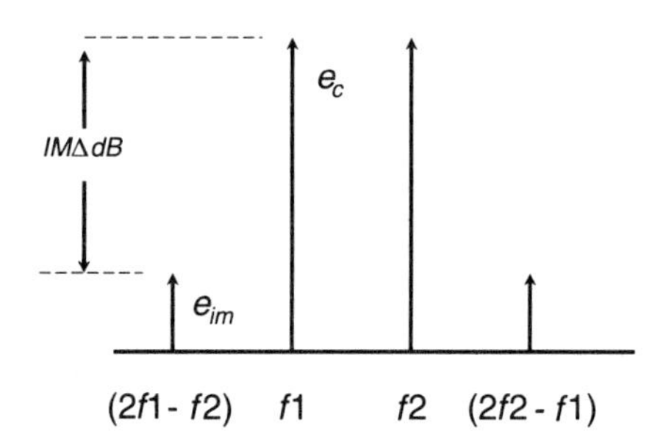

Figure 4.9 Amplifier output spectrum showing two tones and IM side bands.

$$IM\Delta dB = 20 * \log_{10}\left[\frac{e_m}{e_c}\right] \quad (4.10)$$

The difference in IM side band amplitude e_{im} and tone amplitude e_c expressed in dB shown in Figure 4.9 and defined by (4.10) is equal to the ratio derived in (4.9). Equation (4.9) describes third-order intermodulation side band level as a function of total two-tone power input P_{in}, amplifier small signal gain G_{ss}, saturated power input P_{sat}, and compression coefficient K. This equation is a significant tool for understanding why amplifier third-order IM side bands do not always increase at a rate of 3 dB per each dB increase of power input. Consider the value of (4.9) where combined power input of the two tones causes 1 dB compression. The compression term of (4.4) has value

$$-1 = -K\log_{10}\left[1 + 10^{\left(\frac{P_{in} + G_{ss} - P_{sat}}{K}\right)}\right]$$

which when divided on both sides by 20 can be expressed as

$$-\frac{1}{20} = \log_{10}\left\{\left[1 + 10^{\left(\frac{P_{in} + G_{ss} - P_{sat}}{K}\right)}\right]^{\frac{-K}{20}}\right\}$$

When the anti-logarithm of this equality is taken, a value for the logarithm's argument is found to be

$$10^{\frac{-1}{20}} = \left[1 + 10^{\left(\frac{P_{in} + G_{ss} - P_{sat}}{K}\right)}\right]^{\frac{-K}{20}} = 0.8913 \quad (4.11)$$

Equation (4.11) is extremely significant, for it says that regardless of the value of compression coefficient K, the value of the expression in the brackets raised to the $\frac{-K}{20}$ power is 0.8193 at 1 dB compression. This expression is embedded in (4.9). When power input to any amplifier creates 1 dB compression, the expression in (4.11) equals 0.8913, and (4.9) becomes

$$IMdB = 20 * \log10[1 - 0.8913]$$

$$IMdB = -19.27 \text{ dB}$$

Figure 4.10 illustrates the significance of this result. The third-order intercept point is defined to be the projected intersection of linear output power growth at 1 dB per dB of input power and of IM sideband power growth at 3 dB per dB of input power. When the amplifier is compressed 1 dB, this intersection point lies at a projected linear power output that is 9.635 dB higher than the calculated uncompressed power output at P_{1dB}, see Figure 4.10. The difference of 9.635 dB between uncompressed power output at P_{1dB} and the 3rd OIP is half the decibel difference between the uncompressed power output at P_{1dB} and the computed IM sideband, which is 19.27 dB less than the uncompressed calculated power output.

The 3rd OIP then is 10.635 dB greater than power output at P_{1dB}. This is a significant result because at the 1 dB compression point, the compression coefficient K is not a determining factor in the calculation of the IM sideband level as indicated by (4.11).

The only way 3rd OIP of amplifiers having different compression coefficient K values can be compared is to find the P_{1dB} power output and add 10.635 dB. Additional support for this concept is realized by considering the magnitude of the IM component that is capable of compressing carriers $f1$ and $f2$ by 1 dB. A unity vector reduced 1 dB has amplitude 0.8913. The magnitude of an opposing vector that reduces the unity vector by 1 dB is 0.1087. Magnitude of this opposing vector relative to the unity vector is −19.275 dB. These dynamics are true for every amplifier that is compressed P_{1dB} regardless of amplifier compression coefficient K. Projection of linear power output and IM side band power output from the IM magnitude at P_{1dB} gives a 3rd OIP exactly 10.63 dB greater than the P_{1dB} power output as shown in Figure 4.10.

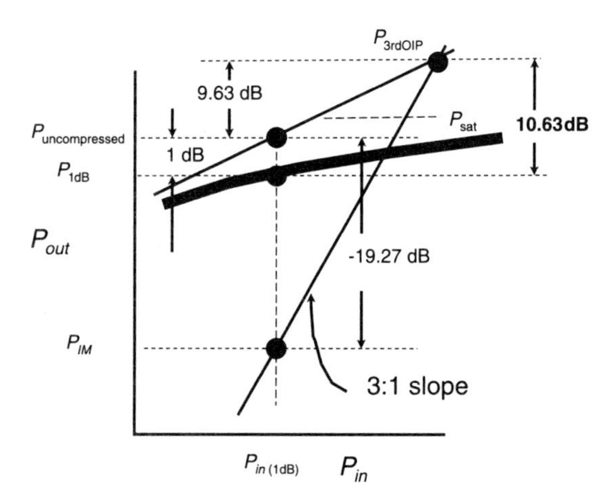

Figure 4.10 The relationship between P_{1dB} and the 3rd OIP.

Equation (4.9) gives a powerful convincing argument that an amplifier's 3rd OIP should only be defined relative to the amplifier's P_{1dB} point. An amplifier's third-order intercept can be quantified as

$$P_{3rdOIP} = P_{1dB} + 10.63 \text{ dB}$$

Substituting Equation (4.7) for P_{1dB} into this relationship gives an equation for amplifier output third-order intercept

$$P_{OIP} = P_{sat} + 9.63 + K\log_{10}\left[10^{\left(\frac{1}{K}\right)} - 1\right] \quad (4.12)$$

as a function of saturated power output and compression coefficient K. Amplifier input third-order intercept P_{IIP} is a useful quantity in calculations where cascaded amplifiers are involved. Input intercept point is related to output intercept by amplifier small signal gain

$$P_{IIP} = P_{sat} + 9.63 + K\log_{10}\left[10^{\left(\frac{1}{K}\right)} - 1\right] - G_{ss} \quad (4.13)$$

Third-order intermodulation side band growth rate is not always 3 dB per 1 dB increase of input power. This is illustrated by using (4.9) to calculate IM sideband power level as a function of power input P_{in} for different compression coefficient K values.

Realize that (4.9) calculates intermodulation side band to carrier ratio. Intermodulation side band absolute power level is the power level of the amplifier's output power plus IM ratio as defined by (4.9). If an increase in IM side band absolute power level of 3.0 dB is observed (as seen in a spectrum analyzer display) for a 1.0 dB increase in amplifier output power level, the IM ratio value as calculated by (4.9) has increased only 2.0 dB.

Use a Spreadsheet to Experiment with Behavioral Models

Construct a spreadsheet that calculates amplifier power output as a function of power input using (4.4). Select arbitrary G_{ss} and P_{sat} values. Arrange P_{in} and P_{out} values in columns. Increment P_{in} values 1.0 dB per spreadsheet cell with values ranging from small signal linear to saturated power output. Design the spreadsheet such that compression coefficient K value is easily varied and is automatically copied into all equations that use K. Add a column that calculates amplifier compression depth using (4.5). Add a column that calculates IM sideband ratio using (4.9). Add another column that calculates IM sideband absolute power level by adding P_{out} and IM sideband ratio. Add yet another column that calculates the difference in IM sideband absolute power level between each two contiguous cells. This last column calculates the IM sideband power level growth rate per dB of increase in power output. Now enter various compression coefficient K values ranging from 2.5 to 12 and note the IM sideband power level growth rate per decibel of power output increase. Table 4.1 illustrates results of such an experiment.

The expected IM side band rate of increase of 3 dB per dB input power is realized only when $K = 10$. Rate of IM side band increase is greater than 3 dB per dB where $K < 10$. Figure 4.11 shows the result of plotting IM level as a function of power in P_{in} for different K values for an amplifier having 25 dB small signal gain and +20 dBm saturated power output. Actual amplifier measured data supports the results shown in Figure 4.11 where amplifier compression coefficient K is quantified by using the technique described by (4.6).

4.1.4 Amplifier Stage Phase Shift as a Function of Power Input

The fundamental component in the amplifier's output experiences phase shift as the amplifier stage is driven into compression. This behavior is evident as AM to PM conversion where an amplitude modulated signal is applied to an amplifier stage that is driven into compression. The phase sensitivity θ of a single amplifier stage is typically five degrees per decibel of compression. Phase shift φ develops rapidly if the compression coefficient K is small, and slowly if the compression coefficient K is large. Phase sensitivity θ remains constant

Table 4.1
IM Sideband Power Growth Rate Versus Compression Coefficient K Value

K	Small Signal IM Growth Rate Per Decibel Increase In P_{out}
2.5	9.00
3.5	6.71
4.5	5.44
5.5	4.64
6.5	4.08
7.5	3.67
8.5	3.35
9.5	3.10
10.5	2.90
11.5	2.74

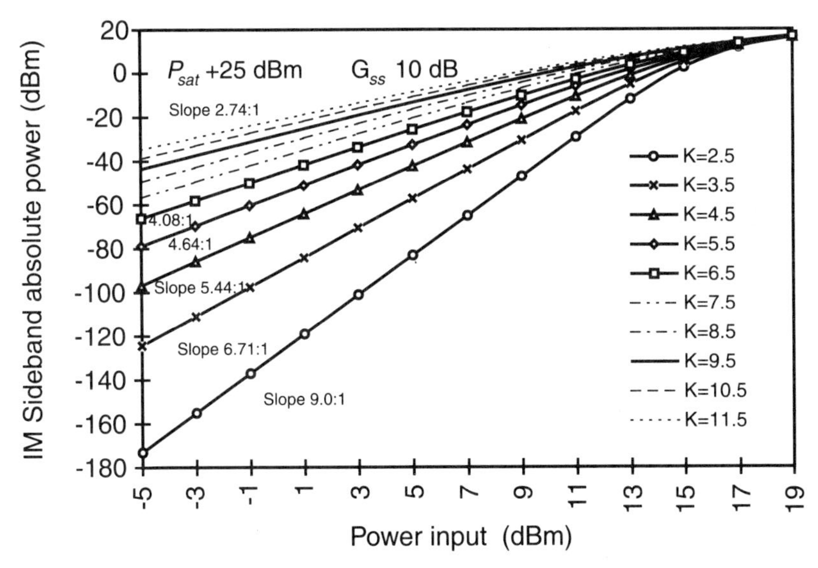

Figure 4.11 Growth rate of third-order IM side bands depends on compression coefficient K value.

through moderate compression depth (up to 8 dB) regardless of the value of K. A behavioral model that gives phase shift as a function of power input is

$$\varphi = \theta K \log_{10}\left[1 + 10^{\left(\frac{P_{in} + G_{ss} - P_{sat}}{K}\right)}\right] \tag{4.14}$$

4.1.5 Defining Power-Added Efficiency

Amplifier power-added efficiency (PAE) is defined as the ratio of difference in signal power output minus signal power input, to power consumed from a prime power source

$$PAE = \frac{p_{out} - p_{in}}{p_{dc}} * 100\% \tag{4.15}$$

Where signal power is expressed in dBm, (4.15) can be expressed as

$$\text{PAE} = \frac{10^{\left(\frac{P_{out} - 30}{10}\right)} - 10^{\left(\frac{P_{in} - 30}{10}\right)}}{v_{dc} * i_{dc}} * 100\% \tag{4.16}$$

prime power consumption by class A amplifiers and is nearly constant as a function of signal power input. As the amplifier goes into compression, power output ceases to increase as rapidly as power input and power-added efficiency peaks as shown in Figure 4.12.

4.1.6 Estimating Amplifier Noise Figure

The noise figure of an amplifier stage is dependent on many variables too complex to treat with a rigorous behavioral model. Observation of measured data, however, leads to the conclusion that there is an empirical relationship between current drawn from a prime power source by the amplifier's transistor and the transistor's noise figure expressed in decibels. That rule of thumb relationship is

$$NF \approx 10\sqrt{i_{\mathrm{dc}}} + 0.5 \tag{4.17}$$

where i_{dc} is expressed in amperes. An amplifier noise figure is the transistor's noise figure plus loss in decibels of the input impedance match network modified by noise figure of circuitry on the output side of the amplifier. Amplifier noise factor is defined to be the power ratio of noise figure

$$F = 10^{\frac{NF}{10}} \approx 10^{(\sqrt{i_{\mathrm{ds}}}+0.05)} \tag{4.18}$$

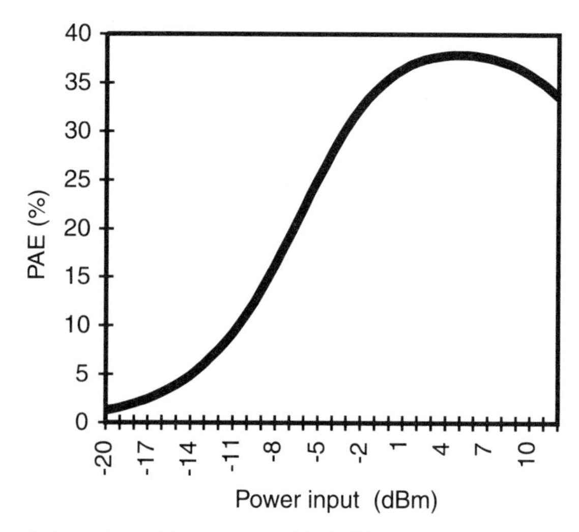

Figure 4.12 Typical class A amplifier power-added efficiency.

Example 4.1 Determine a Relationship Between Compression Coefficient K and $(P_{sat} - P_{1dB})$.

Equation (4.7) defines P_{1dB} in terms of P_{sat} and compression coefficient K. Here it is rearranged to give a relationship between $(P_{sat} - P_{1dB})$ and compression coefficient K

$$(P_{sat} - P_{1dB}) = 1 - K\log_{10}\left[10^{\left(\frac{1}{K}\right)} - 1\right] \tag{4.19}$$

A direct solution of (4.19) for K is not possible, but values of $(P_{sat} - P_{1dB})$ that result from specific values of compression coefficient K can be determined and plotted. A spread sheet is created to calculate differences between P_{sat} and P_{1dB} for various values of compression coefficient K. The computed values are shown in Table 4.2 and are plotted in Figure 4.13.

Example 4.2 A Class A Amplifier

Given: A class A amplifier has a transistor that provides 18 dB small signal gain, operates from a prime power source of 4 volts, and draws 0.025 amperes. One

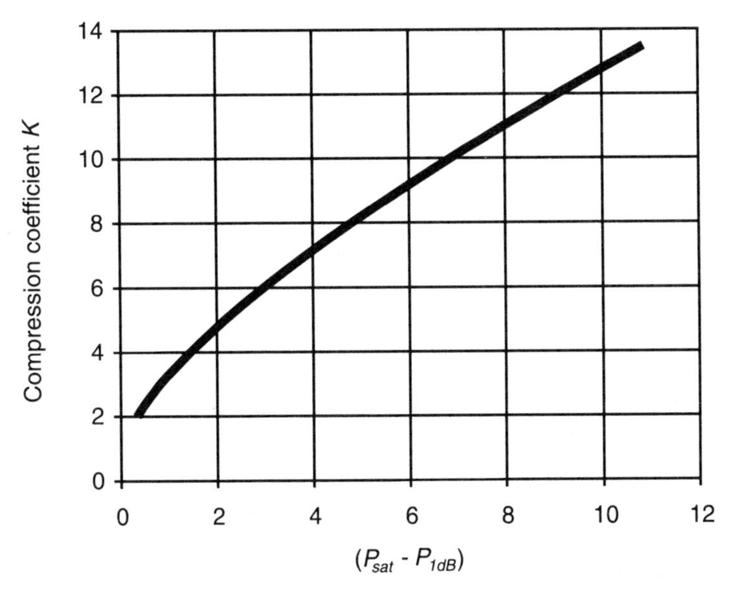

Figure 4.13 Compression coefficient K as a function of $(P_{sat} - P_{1dB})$.

Table 4.2
Calculated Values

$(P_{sat} - P_{1dB})$	K
0.33	2.00
0.55	2.50
0.81	3.00
1.11	3.50
1.44	4.00
1.79	4.50
2.16	5.00
2.56	5.50
2.98	6.00
3.41	6.50
3.87	7.00
4.33	7.50
4.82	8.00
5.31	8.50
5.82	9.00
6.34	9.50
6.87	10.00
7.41	10.50
7.96	11.00
8.52	11.50
9.10	12.00
9.68	12.50
10.27	13.00
10.86	13.50

dB compression occurs at a power output $P_{1dB} = +12.7$ dBm. Noise figure of stages following the amplifier is 3.0 dB. Amplifier input impedance match network loss is -0.7 dB.

Determine: Saturated power output, output 3rd OIP, input 3rd OIP, maximum power-added efficiency, noise figure, and compression coefficient.

Procedure: Saturated power output is determined by assuming the transistor has a knee voltage of 1 volt, then calculating output signal peak-to-peak current and peak-to-peak voltage swings.

$$p_{sat} = \frac{2(v_{dc} - 1)2i_{dc}}{8} = \frac{(4 - 1)0.025}{2} = 0.0375 \text{ watts}$$

Converted to decibels relative to 1 milliwatt

$$P_{sat} = 15.74 \text{ dBm}$$

Compression coefficient K is determined from Figure 4.13 in Example 4.1. The difference between P_{sat} and P_{1dB} for this amplifier is 3.0 dB, which gives a value of $K = 6.0$ from Figure 4.13.

Output 3rd OIP is determined using (4.12)

$$P_{OIP} = P_{sat} + 9.63 + K\log_{10}\left[10^{\left(\frac{1}{K}\right)} - 1\right]$$

$$= 15.74 + 9.63 + 6\log_{10}\left[10^{\left(\frac{1}{6}\right)} - 1\right] = +23.39 \text{ dBm}$$

Input 3rd OIP, $P_{IIP} = P_{OIP} - G_{ss} = 23.39 - 18 = +5.39$ dBm.

Noise figure of the amplifier's transistor is estimated to be $NF = 10\sqrt{0.025} + 0.5 = 2.08$ dB. Noise figure of following stages adds to the transistor noise figure as calculated by

$$NF = 10\log_{10}\left[10^{\frac{2.08}{10}} + \frac{10^{\frac{3.0}{10}} - 1}{10^{\frac{18}{10}}}\right] + 0.7 = 2.82 \text{ dB}$$

Amplifier power-added efficiency as a function of power input is calculated using (4.15) to develop a spreadsheet table. The data is plotted in Figure 4.14 to show maximum PAE and the power input level at which it occurs.

4.2 Summary

New curve-fitting techniques developed in Chapter 1 and results obtained from transistor behavioral models developed in Chapter 2 have been used to create new behavioral models of class A amplifiers. Three key model equations, one describing power output as a function of power input, another describing phase shift as a function of power input, and a rule of thumb for estimating amplifier noise figure as a function of average bias current, are the basis for all of the relationships developed for the class A amplifier. A unique compression coefficient

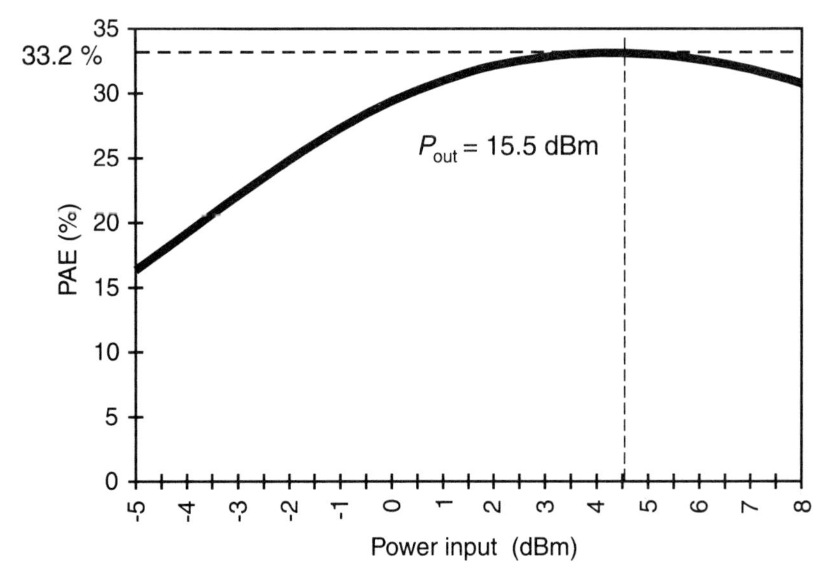

Figure 4.14 Amplifier power-added efficiency peaks at 33.2 percent.

K, was defined. This compression coefficient is needed to model differences in amplifier compression that are observed in actual circuitry and are predicted by the transistor behavioral models developed in Chapter 3. Compression coefficient *K* is found to be a key parameter that allows relationships to be developed between all of the nonlinear parameters P_{1dB}, P_{sat}, P_{OIP} and third-order intermodulation side band levels. Methods for determining the value of *K* have been described. An extremely important equation was developed that calculates IM side band level as a function of two-tone power input. The relationship was shown of IM side band growth rate and the value of compression coefficient *K* as power input increases. Amplifier power-added efficiency was defined, and the dynamic relationship between *PAE* and power input was illustrated.

The nonlinear amplifier relationships developed in this chapter demonstrate the power of behavioral modeling to tie together system, subsystem, circuit, and device parameters with closed-form equations providing a means of communication and parameter flow down that heretofore did not exist. With these tools, the systems engineer can find assurance that requirements developed at the system level can be satisfied by subsystem hardware developers, subsystem engineers can be assured that their flow-down requirements can be satisfied by circuit designers and that real devices can be found to satisfy circuit developer's needs.

4.3 Problems

Problem 4.1

What is the typical compression coefficient K for transistors biased at 50 percent of i_{dss}, or i_{max}?

Problem 4.2

What bias level as a percent of i_{dss}, or i_{max} is expected to give compression coefficient $K = 6.66$ for MESFETs, for spike-doped MESFETs, for bipolar transistors?

Problem 4.3

What is the expected growth rate of IM side bands for the amplifier in Problem 4.2 above as power input is increased 1 dB?

Problem 4.4

A prime power source of 8 volts is available. What current is required of an 18 dB small signal gain class A MESFET amplifier required to have $P_{1dB} = +20$ dBm? What noise figure will it be expected to have? What P_{OIP}? What P_{sat}?

Problem 4.5

A prime power source of 8 volts is available. What current is required of an 18 dB small signal gain class A step-doped MESFET amplifier required to have $P_{1dB} = +20$ dBm? What noise figure will it be expected to have? What P_{OIP}? What P_{sat}?

Problem 4.6

A prime power source of 8 volts is available. What current is required of an 18 dB small signal gain class A bipolar amplifier required to have $P_{1dB} = +20$ dBm? What noise figure will it be expected to have? What POIP? What P_{sat}?

5

Power Amplifier Behavior

5.1 Class AB Amplifiers

The class A amplifier model developed in Chapter 4 assumes characteristic class A amplifier behavior by exhibiting a relatively stable average bias current, and stable RF gain as signal level is varied from small signal, linear, to large signal, and into saturation. Class A amplifiers are biased for quiescent current approximately 50 percent of i_{max} or i_{dss}. The class A amplifier does not normally exhibit gain expansion as RF input signal level is increased. Gain expansion is defined to be an increase in small signal gain resulting from an increase in average bias current.

Class AB amplifiers, the topic of this chapter, do not exhibit the same stable average DC current and gain as a function of input signal level. The class AB amplifier is typically biased for quiescent current of 10 percent to 30 percent of i_{max} or i_{dss} to improve power-added efficiency, to reduce heat dissipation when signal levels are small, and in some cases, where load impedance is lower, to obtain higher-power output with lower heat dissipation. Under certain quiescent bias conditions, the class AB amplifier does exhibit gain expansion as average bias current changes when RF signal input power is increased.

Data developed from DFT analysis of transistor output current waveforms (see behavioral model, Examples 3.1 through 3.3 in Chapter 3) show that average DC current increases as input signal ranges from small signal, linear, to large signal, and into saturation when quiescent bias is between 10 percent and 40 percent i_{max} or i_{dss}. The class AB amplifier average DC current approaches 50 percent of i_{max} in the limit where the transistor is driven to hard saturation. This causes the output current waveform to approach a square wave. Output current waveform DFT analysis also indicates moderate gain expansion in

MESFET devices biased at 10 percent and 20 percent of i_{dss} as RF input power is increased and output power approaches saturation. Bipolar transistors exhibit moderate gain expansion where quiescent bias is 40 percent to 60 percent of i_{max} and large gain expansion where quiescent bias is 20 percent of i_{max}. Bipolar transistor gain is directly proportional to average bias current.

Class AB amplifier heat dissipation is a dynamic function of power input as average bias current increases and power output increases. Class AB amplifier power-added efficiency behavior is more dynamic than class A *PAE* because DC power consumption and gain are both changing as power input increases. Class AB amplifier noise figure, third-order intercept, and intermodulation side band levels are also more dynamic than that of class A amplifiers due to change in average bias current.

Techniques for modeling these class AB amplifier characteristics are developed in this chapter. The class AB amplifier behavioral model is developed with power input normalized to saturated power output and average DC current normalized to i_{dss} or i_{max} in order to be universally applicable.

5.2 Basis for the Class AB Amplifier Behavioral Model

Even though class AB amplifiers differ significantly from class A amplifiers in the way average DC bias current and small signal gain vary as a function of signal power input, class AB amplifier behavior of power-out versus power-in and phase-out versus phase-in can be modeled by the general amplifier behavioral model equations for power output

$$P_{out} = P_{in} + G_{ss} - K\log_{10}\left[1 + 10^{\left(\frac{P_{in} + G_{ss} - P_{sat}}{K}\right)}\right] \tag{5.1}$$

and phase shift

$$\varphi_{out} = \theta K\log_{10}\left[1 + 10^{\left(\frac{P_{in} + G_{ss} - P_{sat}}{K}\right)}\right] + \varphi_{in} \tag{5.2}$$

defined in Chapter 4. Compression coefficient K is a dynamic parameter in class AB amplifiers. A behavioral model of compression coefficient K as it varies as a function of quiescent bias coefficient will be developed. Small signal gain G_{ss} in (5.1) and (5.2) is a dynamic parameter for the class AB amplifier where gain is a function of average bias current. A complete class AB amplifier behavioral model needs an equation to describe average DC bias current $i_{ds}(b, P_{in})$ as a function of RF signal power input and bias condition, and an equation to relate gain to average bias current $G(b, i_{ds})$. A new dynamic small signal gain equation $G(b, P_{in})$, the product of $i_{ds}(b, P_{in})$ and $G(b, i_{ds})$, will be developed and substituted for the parameter G_{ss} to account for gain variation as a function of RF power input in the above amplifier model equations. These new equations will be used to calculate the dynamic parameters of heat dissipation and power-added efficiency. Several new parameters useful in describing and modeling class AB amplifier behavior are now developed.

5.2.1 New Parameters for Class AB Amplifier Behavioral Models

Useful parameters in discussing class AB amplifiers are the RF load impedance and load line drawn on the transistor's I-V curves and a definition of average bias coefficient b_{avg}. Additional useful parameters are normalized power input $(P_{in} - P_{sat})$, normalized average current $\frac{i_{dc}}{i_{dss}}$, and output power P_K where compression begins to occur. These parameters are defined and developed in this section.

5.2.1.1 Class AB Amplifier Quiescent Bias and Load Line

Amplifier load line impedance value can vary by design from high impedance for highest possible gain to low impedance for highest possible power output. Amplifier-saturated power output, average bias current at saturated power output, quiescent bias current, and bias voltage are measurable amplifier parameters that can reveal much about the load line impedance value. If an amplifier develops saturated power output P_{sat} with v_{dc} applied, it can be assumed that the transistor has knee voltage $v_k = 1$ volt, and that peak-to-peak RF output current is

$$i_{p-p} = \frac{4 * 10^{\left(\frac{P_{sat} - 30}{10}\right)}}{(v_{dc} - v_k)} \tag{5.3}$$

Average amplifier current drawn from the voltage source v_{dc} with no RF power input is the quiescent bias current i_q. When the load line impedance value is high, saturated power output is low, i_{p-p} is small and average bias current does not vary from value i_q. As load line impedance value decreases, i_{p-p} increases, and saturated power output increases. When i_{p-p} becomes greater than $2i_q$, average bias current begins to increase. Maximum average bias current obtainable at saturated power output is $i_{dc} = 0.5 i_{p-p}$.

A load line selected for power output has nominal impedance

$$R_l = \frac{2(v_{ds} - v_k)}{i_{dss}} \quad \text{or} \quad R_l = \frac{2(v_{ce} - v_k)}{i_{max}} \tag{5.4}$$

where dividing by i_{dss} applies to MESFET devices and dividing by i_{max} applies to bipolar transistors. Figure 5.1 illustrates typical class AB amplifier quiescent bias conditions. The value b illustrated in Figure 5.1 is the ratio i_q/i_{dss} or i_q/i_{max}.

5.2.1.2 Quiescent Bias Coefficient

Quiescent bias coefficient

$$b_q = \frac{i_q}{i_{dss}} \tag{5.5}$$

is the ratio of quiescent bias current i_q (no signal input) to the peak saturation current i_{dss} or i_{max} that flows when the loaded transistor is driven into saturation. Selection of the best value for quiescent bias coefficient b_q depends on specific transistor characteristics. A good starting value is $b_q = 0.2$. An optimum value of b_q will be discovered as the class AB amplifier behavioral model develops and advantages of specific values become evident. For instance, power-added efficiency of MESFET transistors biased class AB tends to peak when $0.1 \leq b_q \leq 0.3$.

5.2.1.3 Average Bias Coefficient

Average bias coefficient

$$b_{avg} = \frac{i_{avg}}{i_{dss}} \tag{5.6}$$

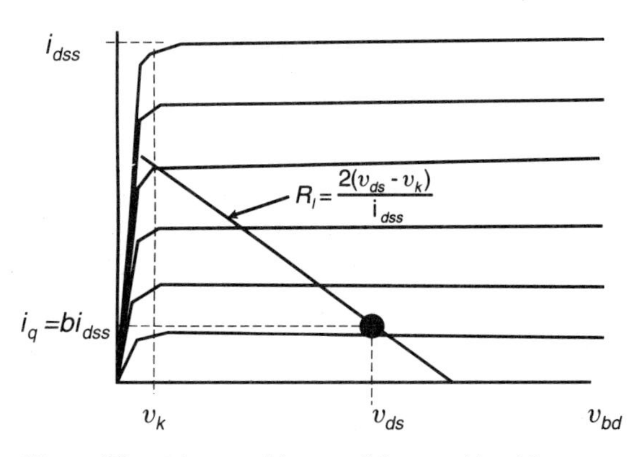

Figure 5.1 Class AB amplifier quiescent bias conditions and load line.

is the bias coefficient that results when RF power input causes average DC current through the class AB biased transistor to change from the quiescent value b_q. Average bias coefficient value always equals quiescent bias coefficient value when there is no RF present. Maximum value of average bias coefficient b_{avg} = 0.5 occurs at saturated power output in power amplifiers where load line impedance value is low and peak instantaneous RF current is i_{dss} or i_{max}. Average bias coefficient is a function of power input and the load line impedance.

5.2.1.4 Saturated Power Output

Saturated power output at the fundamental frequency for the class AB biased transistor is defined at the point where output current swing is maximum along the RF load line as illustrated in Figure 5.2. Its value in units of watts is the product of peak-to-peak voltage and peak-to-peak current divided by 8

$$p_{sat} = \frac{2(v_{ds} - v_k)i_{p-p}}{8} \quad \text{or} \quad p_{sat} = \frac{(v_{ce} - v_k)i_{dss}}{4} \text{ watts} \tag{5.7}$$

For behavioral model use, express the value of saturated power output in decibels relative to one milliwatt

$$P_{sat} = 10 * \log_{10}\left[\frac{(v_{ds} - v_k)i_{dss}}{4}\right] + 30 \text{ dBm} \tag{5.8}$$

DFT analysis of bipolar transistor output current waveform at saturated power output indicates saturated power output is 3 dB to 4 dB less than that

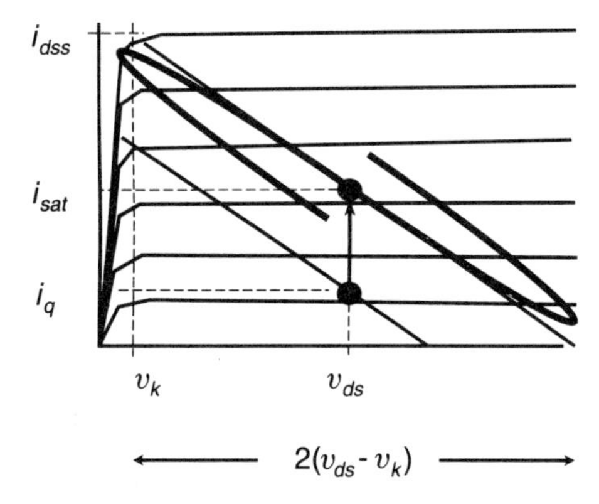

Figure 5.2 Output RF current swing at saturated power output.

calculated by (5.8). The shortfall in bipolar saturated power output is due to the exponential curvature of the output RF current waveform (see Figure 3.23). The saturated bipolar transistor power output waveform just does not have the same peak-to-peak to RMS voltage ratio for fundamental and harmonic components as is found in the saturated MESFET waveform.

5.2.1.5 Defining P_K, Power Output as Compression Begins

A useful definition in the discussion of class AB amplifiers is the output power level P_K at which compression of power output begins. Compression coefficient K has been defined as a parameter that expresses the softness or sharpness with which an amplifier goes into compression. Figure 5.3 further clarifies the definition of K.

The parameter K as used here is the RHF transition range as defined in Chapter 2, Section 2.1.1, and Figure 2.4. Referring to Figure 5.5, it is clear that the power input value P_{in} at the intersection of linear gain G and P_{sat} is $P_{in} = (P_{sat} - G)$. Compression begins at power input $P_{in} = (P_{sat} - G - K)$. Power output at the point compression begins is determined by using the amplifier P_{out} versus P_{in} behavioral model (5.1) and setting power input

$$P_{in} = (P_{sat} - G - K) \tag{5.9}$$

$$P_K = P_{sat} - G - K + G - K\log_{10}\left[1 + 10^{\left(\frac{P_{sat} - G - K + G - P_{sat}}{K}\right)}\right]$$

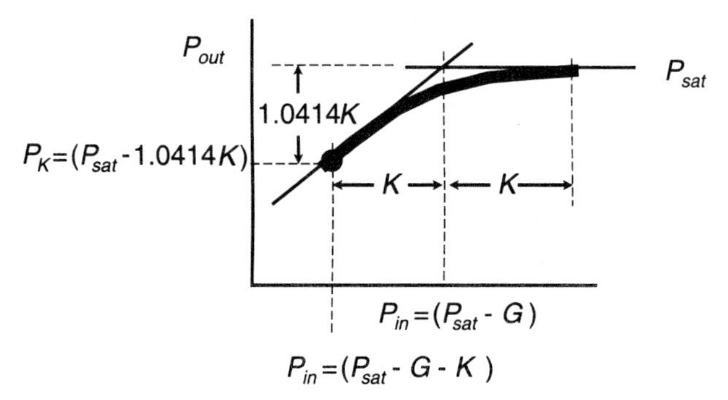

Figure 5.3 Compression coefficient K is the RHF transition range in the amplifier P_{out} versus the P_{in} behavioral model.

which reduces to

$$P_K = P_{sat} - 1.0414K \tag{5.10}$$

This is a significant power output point to keep in mind. P_K is the power output level in decibels that saturation effects such as output power compression begin to appear in the nonlinear output of any class A or class AB amplifier.

If there is no gain expansion, the gain in (5.9) is just small signal gain G_{ss}. If there is gain expansion, the gain is small signal gain plus the gain expansion $G_{ss} + \Delta G$.

5.2.1.6 Compression Coefficient K as a Function of Quiescent Bias Coefficient b_q

Compression coefficient K varies as a function of quiescent bias coefficient b_q and varies for transistor type as is shown in Section 4.1.1.2 and in Figures 4.4, 4.5, and 4.6. Compression coefficient reaches a maximum of approximately 16 as quiescent bias coefficient b_q value approaches unity. The class AB amplifier behavioral model needs a dynamic equation $K(b)$ for compression coefficient as a function of quiescent bias coefficient over the range $0.1 \le b_q \le 0.4$. The class A behavioral model was only concerned with quiescent bias at 50% I_{dss}, and a fixed value for K was sufficient. When class A and class AB amplifier behavioral models are expanded into the frequency domain in Chapter 6, quiescent bias coefficient b_q values will become a function of frequency and will be found to range from $0.1 \le b_q \le 1.0$. This simply says that although quiescent bias current i_q (no RF present) remains constant across the frequency domain, the saturation current at the peak of the RF current waveform can become equal to the quiescent bias current. This occurs when RF frequency is outside of the amplifier's operating bandpass. The RF load line impedance becomes very high outside the operating band, and there is very little RF output current flowing.

Compression coefficient $K(b)$ exhibits sufficient uniformity within the MESFET type of transistor so that a single model is usually sufficient for either square law or step-doped devices. Compression coefficients for square law and step-doped MESFETs are plotted in Figure 5.4 where the average of the two values is also plotted. Curve-fit techniques developed in Chapter 2 are used to formulate a model for the average $K(b)_{MESFET}$ of the two types of MESFETs over the quiescent bias range of interest.

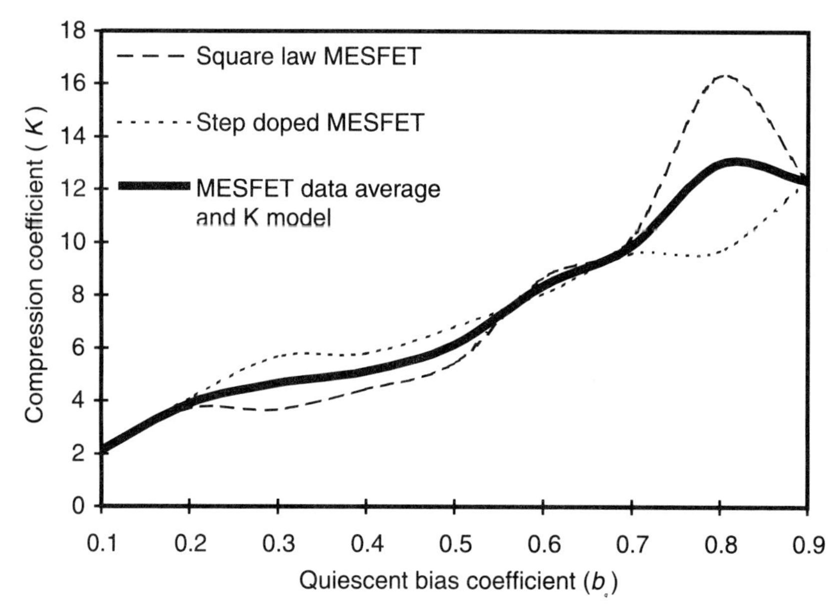

Figure 5.4 MESFET transistor compression coefficient model compared to DFT source data average.

$$K(b)_{\text{MESFET}} = -6.46 + 24.8b - 0.244 \log_{10}\left[1 + 10^{\left(\frac{b-0.597}{0.02}\right)}\right] + 0.471 \log_{10}\left[1 + 10^{\left(\frac{b-0.696}{0.02}\right)}\right] - 0.947 \log_{10}\left[1 + 10^{\left(\frac{b-0.8}{0.02}\right)}\right] + 0.769 \log_{10}\left[\frac{10^{\left(\frac{b-0.217}{0.05}\right)}}{1 + 10^{\left(\frac{b-0.217}{0.05}\right)}}\right] - 1.042 \log_{10}\left[\frac{10^{\left(\frac{b-0.478}{0.05}\right)}}{1 + 10^{\left(\frac{b-0.478}{0.05}\right)}}\right] \quad (5.11)$$

Figure 5.4 shows the MESFET compression coefficient model equation plotted over the average coefficient value curve. There is so little difference between the model and the original data trace that they appear as one curve.

Bipolar transistor compression coefficient $K(b)$ is significantly different from that of the MESFET at low values of quiescent bias. A separate model is needed. Bipolar transistor compression coefficient is also modeled by the curve-fit methods described in Chapter 2 to obtain

$$K(b)_{\text{Bipolar}} = 0.05 + 12.25b + 3.46\log_{10}\left[1 + 10^{\left(\frac{b-0.73}{0.08}\right)}\right] - 1.71\log_{10}\left[1 + 10^{\left(\frac{b-0.897}{0.03}\right)}\right] - 0.505\log_{10}\left[\frac{10^{\left(\frac{b-0.37}{0.05}\right)}}{1 + 10^{\left(\frac{b-0.37}{0.05}\right)}}\right] + 0.113\log_{10}\left[\frac{10^{\left(\frac{b-0.5}{0.01}\right)}}{1 + 10^{\left(\frac{b-0.5}{0.01}\right)}}\right] \quad (5.12)$$

which is shown in Figure 5.5 in comparison to data derived from DFT analysis of bipolar transistor output current waveforms. The model reproduces the original data with little error.

5.2.2 Average DC Current as a Function of Power Input

Transistor gain is a function of average DC bias current. Average DC bias current is a function of RF power input. These facts hold true for bipolar and

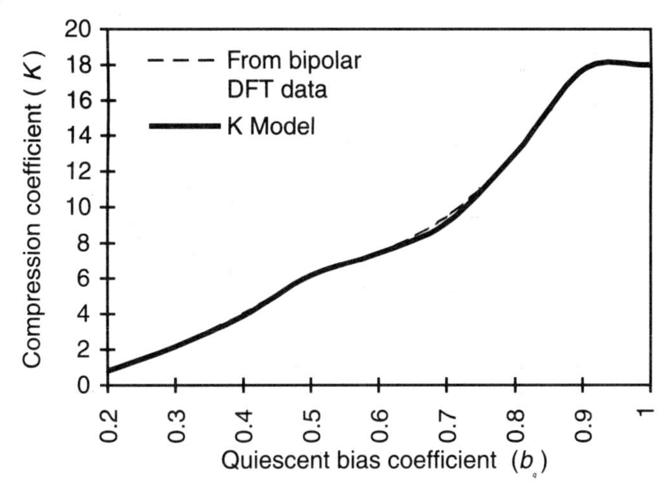

Figure 5.5 Bipolar transistor compression coefficient model shown in comparison to DFT source data.

field-effect transistors. Behavioral models for square law MESFETs, step-doped MESFETs, and bipolar transistors in Examples 3.1, 3.2, and 3.3 generate data showing average DC current as a function of RF power input. The data of interest appears in Figures 3.13, 3.18, and 3.22. Normalized average DC current $\left(\frac{i_{dc}}{i_{dss}}\right)$ as a function of normalized RF power input $(P_{in} - P_{sat})$ for class AB amplifiers is similar for all three transistor types and has the appearance of the curves shown in Figure 5.6. The data shown in Figure 5.6 is for the square law MESFET of Example 3.1.

The objective is to develop a unique equation that gives normalized average DC current as a function of normalized RF power input and quiescent bias coefficient b_q for each transistor type. Characteristic of all three transistor types is the quiescent DC current $i_q = bi_{dss}$ with no signal input. Also characteristic is the tendency to approach 50 percent i_{dss} at saturation regardless of the value of quiescent bias coefficient b_q. Three asymptotes and two right-hand functions are used to model bias current as a function of power input for a specific quiescent bias coefficient value as illustrated in Figure 5.7.

The general equation for the model described in Figure 5.7 is of the form of the step-up function (2.34) where change in slope

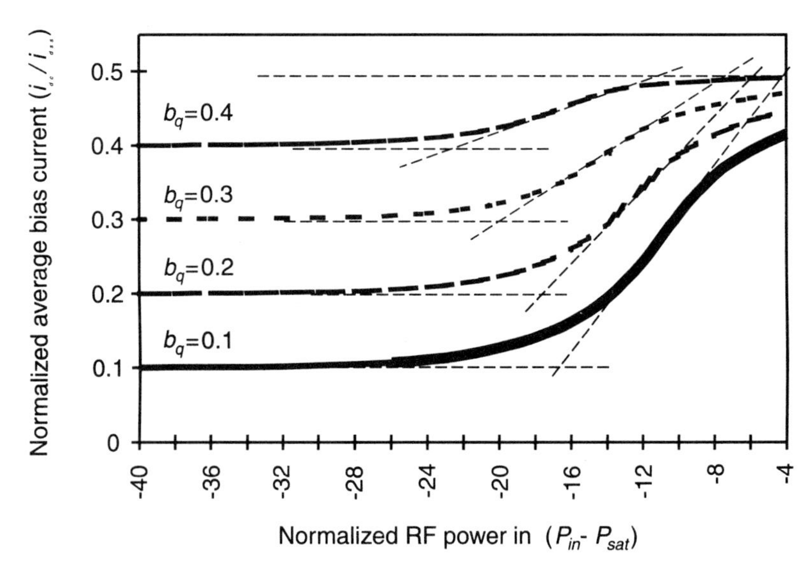

Figure 5.6 Normalized average bias current as a function of normalized RF power input for the square law MESFET.

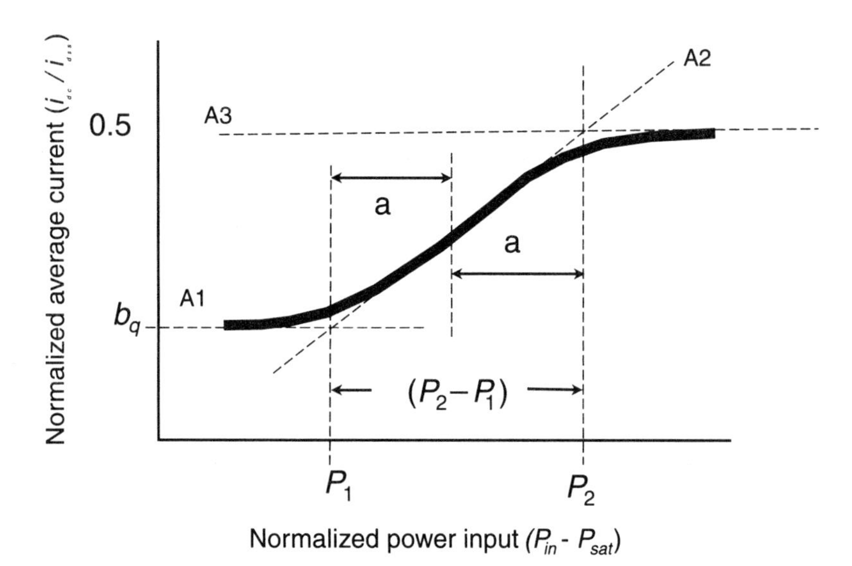

Figure 5.7 Three asymptotes and two right-hand functions model class AB amplifier normalized bias current as a function of normalized input power.

$$\Delta S = \frac{(0.5 - b)i_{dss}}{(P_2 - P_1)} \tag{5.13}$$

and transition range

$$a = \frac{(P_2 - P_1)}{2} \tag{5.14}$$

The resulting equation

$$\frac{i_{dc}}{i_{dss}} = b_q + \frac{(0.5 - b_q)}{2} \log_{10}\left[\frac{1 + 10^{\frac{2(P_{in} - P_{sat} - P_1)}{(P_2 - P_1)}}}{1 + 10^{\frac{2(P_{in} - P_{sat} - P_2)}{(P_2 - P_1)}}}\right] \tag{5.15}$$

applies to each quiescent bias coefficient for each of the transistor types.

An equation for each quiescent bias coefficient b_q is developed using (5.15) as a model. The equations all have the same form; only the coefficients P_1 and P_2 in the equations differ. The value of $(P_2 - P_1)$ in each equation is usually a constant for all values of b_q. Polynomial expressions for each of the coefficients $P_1(b)$ and $P_2(b)$ are developed giving the ability to interpolate values for each of the coefficients for any desired value of quiescent bias coefficient b_q. Procedures for this model development are given in Example 5.1.

5.2.3 Gain as a Function of Average DC Current

Transistor small signal gain as a function of quiescent bias coefficient b_q is easily obtained from transistor behavioral models as illustrated in Examples 3.1, 3.2, and 3.3. Results from discrete Fourier transform (DFT) analysis of transistor RF output current waveforms for each of the three transistor types, square law MESFETs, step-doped MESFETs, and bipolar transistors, are plotted in Figure 5.8. Notice that step-doped MESFET gain is relatively constant where $b_q > 0.4$. This is the region where the transistor is most linear due to the doping applied (see Section 3.5 and Example 3.2). Also note in Figure 5.8 (c) that bipolar transistor gain increases monotonically as average bias current increases.

Transistor gain as a function of power input is also obtainable from DFT analysis of output waveforms. Gain at small signal has the value shown in Figures 5.8 (a), (b), and (c). As signal level increases and average bias current increases, gain changes in a unique way for each transistor type and for each quiescent bias coefficient within each type. Figure 5.9 shows gain as a function of normalized power input for each of the three transistor types. Gain as a function of power input is relatively constant for MESFET amplifiers until power output approaches P_K where compression begins to occur. Power output ceases to increase dB for dB, causing reduction of gain. Bipolar transistor small signal gain is also relatively stable at small signal levels but increases exponentially as power input increases, resulting in significant gain expansion. Notice that the MESFET devices exhibit only moderate gain expansion as input power increases.

The appearance of curves plotted in Figure 5.9 leads to the conclusion that modeling MESFET device gain behavior as a function of power input is easier than modeling bipolar gain behavior. However, amplifier gain is a direct function of average bias current, not of power input. Bias current is a complex function of power input as shown in Figure 5.6 and modeled by (5.15). Data from DFT analysis of output current waveforms in Examples 3.1, 3.2, and 3.3 is rearranged in order to obtain gain as a function of normalized average bias current. This is the parameter that needs to be modeled. The characteristic of gain versus normalized average bias current for each transistor type is illustrated in Figure 5.10. Here the difference in gain behavior within the MESFET type becomes more pronounced, and gain as a function of average current for bipolar devices appears to be more uniform. Behavioral models for each transistor type are developed separately in the following sections.

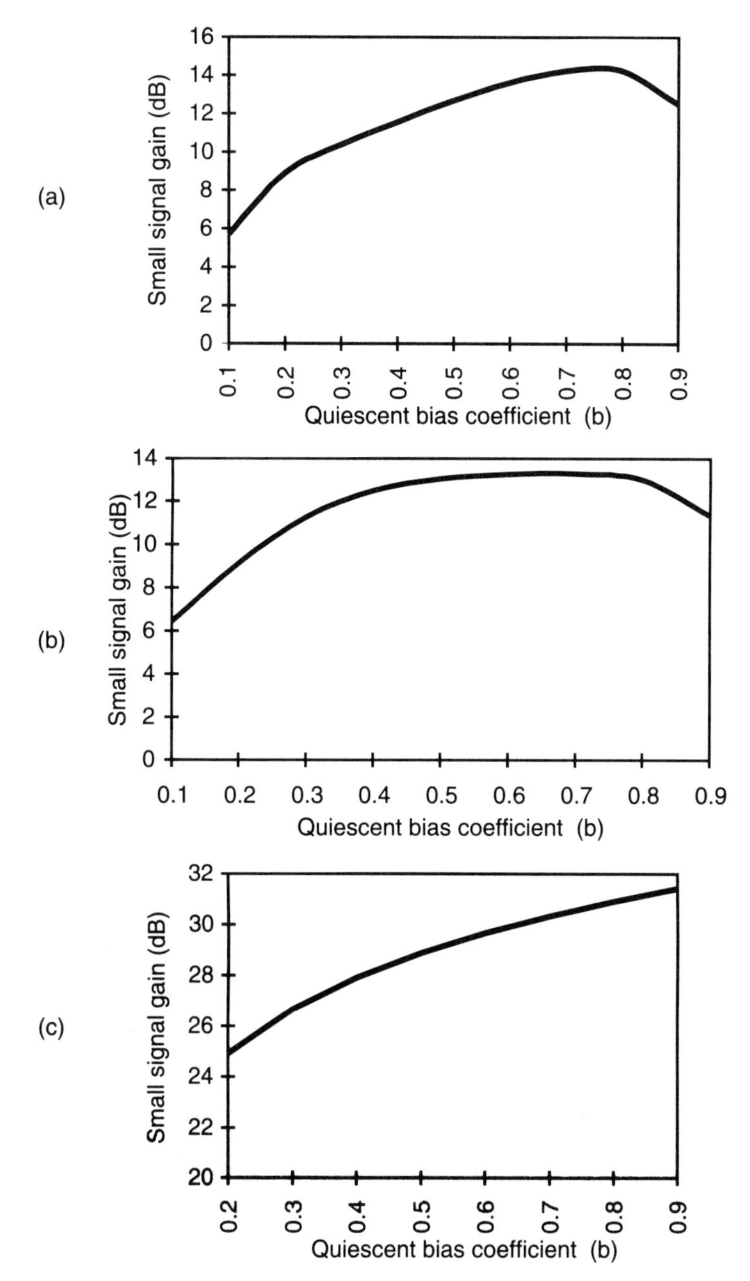

Figure 5.8 Transistor small signal gain as a function of quiescent bias coefficient *b*; (a) square law; (b) step-doped; (c) bipolar.

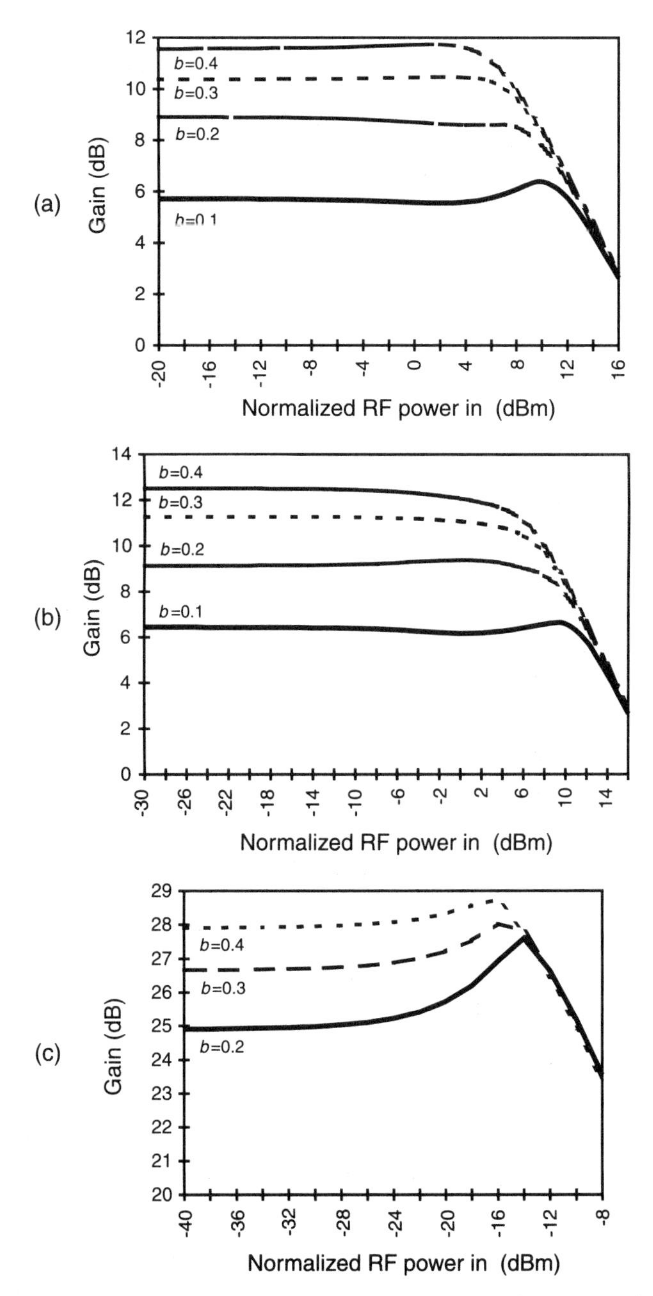

Figure 5.9 Transistor gain as a function of normalized RF power input is unique to the transistor type and bias coefficient; (a) square law; (b) step-doped; (c) bipolar.

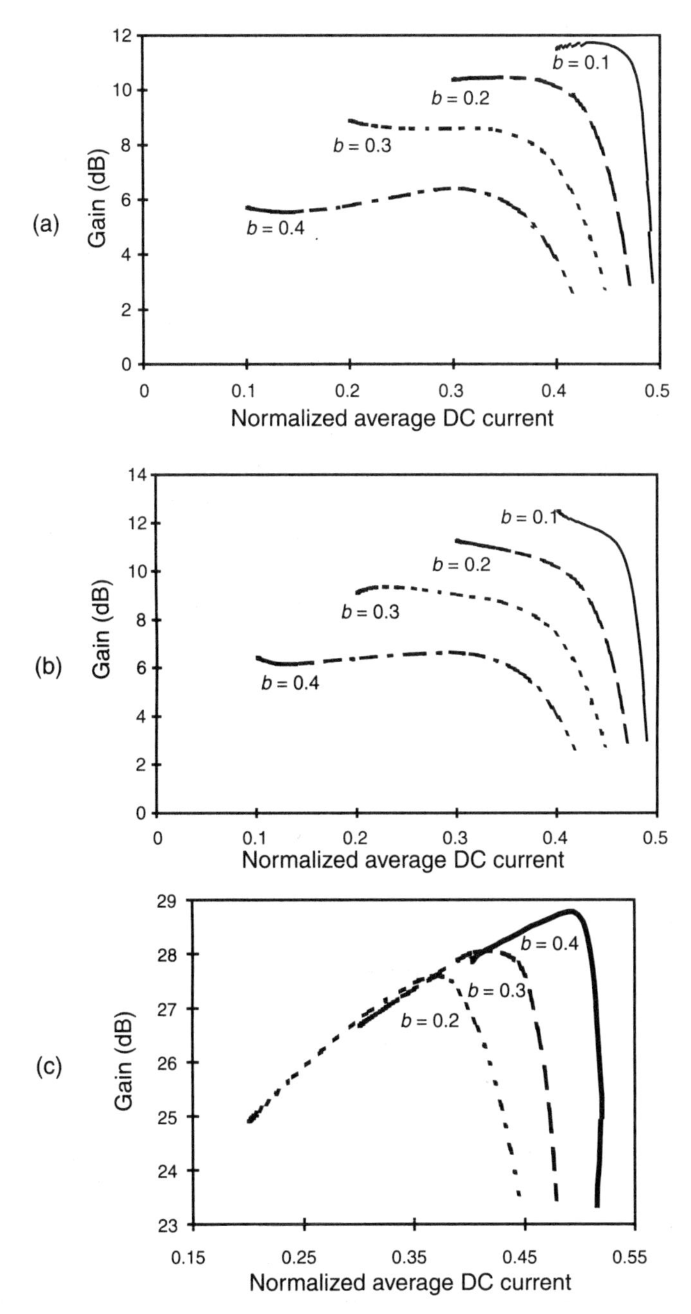

Figure 5.10 Gain as a function of normalized average bias current; (a) square law; (b) step-doped; (c) bipolar.

5.2.3.1 MESFET Gain Versus Average Bias Current

All of the curves shown in Figure 5.10 include the affect of gain compression as power output exceeds P_K. Ignore gain compression when modeling gain as a function of average bias current. Gain compression is accounted for in amplifier behavioral models (5.1) and (5.2). Transistor gain as a function of current is assumed to be a smoothly varying parameter if compression effects are ignored. Each curve in the two MESFET gain plots Figures 5.10(a) and 5.10(b), can be modeled using two asymptotes and a right-hand function (RHF). The range of interest of curve fit for each gain curve is from quiescent bias current i_q (the current at which the curve trace begins) to a linearly extended point into the region of compression. In cases where $b < 0.3$ where a single asymptote would suffice as a model, insert a second phantom asymptote having a small slope change with intercept point at greater than 50 percent i_{dss} to create a two asymptote model. Use the same transition range from one asymptote to the next for all curves. Figure 5.11 illustrates the selection of asymptotes for the modeling of MESFET gain as a function of average bias current.

The model equation for gain as a function of average bias current is

$$G\left(b, \frac{i_{ds}}{i_{dss}}\right) = C(b) + D(b)\frac{i_{ds}}{i_{dss}} + E(b)\log_{10}\left[1 + 10^{\frac{\left(\frac{i_{ds}}{i_{dss}} - F(b)\right)}{a}}\right] \quad (5.16)$$

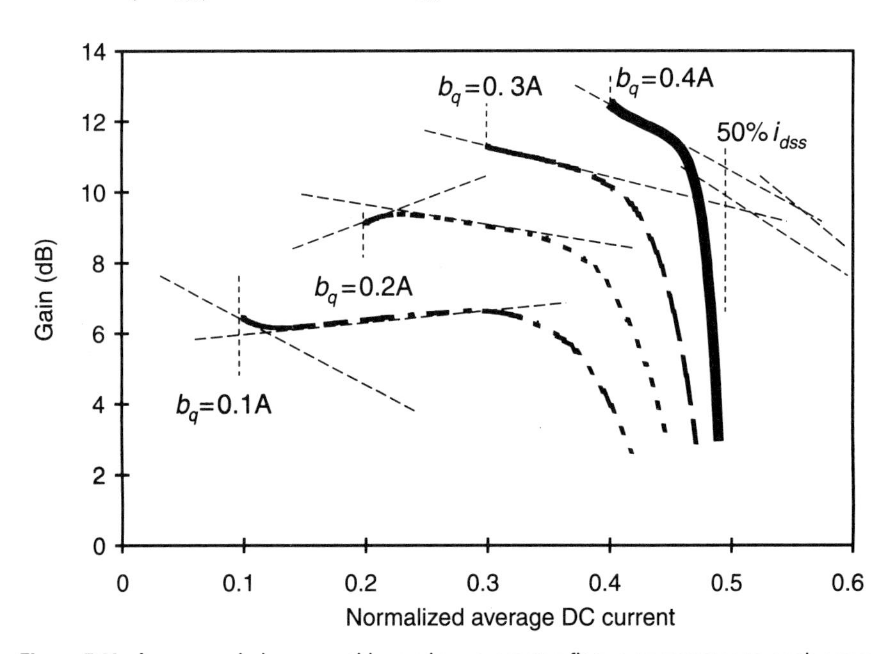

Figure 5.11 Assume gain is a smoothly varying parameter, fit two asymptotes to each curve.

where $C(b)$, $D(b)$, $E(b)$, and $F(b)$ are polynomials that give coefficients the appropriate value for each quiescent bias coefficient b_q. Polynomial equations are developed by first creating a table of values of the coefficients $C(b)$, $D(b)$, $E(b)$, and $F(b)$ as a function of b from the set of four curve-fit equations modeled using (5.16). Four simultaneous cubic equations with variable b expanded around $b = 0.25$ are then written using the values entered into the table. The simultaneous equations are solved using matrix methods shown in Example 1.1, resulting in polynomial equations for $C(b)$, $D(b)$, $E(b)$, and $F(b)$.

Example 5.1 Modeling Square Law MESFET Gain as a Function of Power Input

The square law MESFET transistor described in Example 3.1 of Chapter 3 is used here to illustrate the method of developing a class AB amplifier model. Data generated in Example 3.1 by discrete Fourier transform analysis of the transistor's output current waveform is reformated to be useful for class AB amplifier model development. Only data for quiescent bias coefficient values $0.1 \leq b_q \leq 0.4$ applies to class AB amplifiers.

Current Versus Power Input

The first model to be developed for square law MESFET amplifier behavior is normalized average bias current as a function of normalized power input for different quiescent bias coefficient values. Data showing average bias current as a function of normalized power input for the square law MESFET from Example 3.1 DFT analysis is reproduced in Figure 5.12.

Notice that the difference $[P_2(b) - P_1(b)]$ is approximately 11.6 dB for all curves tracing current versus power-in. This allows the transition range $a = 5.8$ to apply to all four curve equations. Since $[P_2(b) - P_1(b)] = 11.6$ is allowed for all four curves, only two polynomial equations need be developed: one for $P_1(b)$ and one for the product of slope change and transition range. The second intersect point is $P_2(b) = 11.6 + P_1(b)$. The general form of the model is from (5.15)

$$\frac{i_{ds}(b, P_{in})}{i_{dss}} = b_q + \frac{(0.5 - b_q)}{2} * \log_{10}\left[\frac{1 + 10^{\frac{(P_{in} - P_{sat} - P_1(b))}{5.8}}}{1 + 10^{\frac{(P_{in} = P_{sat} - P_2(b))}{5.8}}}\right] \quad (5.17)$$

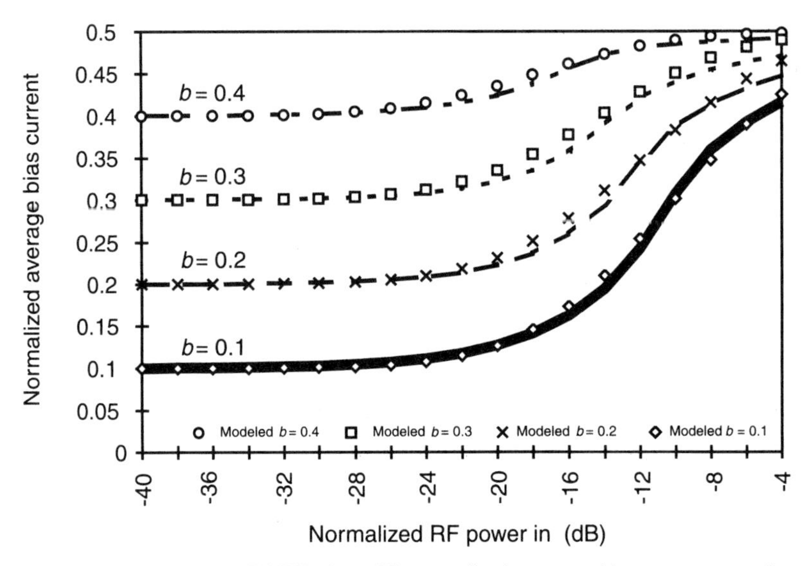

Figure 5.12 Square law MESFET class AB normalized average bias current as a function of normalized power-in and quiescent bias coefficient.

Equations for each of the bias curves are developed resulting in the following:

for $b = 0.1$

$$\frac{i_{ds}}{i_{dss}} = 0.1 + 0.2 * \log_{10}\left[\frac{1 + 10^{\frac{(P_{in} - P_{sat} + 16.1)}{5.8}}}{1 + 10^{\frac{(P_{in} - P_{sat} + 4.2)}{5.8}}}\right]$$

for $b = 0.2$

$$\frac{i_{ds}}{i_{dss}} = 0.2 + 0.15 * \log_{10}\left[\frac{1 + 10^{\frac{(P_{in} - P_{sat} + 17.5)}{5.8}}}{1 + 10^{\frac{(P_{in} - P_{sat} + 5.8)}{5.8}}}\right]$$

for $b = 0.3$

$$\frac{i_{ds}}{i_{dss}} = 0.3 + 0.1 * \log_{10}\left[\frac{1 + 10^{\frac{(P_{in} - P_{sat} + 19.2)}{5.8}}}{1 + 10^{\frac{(P_{in} - P_{sat} + 7.4)}{5.8}}}\right]$$

and for $b = 0.4$

$$\frac{i_{ds}}{i_{dss}} = 0.4 + 0.05 * \log_{10}\left[\frac{1 + 10^{\frac{(P_{in} - P_{sat} + 22)}{5.8}}}{1 + 10^{\frac{(P_{in} - P_{sat} + 10.7)}{5.8}}}\right]$$

A third-order polynomial equation for coefficient $P_2(b)$ is derived given the data from the above equations, which are tabulated in Table 5.1.

Third-order polynomial expansion about quiescent bias coefficient value $b_q = 0.15$ is used to obtain the following equation

$$P_2(b) = -5.106 - 15.2916(b - 0.15) + 42.5(b - 0.15)^2 - 283.333(b - 0.15)^3 \quad (5.18)$$

and

$$P_1(b) = P_2(b) - 11.8 \quad (5.19)$$

These results are substituted back into (5.17) to complete the behavioral model for normalized average current as a function of normalized power input. Normalized average bias current versus normalized RF power input behavioral model results compare favorably to data obtained from MESFET transistor output waveform DFT analysis. Model results are plotted over original data in Figure 5.12.

Now develop a model for gain as a function of normalized average bias current.

Gain Versus Normalized Average Bias Current

Data from DFT analysis of transistor output current waveform for quiescent bias coefficients $0.1 \leq b_q \leq 0.4$ are formated to show transistor gain as a function of normalized average bias current in Figure 5.13.

Following the procedure described in Section 5.2.3.1 above, two asymptotes are fitted to each curve in Figure 5.13, creating four equations, each equation having different coefficient values. These equations for gain as a function of average bias current are all of the form

Table 5.1
Square Law MESFET Normalized Average Bias Current Model Coefficients as Function of b_q

b_q	$P_1(b)$	$P_2(b)$	$[P_2(b) - P_1(b)]$
0.1	−16.1	−4.2	11.9
0.2	−17.5	−5.8	11.7
0.3	−19.2	−7.4	11.8
0.4	−22	−10.7	11.3

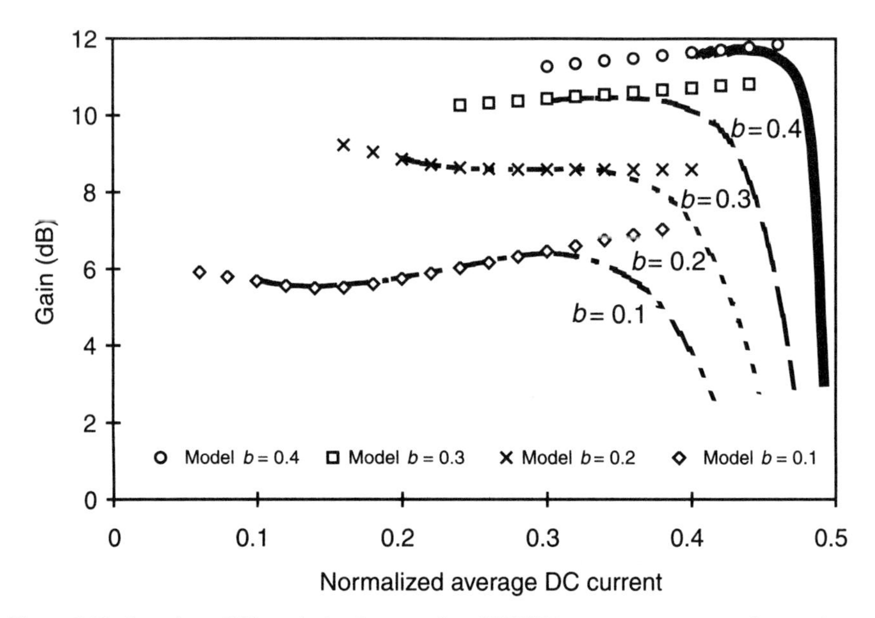

Figure 5.13 Data from DFT analysis of square law MESFET output current waveforms shows gain as function of normalized average bias current.

$$G\left(b, \frac{i_{\text{ds}}}{i_{\text{dss}}}\right) = H(b) + J(b)\frac{i_{\text{ds}}}{i_{\text{dss}}} + L(b)\log_{10}\left[1 + 10^{\frac{\left(\frac{i_{\text{ds}}}{i_{\text{dss}}} - M(b)\right)}{a}}\right] \tag{5.20}$$

Transition range is arbitrarily set constant at value $a = 0.05$ for all four equations to reduce the number of variables. Phantom asymptotes are drawn for curves of $b_q = 0.3$ and $b_q = 0.4$ to give their models the form of (5.20). Asymptote intersect values for these two curves are arbitrarily set at $\frac{i_{\text{ds}}}{i_{\text{dss}}} =$ 0.34 and 0.45 respectively. Values for coefficients of the four curve-fit equations are developed using techniques described in Chapter 2.

Four equations that model each of the curves for gain at different quiescent bias coefficients are

$$b_q = 0.1 \qquad G\left(\frac{i_{\text{ds}}}{i_{\text{dss}}}\right) = 6.2 - 4.9\frac{i_{\text{ds}}}{i_{\text{dss}}} + 0.62\log_{10}\left[1 + 10^{\frac{\left(\frac{i_{\text{ds}}}{i_{\text{dss}}} - 0.1573\right)}{0.05}}\right]$$

$b_q = 0.2$ $$G\left(\frac{i_{ds}}{i_{dss}}\right) = 10 - 6\frac{i_{ds}}{i_{dss}} + 0.3\log_{10}\left[1 + 10^{\frac{\left(\frac{i_{ds}}{i_{dss}} - 0.25\right)}{0.05}}\right]$$

$b_q = 0.3$ $$G\left(\frac{i_{ds}}{i_{dss}}\right) = 9.3 + 3.6\frac{i_{ds}}{i_{dss}} - 0.32\log_{10}\left[1 + 10^{\frac{\left(\frac{i_{ds}}{i_{dss}} - 0.34\right)}{0.05}}\right]$$

and for $b_q = 0.4$ $$G\left(\frac{i_{ds}}{i_{dss}}\right) = 10 + 4\frac{i_{ds}}{i_{dss}} - 0.2\log_{10}\left[1 + 10^{\frac{\left(\frac{i_{ds}}{i_{dss}} - 0.45\right)}{0.05}}\right]$$

The coefficient values from these equations are tabulated in Table 5.2 as functions of quiescent bias coefficient b, and polynomial expansion is performed about the value $b = 0.25$.

Third-order polynomial solutions obtained for $H(b)$, $J(b)$, $L(b)$, and $M(b)$ are

$$H(b) = 9.8437 - 9.4583(b - 0.25) - 77.5(b - 0.25)^2 + 983.33(b - 0.25)^3$$

$$J(b) = -1.2937 + 104.2916(b - 0.25) + 37.5(b - 0.25)^2 - 3316.66(b - 0.25)^3$$

$$L(b) = -0.0375 - 6.6333(b - 0.25) + 0.3325(b - 0.25)^2 + 173.333(b - 0.25)^3$$

Table 5.2
Coefficients of Gain as a Function of Quiescent Bias Coefficient b

b	***H*(*b*)**	***J*(*b*)**	***L*(*b*)**	***M*(*b*)**
0.1	6.2	−4.9	0.62	0.1573
0.2	10	−6	0.3	0.25
0.3	9.3	3.6	−0.32	0.344
0.4	10	4	−0.2	0.45

$$M(b) = 0.29617 + 0.93554(b - 0.25) + 0.3325(b - 0.25)^2 + 1.78333(b - 0.25)^3$$

These polynomial expressions for the coefficients are substituted back into (5.20), where the value for transition range $a = 0.05$. Gain is calculated as a function of normalized average bias current for the four values of quiescent bias coefficient $b_q = 0.1$, $b_q = 0.2$, $b_q = 0.3$, and $b_q = 0.4$. Calculated results from the gain versus normalized average bias current behavioral model are plotted in Figure 5.13 in comparison with data obtained from DFT analysis of RF output current waveforms. Modeled results compare favorably when the gain compression effects included in the DFT data are discounted.

Gain as a Function of RF Power Input

The product of gain as a function of normalized average bias current $G\left(b, \frac{i_{ds}}{i_{dss}}\right)$, (5.20), and normalized average bias current as a function of normalized RF power input $\frac{i_{ds}[b, (P_{in} - P_{sat})]}{i_{dss}}$, (5.17), gives gain as a function of normalized RF power input $G[b, (P_{in} - P_{sat})]$. This is the parameter needed for insertion into the class AB amplifier power output versus power input and phase out versus power input behavioral model (see (5.1) and (5.2)). Gain compression is accounted for when gain as a function of normalized power input $G[b, (P_{in} - P_{sat})]$ is calculated and inserted into the general amplifier behavioral model with the appropriate dynamic compression coefficient $K(b)$ (see Section 5.2.1.6, (5.11)). Gain contraction and expansion are modeled correctly as is shown in Figure 5.14. Phase shift as a function of normalized RF power input is calculated using the same modeled parameters for gain.

Amplifier Transfer Phase as a Function of Power Input

The product of gain as a function of normalized average bias current (5.20) and current as a function of normalized RF power input (5.17) is also used to determine gain as a function of normalized power input in the behavioral model for amplifier phase shift as a function of power input (5.2). The model for MESFET compression coefficient as a function of quiescent bias coefficient (5.11) is also used in the phase shift behavioral model. Phase shift as a function of power input is calculated and plotted in Figure 5.15 assuming a typical

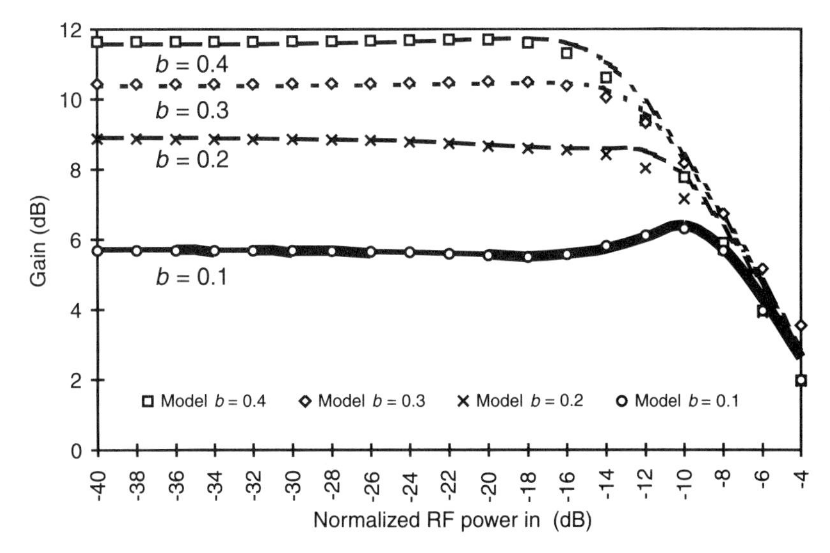

Figure 5.14 Modeled MESFET gain as a function of normalized RF power input compared to data from DFT analysis of output current waveforms.

AM/PM conversion sensitivity of five degrees per dB and including all of the dynamic parameters developed for the class AB amplifier.

Power-Added Efficiency

The definition of power-added efficiency is the same for all amplifier classes and all transistor types (see Section 4.1.5 and (4.16)). When the dynamic parameters developed here for class AB MESFET amplifiers are inserted into (4.16), and *PAE* is calculated as a function of normalized power input and quiescent bias coefficient, a peak efficiency is found for quiescent bias coefficient $0.2 \leq b_q \leq 0.3$ as indicated in Figure 5.16.

The peak in *PAE* illustrated in Figure 5.16 is also sensitive to power input. Data used to generate Figure 5.16 are replotted in Figure 5.17 in a different format showing power-added efficiency as a function of compression depth. The maximum in *PAE* clearly occurs where the amplifier is compressed by 1.5 to 2.0 dB. This same peak in efficiency is observed for step-doped MESFET transistors.

5.2.3.2 Bipolar Transistor Gain Versus Average Bias Current

Bipolar transistor gain as a function of average bias current is much easier to model than that of the MESFET. A single equation consisting of two asymptotes is used to model gain for all values of quiescent bias coefficient over the entire average current range from i_q = 10 percent i_{max} to i_q = 50 percent i_{max}.

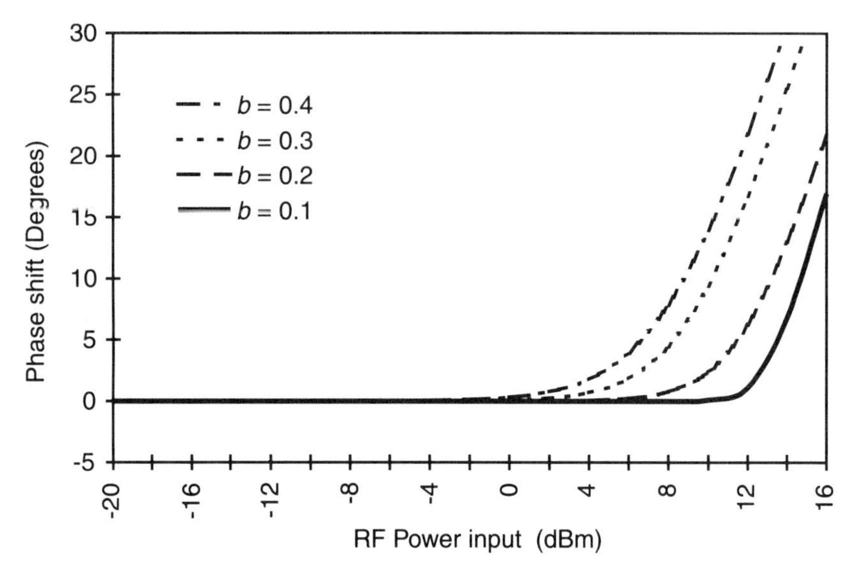

Figure 5.15 Class AB MESFET amplifier phase shift as a function of RF power input.

Gain is assumed to be a smoothly varying function of average bias current, and compression effects at power output greater than P_K are discounted. Figure 5.18 illustrates the selection of two asymptotes for modeling bipolar gain as a function of average bias current.

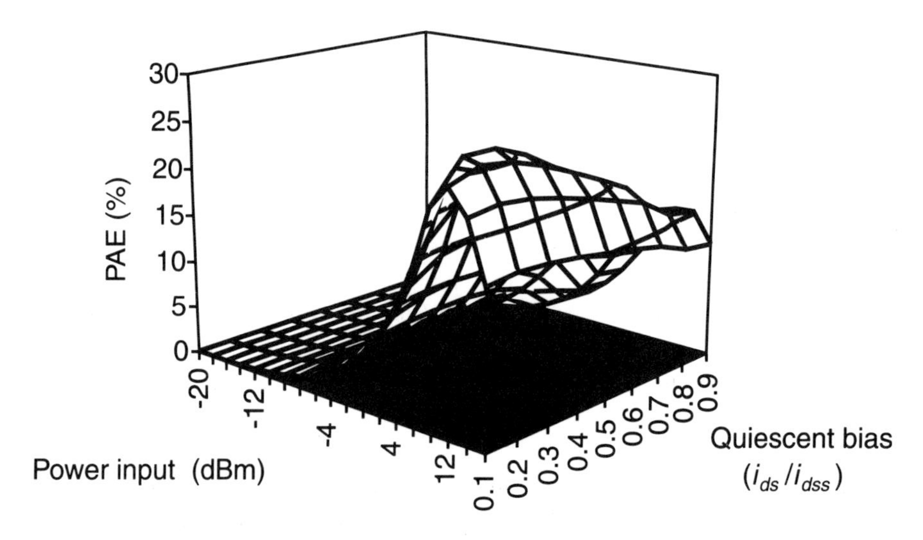

Figure 5.16 Peak power-added efficiency is achieved where quiescent bias coefficient $0.2 \le b_q \le 0.3$.

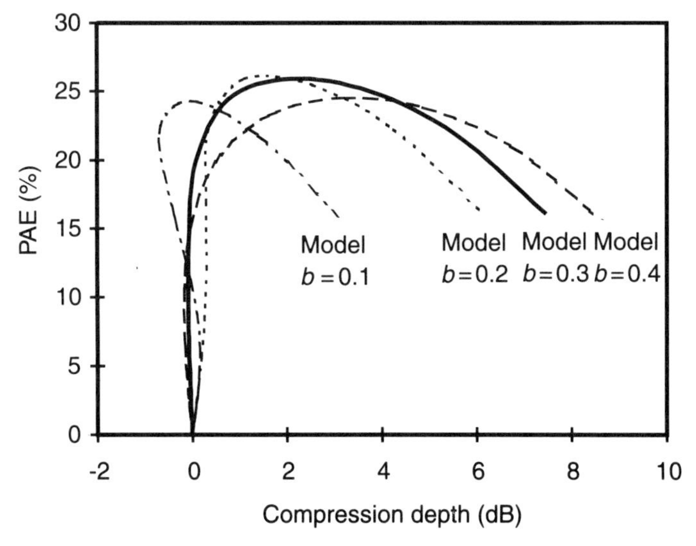

Figure 5.17 Peak *PAE* occurs at 1.5 dB to 2.0 dB compression depth.

The form of the bipolar gain versus normalized average bias current model equation is identical to that of the MESFET gain (5.16) with the exception that all coefficients are constant as a function of quiescent bias coefficient b_q.

$$G\left(\frac{i_{ds}}{i_{dss}}\right) = C + D\frac{i_{ds}}{i_{dss}} + E\log_{10}\left[1 + 10^{\frac{\left(\frac{i_{ds}}{i_{dss}} - F\right)}{a}}\right] \tag{5.21}$$

5.3 Summary

A behavioral model for class AB amplifiers has been developed. The model expands the general nonlinear amplifier model equations for power output and phase as a function of RF power input by adding new equations that describe $i_{ds}(b, P_{in})$, average bias current as a function of quiescent bias and RF power input, $G(b, i_{ds})$, gain as a function of quiescent bias and average current, and $K(b)$ compression coefficient as a function of quiescent bias b. Normalized equations that were developed are easily converted to specific applications by multiplying them by the transistor's actual i_{dss} or adding in P_{sat} where appropriate. The resulting class AB behavioral model is described by the equations

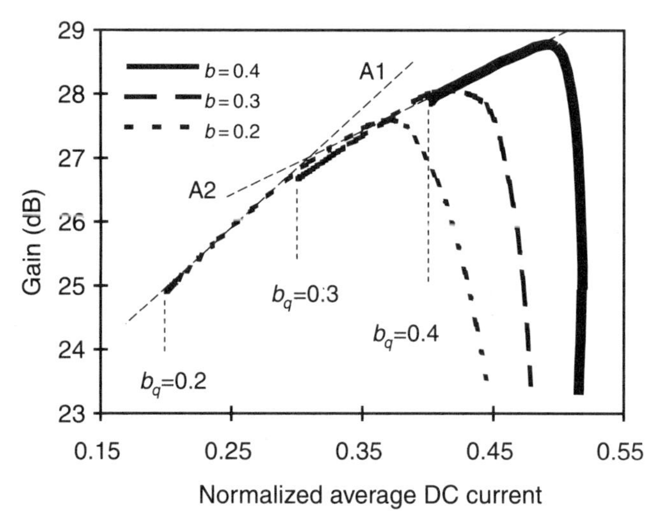

Figure 5.18 Two asymptotes are used to model bipolar gain as a function of normalized average bias current.

$$P_{out} = P_{in} + G(b, P_{in}) - K(b) \log_{10}\left[1 + 10^{\left(\frac{P_{in} + G(b, P_{in}) - P_{sat}}{K(b)}\right)}\right] \tag{5.22}$$

$$\varphi_{out} = \theta K(b) \log_{10}\left[1 + 10^{\left(\frac{P_{in} + G(b, P_{in}) - P_{sat}}{K(b)}\right)}\right] + \varphi_{in} \tag{5.23}$$

where

$$G(b, P_{in}) = G(b, i_{ds}) * i_{ds}(b, P_{in})$$

Equations for $G(b, i_{ds})$, $i_{ds}(b, P_{in})$, and $K(b)$ are specific to the transistor type FET, or bipolar, and to the actual values of P_{sat} and i_{dss} of the transistor. The general form of average current as a function of quiescent bias coefficient and power input is

$$i_{ds}(b, P_{in}) = bi_{dss} + \frac{(0.5 - b)i_{dss}}{2} \log_{10}\left[\frac{1 + 10^{\frac{2(P_{in} - P_1(b))}{(P_2(b) - P_1(b))}}}{1 + 10^{\frac{2(P_{in} - P_2(b))}{(P_2(b) - P_1(b))}}}\right] \tag{5.24}$$

where $P_1(b)$ and $P_2(b)$ are specific to the transistor. The difference $[P_2(b) - P_1(b)]$ is nearly constant for all quiescent bias coefficient b_q values and has approximate values of 11.8 dB for MESFETs and 10 dB for bipolar transistors.

The general form of gain as a function of quiescent bias coefficient and average current is

$$G(b, i_{ds}) = C(b) + D(b)i_{ds} + E(b)\log_{10}\left[1 + 10^{\frac{(i_{ds} - F(b))}{a}}\right] \tag{5.25}$$

where $C(b)$, $D(b)$, $E(b)$, and $F(b)$ are polynomials that give coefficients the appropriate value for each quiescent bias coefficient b_q. Coefficients $C(b)$, $D(b)$, $E(b)$, and $F(b)$ are unique to the transistor type, (FET or bipolar), and to variations of design within each type. Normalized polynomial equations for $C(b)$, $D(b)$, $E(b)$, and $F(b)$ were developed in Example 5.1 for the square law MESFET.

Compression coefficient $K(b)$ is also unique to transistor type. Average value of $K(b)$ is used within a specific type without introducing significant error. The general form of compression coefficient

$$K(b) = k_0 + k_1 bq + k_2 \log_{10}\left[1 + 10^{\frac{b_q - k_3}{k_4}}\right] + k_5 \log_{10}\left[1 + 10^{\frac{b_q - k_6}{k_7}}\right]$$
$$+ k_8 \log_{10}\left[1 + 10^{\frac{b_q - k_9}{k_{10}}}\right] + k_{11} \log_{10}\left[\frac{10^{\frac{b_q - k_{12}}{k_{13}}}}{1 + 10^{\frac{b_q - k_{12}}{k_{13}}}}\right]$$
$$+ k_{14} \log_{10}\left[\frac{10^{\frac{b_q - k_{15}}{k_{16}}}}{1 + 10^{\frac{b_q - k_{15}}{k_{16}}}}\right] \tag{5.26}$$

is used where coefficient values k_j are specific to transistor type.

5.4 Problems

Problem 5.1

Using the transistor data given in Example 3.2 for a step-doped MESFET transistor, develop a class AB amplifier model and show that peak power-added efficiency occurs where quiescent bias coefficient $0.2 \leq b_q \leq 0.3$ and compression depth is between 1.5 dB and 2.0 dB.

Problem 5.2

Using transistor data given in Example 3.3 for a bipolar transistor, develop a class AB amplifier model and show that *PAE* continues to rise as power input increases and that the *PAE* is less than that of the MESFET devices used in Examples 3.1 and 3.2.

6

Modeling It With Frequency

6.1 Adding Frequency as a Behavioral Model Variable

Behavioral models developed in Chapters 3, 4, and 5 describe transistor and amplifier characteristics at a single frequency. Expanding models to work over the frequency domain adds significant utility. Basic parameters used in amplifier behavioral models are small signal gain, saturated power output, and compression coefficient. Equations for these parameters as a function of frequency can be obtained directly by curve fitting measured or published data or can be simulated using techniques developed in this chapter. When equations for $G_{ss}(f)$, $P_{sat}(f)$ and $K(f)$ are known, power output $P_{out}(f)$, compression depth, phase shift $\varphi(f)$, one dB compression point P_{1dB}, third-order intercept P_{3rdOIP}, can all be calculated. Variations in $i_{max}(f)$ and load line impedance $R_l(f)$ can be estimated from knowledge of $P_{sat}(f)$, b_q, and $i_{ds}(f)$.

Techniques presented in this chapter for modeling the behavior of filter and amplifier response as a function of frequency give good approximations where physical elements that control poles and zeros in the networks are unknown. Rigorous modeling of network behavior where physical elements in the networks are known is covered by Milton Dishal in Section 12 of the *Electronic Engineer's Handbook* [1], and in Chapters 7-9 of *Reference Data for Radio Engineers* [2].

6.2 Amplifier Gain as a Function of Frequency

Transistor small signal gain values determined by DFT analysis of output current waveforms in Chapter 3 are low-frequency gain values. Variation of this small signal gain $G(b, i_{ds})$ with quiescent bias coefficient b_q and average bias current i_{ds} was described in Chapter 5, Section 5.2.3. All transistors exhibit an upper frequency limit f_t above which small signal power gain is less than unity. Small signal gain decreases at a 6 dB per octave rate from maximum low frequency gain to unity power gain at f_t. Typical f_t values for various transistor types are shown in Table 6.1. Figure 6.1 illustrates the roll-off of small signal gain in transistors as a function of frequency.

The low pass frequency response illustrated in Figure 6.1 is easily modeled by

$$G_{ss}(f) = G_{ss}(b, i_{ds}) - 10 \log_{10}\left[1 + \left(\frac{f}{f_0}\right)^2\right] \tag{6.1}$$

where f_o is the frequency at which power gain has diminished by 3 dB, and $G_{ss}(b, i_{ds})$ is very low-frequency small signal gain as determined by DFT analysis of output current waveform performed in examples 3.1, 3.2, and 3.3. The frequency f_0 can easily be determined, given values for $G_{ss}(b, i_{ds})$ and f_t, by

$$f_0 = f_t 10^{\frac{-G_{ss}(b, i_{ds})}{20}} \tag{6.2}$$

Amplifiers designed for a specific bandwidth in a specific frequency range can have no greater small signal gain value than that defined by (6.1) where $f = f_{hi}$, the upper frequency of the required bandwidth. Amplifier input and output impedance matching networks are usually designed to compensate for the transistor's 6 dB per octave roll-off such that the amplifier presents a relatively flat gain over the required bandwidth. The amplifier's input and output impedance matching networks do not do their work without adding insertion

Table 6.1
Typical f_t Unity Power Gain for Different Transistor Types

Transistor Type	Typical Low-Frequency Gain	Typical Unity Gain Frequency f_t
Silicon Bipolar	25 db	8 Ghz
GaAs Heterojunction Bipolar (HBT)	25 dB	25 Ghz
GaAs MESFET	12 dB	40 Ghz
GaAs PHEMT	14 dB	65 GHz

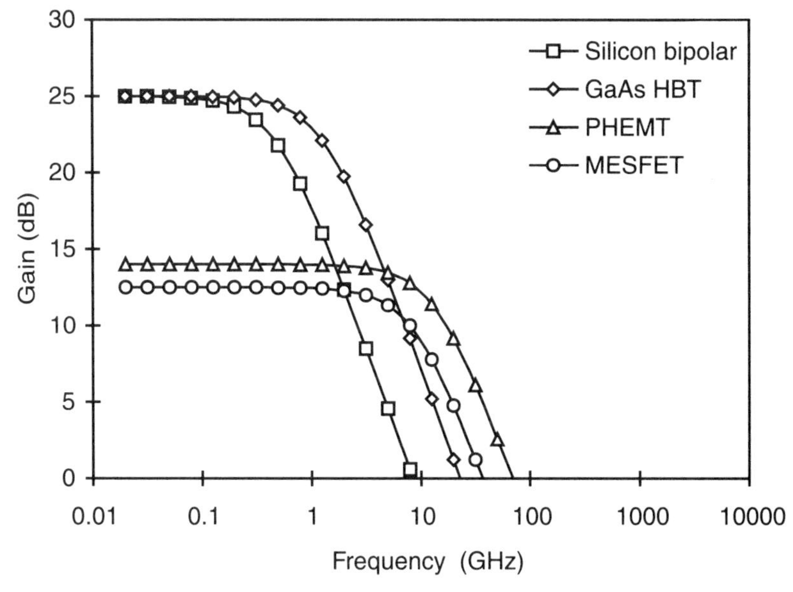

Figure 6.1 Typical transistor small signal gain as a function of frequency.

loss. Insertion loss value for input and output impedance matching networks can range from a few tenths of a decibel to several decibels depending on factors such as the bandwidth being matched, the frequency band in which the amplifier is designed to work, and the type of components used to fabricate the impedance match networks (lumped or distributed). Narrow band amplifiers, defined as amplifiers that are less than or equal to 10 percent bandwidth, are usually tuned with a single pole in the complex plane. Simple, single tuned networks exhibit little or no gain ripple within the operating bandwidth. Gain simply peaks in the center of the band and rolls off monotonically at either side of the center frequency. Broader bandwidth amplifiers, defined as amplifiers with up to 40 percent bandwidth, typically use two and three poles positioned in the complex plane to control load line impedance match and maintain small signal gain over the broader bandwidth. The greater the number of poles used, the steeper the roll-off in gain out of band. Controlled positioning of the poles in the complex plane results in gain ripple ranging from maximally flat (Butterworth design), to designated in band ripple amplitude (Chebishev design). Somc amplifiers are designed to have a gain slope to compensate for frequency response of other components in the subsystem. A method of simulating bandpass amplifier small signal gain as a function of frequency is now developed.

6.2.1 Ideal Bandpass Amplifier Butterworth Frequency Response Simulation

The ideal bandpass amplifier frequency response has a lower frequency −3 dB cutoff f_{lo} and an upper frequency −3 dB cutoff f_{hi}. Small signal gain at the upper frequency of this range can be no greater than that defined by (6.1) where $f = f_{hi}$, the upper frequency of the required bandwidth. The frequency domain

$$\Delta f_{bw} = (f_{hi} - f_{lo}) \tag{6.3}$$

is defined to be the amplifier's bandpass or 3dB bandwidth. The median frequency

$$f_c = \sqrt{f_{hi} * f_{lo}} \tag{6.4}$$

is defined to be the bandpass amplifier's center frequency. Amplifier small signal gain at center frequency is defined as G_c. The number of poles used in the bandpass amplifier's impedance matching network is designated n. Definition of bandpass amplifier percent bandwidth is

$$\%BW = \frac{(f_{hi} - f_{lo})}{f_c} * 100\% \tag{6.5}$$

These definitions are used in a model for maximally flat or Butterworth bandpass amplifier small signal gain

$$G_{bp}(f) = G_c(f_c, b, i_{ds}) - 10\log_{10}\left[1 + \left(\frac{\left|\frac{f_c^2}{f} - f\right|}{(f_{hi} - f_{lo})}\right)^{2n}\right] \tag{6.6}$$

Figure 6.2 compares the results obtained with (6.6) for five objects that have zero gain $G_c(f, b, i_{ds}) = 0$ and a bandpass where $f_{lo} = 1.7$ GHz, $f_{hi} = 2.6$ GHz where the number of poles n varies from 1 to 5.

6.2.2 Adding Ripple to the Bandpass

A Chebishev ripple response is approximated using RHFs to create in-band ripple. Ripple peak-to-peak amplitude R and the number of poles n are variables that need to be included in the model for the Chebishev response. The number of ripples in the bandpass is directly related to the number of poles n used

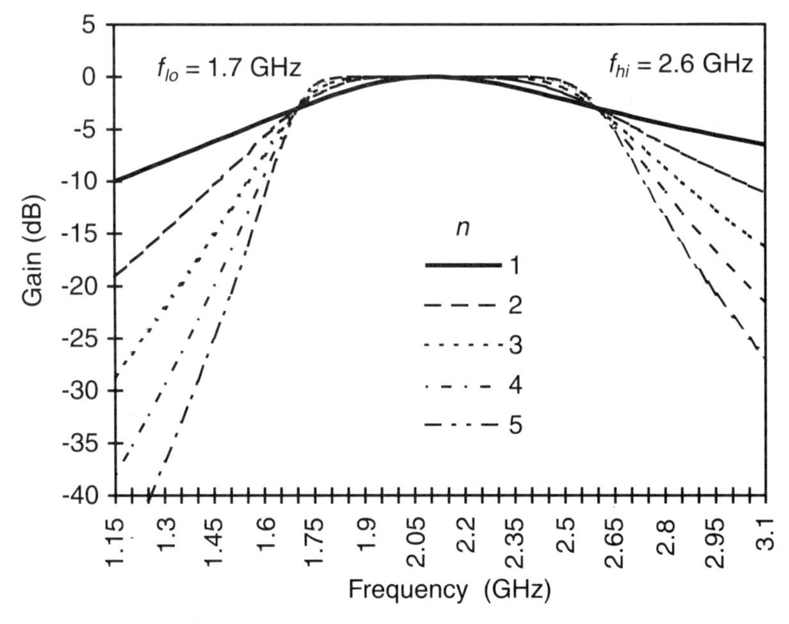

Figure 6.2 Equation (6.6) is illustrated over a 900 MHz bandpass for objects with different numbers of poles.

in the matching network. Equation (6.6) defined above as the maximally flat response model is used as a basis to which the Chebishev ripple model is added. The ripple is approximated by using right-hand functions RHFs added to the fixed gain component G_c in the maximally flat response given by (6.6). Equations for two-pole, and three-pole Chebishev ripple response are developed in the following sections, establishing a procedure and providing guidance for developing equations for higher order networks.

6.2.2.1 The Two-Pole Network Chebishev Ripple Response

The two-pole matching network exhibits two gain peaks in the bandpass with a gain minimum in the center of the band as illustrated in Figure 6.3. Four asymptotes are constructed to represent the two peaks such that the two end asymptotes have a value of zero at frequency points f_{lo} and f_{hi}. The remaining two asymptotes intersect at a point mid-band. The frequency range between f_{lo} and f_{hi} is divided into six equal intervals. The two internal asymptotes intersect with the two external asymptotes at points one-sixth of the way across the range and at five-sixths of the way across the range. Figure 6.3 illustrates the four asymptotes and their locations.

The six intervals each have width $\dfrac{(f_{\text{hi}} - f_{\text{lo}})}{6}$. The intersection of asymptotes

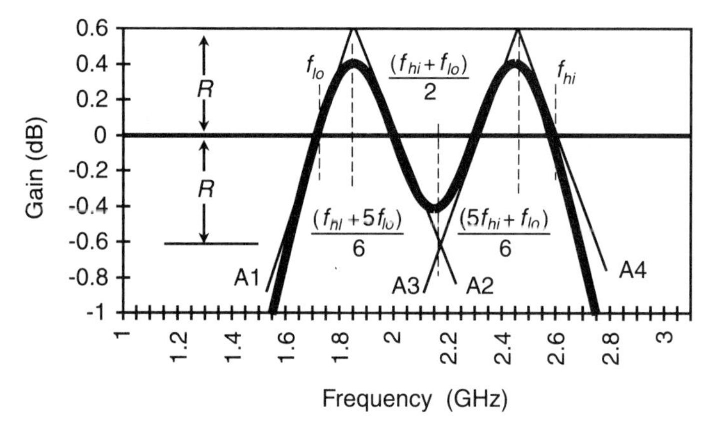

Figure 6.3 Four asymptotes are used to approximate two-pole Chebishev ripple.

A1 and **A2** occurs at frequency $\frac{(f_{hi} + 5f_{lo})}{6}$, the sum of f_{lo} and one interval. The intersection of asymptotes **A2** and **A3** occurs at frequency $\frac{(f_{hi} + f_{lo})}{2}$, mid-band. The intersection of asymptotes **A3** and **A4** occurs at frequency $\frac{(5f_{hi} + f_{lo})}{6}$, the difference of f_{hi} and one interval. The amplitude at which the two end asymptotes intersect is R greater than the amplitudes at the end points f_{lo} and f_{hi}. The amplitude at which the center asymptotes intersect is R below the amplitude at the end points f_{lo} and f_{hi}. The total difference in amplitude between the center intersection point and the end intersections is $2R$. Arbitrarily let the value R be decibels (dB).

Develop a Gain Ripple Equation

A ripple equation is developed by selecting asymptote **A1** as the reference asymptote, then adding the needed RHFs to connect the four asymptotes. The amplitude at the end points is defined to be the gain at band center minus half of the peak-to-peak ripple amplitude ($G_c - 0.5R$). Asymptote slope is either $\pm\frac{6R}{(f_{hi} - f_{lo})}$, **A1** and **A3** are positive, **A2** and **A4** are negative. Transition range for each intersection is arbitrarily set at $\frac{(f_{hi} - f_{lo})}{6}$ to simplify the product of slope change and transition range ΔSa, which is the multiplying coefficient of each RHF.

Asymptote **A1** is the reference asymptote and identifies with the equation

$$G_r = (G_c - 0.5R) + \frac{6R(f - f_{lo})}{(f_{hi} - f_{lo})} \tag{6.7}$$

The first RHF has change of slope $\frac{-12R}{(f_{hi} - f_{lo})}$, transition range $\frac{(f_{hi} - f_{lo})}{6}$, and intersect point $\frac{(f_{hi} + 5f_{lo})}{6}$ giving a value for the RHF of

$$-2R\log_{10}\left[1 + 10^{\frac{(6f - f_{hi} - 5f_{lo})}{(f_{hi} - f_{lo})}}\right] \tag{6.8}$$

The second RHF has change of slope $\frac{12R}{(f_{hi} - f_{lo})}$, transition range $\frac{(f_{hi} - f_{lo})}{6}$, and intersect point $\frac{(f_{hi} + f_{lo})}{2}$ giving a value for the RHF of

$$+2R\log_{10}\left[1 + 10^{\frac{(6f - 3f_{hi} - 3f_{lo})}{(f_{hi} - f_{lo})}}\right] \tag{6.9}$$

The third RHF has change of slope $\frac{-12R}{(f_{hi} - f_{lo})}$, transition range $\frac{(f_{hi} - f_{lo})}{6}$, and intersect point $\frac{(5f_{hi} + f_{lo})}{6}$ giving a value for the RHF of

$$-2R\log_{10}\left[1 + 10^{\frac{(6f - 5f_{hi} - f_{lo})}{(f_{hi} - f_{lo})}}\right] \tag{6.10}$$

Sum the equation for asymptote **A1** and the three RHFs to obtain the model for in-band Chebishev ripple for a two-pole network.

$$G_2 = (G_c - 0.5R)$$

$$+ 2QR\left\{\frac{3(f - f_{lo})}{(f_{hi} - f_{lo})} + \log_{10}\left[\frac{\left(1 + 10^{\frac{(6f - 3f_{hi} - 3f_{lo})}{(f_{hi} - f_{lo})}}\right)}{\left(1 + 10^{\frac{(6f - f_{hi} - 5f_{lo})}{(f_{hi} - f_{lo})}}\right)\left(1 + 10^{\frac{(6f - 5f_{hi} - f_{lo})}{(f_{hi} - f_{lo})}}\right)}\right]\right\} \tag{6.11}$$

The approximation given to transition range causes a slight error in peak-to-peak ripple amplitude that is corrected by factor Q (not to be confused with quality factor normally assigned to resonant LC circuits). The value of Q varies slightly depending on the number of poles n in the model equation. The value of correction factor $Q = 1.22$ for a two-pole Chebishev bandpass model. The value G_2, (6.11), is substituted for the value $G_c(f_c, b, i_{ds})$ in (6.6) to complete the model of a two-pole Chebishev bandpass response.

$$G_{bp}(f) = (G_c - 0.5R)$$

$$+ 2QR\left\{\frac{3(f - f_{lo})}{(f_{hi} - f_{lo})} + \log_{10}\left[\frac{\left(1 + 10^{\frac{(6f - 3f_{hi} - 3f_{lo})}{(f_{hi} - f_{lo})}}\right)}{\left(1 + 10^{\frac{(6f - f_{hi} - 5f_{lo})}{(f_{hi} - f_{lo})}}\right)\left(1 + 10^{\frac{(6f - 5f_{hi} - f_{lo})}{(f_{hi} - f_{lo})}}\right)}\right]\right\}$$

$$- 10\log_{10}\left[1 + \left(\frac{\left|\frac{f_c^2}{f} - f\right|}{(f_{hi} - f_{lo})}\right)^{2n}\right] \tag{6.12}$$

Figure 6.4 illustrates the results for a two-pole, 1 dB ripple Chebishev bandpass response.

Equation (6.12) is inserted into the amplifier behavioral model (4.4) and (4.14) for class A amplifiers and into (5.1) and (5.2) for class AB amplifiers to give variability to gain and phase as a function of frequency.

6.2.2.2 The Three-Pole Network Chebishev Ripple Response

The three-pole matching network exhibits three gain peaks in the bandpass with a gain maximum in the center of the band as illustrated in Figure 6.5. Six asymptotes are constructed to represent the three peaks such that the two end asymptotes **A1** and **A6** pass through zero value at the frequency points f_{lo} and

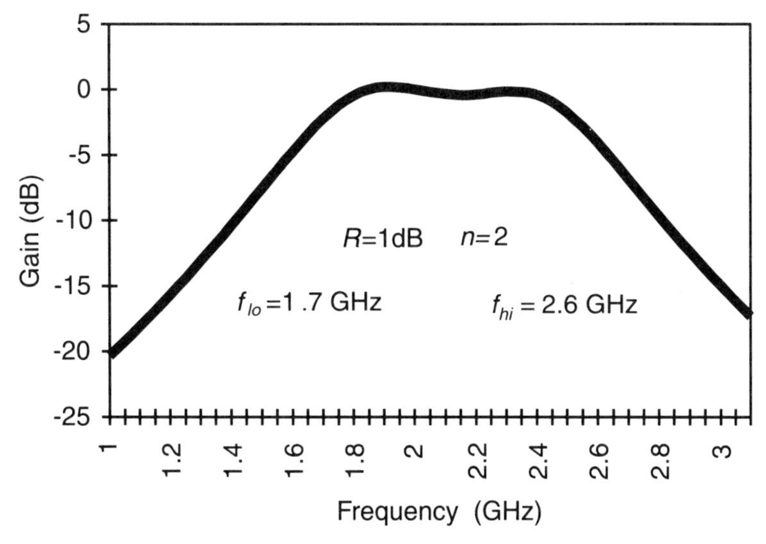

Figure 6.4 Model results for a two-pole, 1 dB ripple, Chebishev bandpass response where $G_c = 0$.

f_{hi}. The two central asymptotes **A3** and **A4** intersect at a point mid-band. The frequency range between points f_{lo} and f_{hi} is divided into ten equal intervals. Asymptotes **A1** and **A2** intersect one-tenth of the way into the band. Asymptotes **A2** and **A3** intersect three-tenths of the way across the band. Asymptotes **A4** and **A5** intersect seven-tenths of the way across the band, and

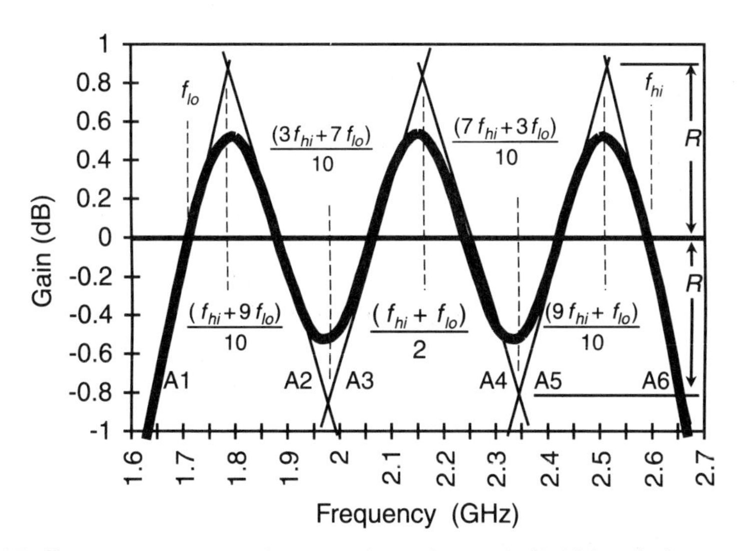

Figure 6.5 Six asymptotes are used to approximate three-pole Chebishev ripple.

asymptotes **A5** and **A6** intersect nine-tenths of the way across the band. The value of asymptote **A1** at f_{lo} and **A6** at f_{hi} is defined to be band center gain minus half of the peak-to-peak ripple amplitude ($G_c - 0.5R$). Figure 6.5 illustrates the six asymptotes and their locations.

The ten intervals each have width $\frac{(f_{hi} - f_{lo})}{10}$. The intersection of asymptotes **A1** and **A2** occurs at frequency $\frac{(f_{hi} + 9f_{lo})}{10}$, the sum of f_{lo} and one interval. The intersection of asymptotes **A2** and **A3** occurs at frequency $\frac{(3f_{hi} + 7f_{lo})}{10}$. The intersection of asymptotes **A3** and **A4** occurs at frequency $\frac{(f_{hi} + f_{lo})}{2}$, mid-band. The intersection of asymptotes **A4** and **A5** occurs at $\frac{(7f_{hi} + 3f_{lo})}{10}$, and the intersection of **A5** and **A6** occurs at $\frac{(9f_{hi} + f_{lo})}{10}$, the upper end of the bandpass f_{hi} minus one interval. The amplitude at which asymptotes **A1** and **A2** intersect is R dB greater than the amplitudes at the end points f_{lo} and f_{hi}. The amplitude at which asymptotes **A4** and **A5** intersect is R dB less than the amplitude at the end points f_{lo} and f_{hi}. The total difference in amplitude between the two intersection point extremes is $2R$. Asymptote slope is either $\pm\frac{10R}{(f_{hi} - f_{lo})}$, **A1**, **A3**, and **A5** have positive slope values, **A2**, **A4**, and **A6** have negative slope values. Transition range for each intersection is arbitrarily set at $\frac{(f_{hi} - f_{lo})}{10}$ to simplify the product of slope change ΔS and transition range a, which is the multiplying coefficient ΔSa of each RHF. Asymptote **A1** is the reference asymptote and is identified with the equation

$$G_r = (G_c - 0.5R) + \frac{10R(f - f_{lo})}{(f_{hi} - f_{lo})} \tag{6.13}$$

The first RHF has change of slope $\frac{-20R}{(f_{hi} - f_{lo})}$, transition range

$\frac{(f_{hi} - f_{lo})}{10}$, and intersect point $\frac{(f_{hi} + 9f_{lo})}{10}$, giving a value for the RHF of

$$-2R\log_{10}\left[1 + 10^{\frac{(10f - f_{hi} - 9f_{lo})}{(f_{hi} - f_{lo})}}\right] \tag{6.14}$$

The second RHF has change of slope $\frac{20R}{(f_{hi} - f_{lo})}$, transition range $\frac{(f_{hi} - f_{lo})}{10}$, and intersect point $\frac{(3f_{hi} + 7f_{lo})}{10}$, giving a value for the RHF of

$$+2R\log_{10}\left[1 + 10^{\frac{(10f - 3f_{hi} - 7f_{lo})}{(f_{hi} - f_{lo})}}\right] \tag{6.15}$$

The third RHF has change of slope $\frac{-20R}{(f_{hi} - f_{lo})}$, transition range $\frac{(f_{hi} - f_{lo})}{10}$, and intersect point $\frac{(f_{hi} + f_{lo})}{2}$, giving a value for the RHF of

$$-2R\log_{10}\left[1 + 10^{\frac{(10f - 5f_{hi} - 5f_{lo})}{(f_{hi} - f_{lo})}}\right] \tag{6.16}$$

The fourth RHF has change of slope $\frac{20R}{(f_{hi} - f_{lo})}$, transition range $\frac{(f_{hi} - f_{lo})}{10}$, and intersect point $\frac{(7f_{hi} + 3f_{lo})}{10}$, giving a value for the RHF of

$$+2R\log_{10}\left[1 + 10^{\frac{(10f - 7f_{hi} - 3f_{lo})}{(f_{hi} - f_{lo})}}\right] \tag{6.17}$$

The fifth and final RHF has change of slope $\frac{-20R}{(f_{hi} - f_{lo})}$, transition range $\frac{(f_{hi} - f_{lo})}{10}$, and intersect point $\frac{(9f_{hi} + f_{lo})}{10}$, giving a value for the RHF of

$$-2R\log_{10}\left[1 + 10^{\frac{(10f-9f_{hi}-f_{lo})}{(f_{hi}-f_{lo})}}\right] \tag{6.18}$$

Sum the reference equation for asymptote **A1** and the five RHFs to obtain the model for approximating three-pole bandpass Chebishev ripple,

$$G_{3p}(f) = (G_c - 0.5R) + 2QR\left\{\frac{5(f-f_{lo})}{(f_{hi}-f_{lo})}\right.$$

$$\left. + \log_{10}\left[\frac{\left(1 + 10^{\frac{(10f-3f_{hi}-7f_{lo})}{(f_{hi}-f_{lo})}}\right)\left(1 + 10^{\frac{(10f-7f_{hi}-3f_{lo})}{(f_{hi}-f_{lo})}}\right)}{\left(1 + 10^{\frac{(10f-f_{hi}-9f_{lo})}{(f_{hi}-f_{lo})}}\right)\left(1 + 10^{\frac{(10f-5f_{hi}-5f_{lo})}{(f_{hi}-f_{lo})}}\right)\left(1 + 10^{\frac{(10f-9f_{hi}-f_{lo})}{(f_{hi}-f_{lo})}}\right)}\right]\right\} \tag{6.19}$$

The approximation of transition range causes a slight error in peak-to-peak ripple amplitude that is corrected by factor Q (not to be confused with quality factor normally assigned to resonant LC circuits). The value of Q varies depending on the number of poles n in the model equation. The value of correction factor Q for a three-pole Chebishev bandpass model is 1.27.

Equation (6.19) $G_{3P}(f)$, describes ripple amplitude in the bandpass. It is substituted for the value $G_c(f_c, b, i_{ds})$ in the maximally flat (6.6) to complete the model of a three-pole Chebishev bandpass response

$$G_{bp}(f) = (G_c - 0.5R) + 2QR\left\{\frac{5(f-f_{lo})}{(f_{hi}-f_{lo})}\right.$$

$$\left. + \log_{10}\left[\frac{\left(1 + 10^{\frac{(10f-3f_{hi}-7f_{lo})}{(f_{hi}-f_{lo})}}\right)\left(1 + 10^{\frac{(10f-7f_{hi}-3f_{lo})}{(f_{hi}-f_{lo})}}\right)}{\left(1 + 10^{\frac{(10f-f_{hi}-9f_{lo})}{(f_{hi}-f_{lo})}}\right)\left(1 + 10^{\frac{(10f-5f_{hi}-5f_{lo})}{(f_{hi}-f_{lo})}}\right)\left(1 + 10^{\frac{(10f-9f_{hi}-f_{lo})}{(f_{hi}-f_{lo})}}\right)}\right]\right\} +$$

$$-10\log_{10}\left[1 + \left(\frac{\left|\frac{f_c^2}{f} - f\right|}{(f_{hi}-f_{lo})}\right)^{2n}\right] \tag{6.20}$$

Figure 6.6 illustrates the results for a three-pole, 1 dB ripple Chebishev bandpass response.

Equation (6.20) is inserted into the amplifier behavioral model (4.4) and (4.14) for class A amplifiers and into (5.1) and (5.2) for class AB amplifiers to give variability to gain and phase as a function of frequency.

6.2.2.3 Higher Order Network Responses

The development procedure and general form of equation for approximating higher order Chebishev ripple response follows from development of (6.12) and (6.20). The number of intervals I into which the bandpass needs to be subdivided is determined by

$$I = 2(2n - 1) \tag{6.21}$$

where n is the number of poles in the network. The number of asymptotes needed to model the function

$$N = 2n \tag{6.22}$$

The number of RHFs required

$$RHF(n) = 2n - 1 \tag{6.23}$$

The reference asymptote's equation

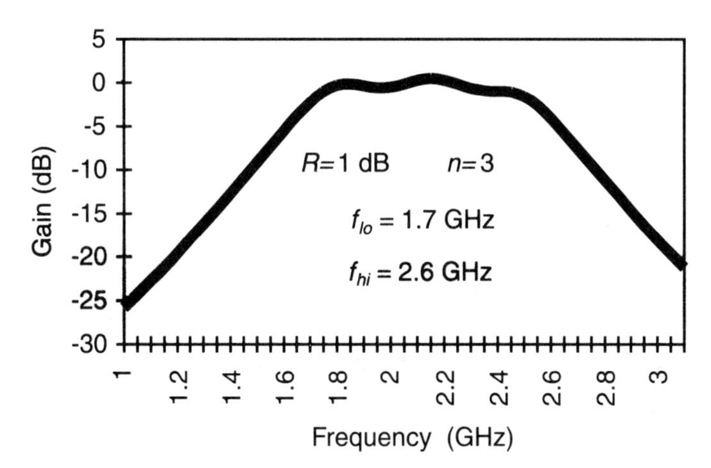

Figure 6.6 Model results for a three-pole, 1 dB ripple, Chebishev bandpass response where $G_c = 0$.

$$G_r = G_c + \frac{2(2n-1)R(f-f_{lo})}{(f_{hi}-f_{lo})} \tag{6.24}$$

Asymptote intersect frequencies f_i are a function of the number of RHFs needed and the number of poles (designated by n). Intersections occur at frequencies

$$f_1 = f_{lo} + \frac{[(2i-1)f_{hi} + (4n-2i-1)f_{lo}]}{2(2n-1)} \tag{6.25}$$

where i is the *ith* RHF and n is the number of poles. The exponent of ten associated with *ith* RHF has the value

$$\frac{2(2n-1)f-(2i-1)f_{hi} - (4n-2i-1)f_{lo}}{(f_{hi}-f_{lo})} \tag{6.26}$$

The change in slope going from one asymptote to the next

$$\Delta S = (-1)^i \frac{4(2n-1)R}{(f_{hi}-f_{lo})} \tag{6.27}$$

The transition range

$$a = \frac{(f_{hi}-f_{lo})}{2(2n-1)} \tag{6.28}$$

The product of slope change and transition range for the *ith* RHF

$$\Delta S a_i = (-1)^i 2R \tag{6.29}$$

Correction coefficient Q is determined empirically for the number of poles in the Chebishev response. A good average value is $Q = 1.244$. Table 6.2 lists correction coefficient Q values for networks having 2 to 6 poles.

6.2.3 Modeling Measured Small Signal Gain Data

Measured small signal gain data rarely obeys the ideal Butterworth or Chebishev response described above. When it does not, a curve-fit equation has to be developed for the data, using techniques described in Chapter 2. Begin with a plot of measured G_{ss} in decibel units over a linear frequency scale and select a point

Table 6.2
Correction Coefficient *Q* Values for *n* Pole Networks

n	*Q*
2	1.22
3	1.27
4	1.28
5	1.22
6	1.23

near the center of the frequency range where an asymptote to the data has a constant value as a function of frequency as shown in Figure 6.7. This constant value of small signal gain will be the reference asymptote.

Construct asymptotes to the data trace proceeding to the left from the reference asymptote as illustrated in Figure 6.8. Use as many asymptotes as is required to capture each major change in data trace first derivative. Define each asymptote with coordinates of two points (x_i, y_i) and (x_j, y_j) on each line. Using (2.4), compute slope values for each of the asymptotes. Using (2.7), compute asymptote intercept points. Compute slope change going from one asymptote to the next and estimate transition ranges at each intersection. Calculate and tabulate coefficients for left-hand functions that model the lower frequency range of data.

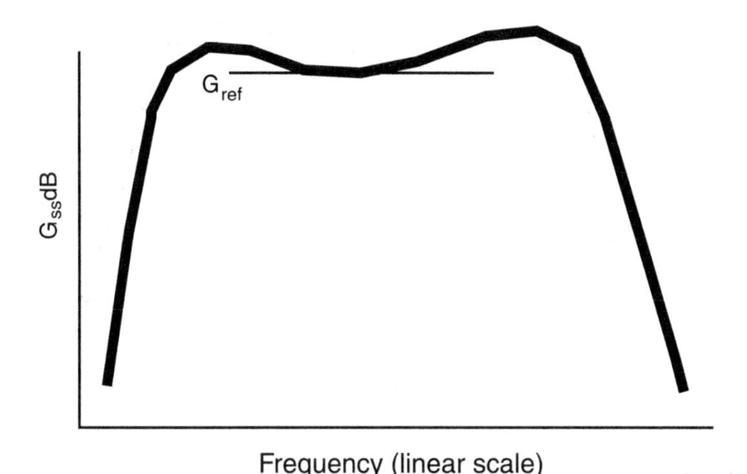

Figure 6.7 Begin curve fitting with a constant value asymptote near the band center.

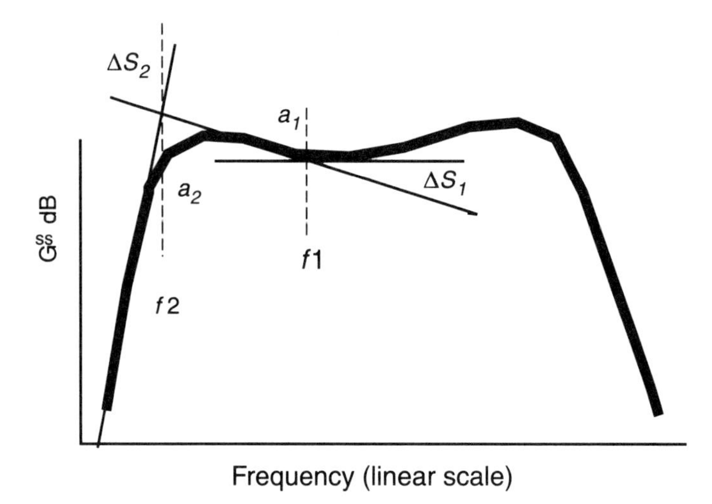

Figure 6.8 Add asymptotes to the left of the reference to capture major data trace first derivative changes.

Add asymptotes to the right of the reference asymptote to capture changes in first derivative for frequencies in the upper end of the data range as illustrated in Figure 6.9. Develop the database needed for calculating right-hand function coefficients.

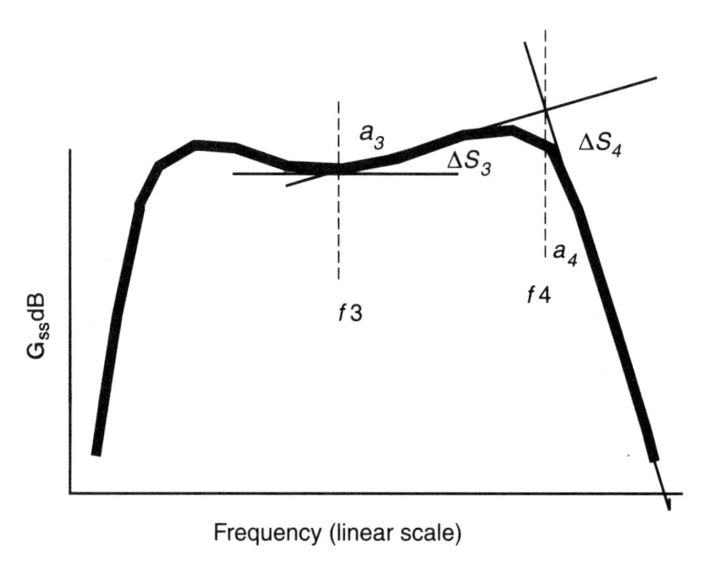

Figure 6.9 Add asymptotes to the right of the reference to capture major data trace first derivative changes.

Form the small signal gain model equation by summing the reference asymptote equation, all of the LHFs and all of the RHFs. The resulting equation will have the form

$$G_{ss}(f) = G_{ss} + \Delta S_1 a_1 \log\left[\frac{10^{\left(\frac{f-f_1}{a_1}\right)}}{1 + 10^{\left(\frac{f-f_1}{a_1}\right)}}\right] + \Delta S_2 a_2 \log\left[\frac{10^{\left(\frac{f-f_2}{a_2}\right)}}{1 + 10^{\left(\frac{f-f_2}{a_2}\right)}}\right] +$$

$$\ldots\ldots\ldots + \Delta S_3 a_3 \log\left[1 + 10^{\left(\frac{f-f_3}{a_3}\right)}\right] + \Delta S_4 a_4 \log\left[1 + 10^{\left(\frac{f-f_4}{a_4}\right)}\right] + \ldots\ldots\ldots \quad (6.30)$$

Equation (6.30) is inserted into the amplifier behavioral models (4.4) and (4.14) for class A amplifiers and into (5.1) and (5.2) for class AB amplifiers to give variability to gain and phase as a function of frequency.

6.3 Saturated Power Output as a Function of Frequency

Saturated power output is realized when RF current swing extends the full length of the load line driving into cut-off at one end and saturation at the other. The RF load line impedance is a dynamic function of frequency. Its value is dependent on matching network design and not directly measurable by the amplifier user. As load line impedance value changes with frequency, the peak-to-peak RF current out of the transistor at saturated power output changes. Transistor maximum current output is limited by either I_{dss} for FETs or I_{max} for bipolar devices when load line impedance is matched for power in the bandpass. The net result is that for load line impedance variation of as much as 1.5:1, saturated power output does not change more than 1.76 dB. Saturated power output over most of the operating frequency band usually falls within ± 0.5 dB of average P_{sat} over the band. Load line impedance value for frequencies out of band increases significantly, causing peak-to-peak current at saturated power output to decrease. For example, consider an amplifier operating at saturated power output at a frequency in band where load line impedance value is R_l. Peak-to-peak current swing is i_{dss}, and saturated power output

$$p_{sat1} = \frac{i_{dss}^{\ 2}}{8} R_l \quad (6.31)$$

When frequency moves out of band and load line impedance increases to $10R_l$, peak-to-peak current at saturated power output drops to $0.1\, i_{dss}$. Peak-to-peak voltage swing at saturated power output is limited to the applied power supply voltage and does not change. Saturated power output drops to

$$p_{\text{sat2}} = \frac{0.01 i_{\text{dss}}^{\;2}}{8} 10 R_1 = 0.1 p_{\text{sat1}} \tag{6.32}$$

which is 10 dB less than in band saturated power output. Load line impedance value continues to increase as frequency moves farther out of band.

6.3.1 Simulate Saturated Power Output Over the Band

Flat, uniform, saturated power output over an amplifier's bandpass can be simulated by developing a P_{sat} model the same way that a small signal gain model was developed (see (6.6)). First postulate a maximally flat saturated power output value at a frequency band center $P_{\text{sat}}(f_c)$, then define a frequency bandpass $(f_{\text{hi}} - f_{\text{lo}})$ with the model equation

$$P_{\text{sat}}(f) = P_{\text{sat}}(f_c) - 10 \log_{10}\left[1 + \left(\frac{\left|\frac{f_c^2}{f} - f\right|}{(f_{\text{hi}} - f_{\text{lo}})}\right)^{2n}\right] \tag{6.33}$$

where n is the number of poles in the impedance matching network.

Saturated power output ripple within the bandpass can be added by developing *nth* order ripple equations like (6.12) and (6.20), and substituting $P_{\text{sat}}(f_c)$ for G_c. An example equation for adding a third-order ripple to saturated power output is

$$P_{\text{sat3}} = (P_{\text{sat}}(f_c) - 0.5R) + 2QR\left\{\frac{5(f - f_{\text{lo}})}{(f_{\text{hi}} - f_{\text{lo}})}\right.$$

$$\left. + \log_{10}\left[\frac{\left(1 + 10^{\frac{(10f - 3f_{\text{hi}} - 7f_{\text{lo}})}{(f_{\text{hi}} - f_{\text{lo}})}}\right)\left(1 + 10^{\frac{(10f - 7f_{\text{hi}} - 3f_{\text{lo}})}{(f_{\text{hi}} - f_{\text{lo}})}}\right)}{\left(1 + 10^{\frac{(10f - f_{\text{hi}} - 9f_{\text{lo}})}{(f_{\text{hi}} - f_{\text{lo}})}}\right)\left(1 + 10^{\frac{(10f - 5f_{\text{hi}} - 5f_{\text{lo}})}{(f_{\text{hi}} - f_{\text{lo}})}}\right)\left(1 + 10^{\frac{(10f - 9f_{\text{hi}} - f_{\text{lo}})}{(f_{\text{hi}} - f_{\text{lo}})}}\right)}\right]\right\} \tag{6.34}$$

which is substituted for $P_{\text{sat}}(f_c)$ in (6.33) giving a final model

$$P_{sat}(f) = (P_{sat}(f_c) - 0.5R) + 2QR\left\{ \frac{5(f-f_{lo})}{(f_{hi}-f_{lo})} \right.$$

$$\left. + \log_{10}\left[\frac{\left(1+10^{\frac{(10f-3f_{hi}-7f_{lo})}{(f_{hi}-f_{lo})}}\right)\left(1+10^{\frac{(10f-7f_{hi}-3f_{lo})}{(f_{hi}-f_{lo})}}\right)}{\left(1+10^{\frac{(10f-f_{hi}-9f_{lo})}{(f_{hi}-f_{lo})}}\right)\left(1+10^{\frac{(10f-5f_{hi}-5f_{lo})}{(f_{hi}-f_{lo})}}\right)\left(1+10^{\frac{(10f-9f_{hi}-f_{lo})}{(f_{hi}-f_{lo})}}\right)} \right] \right\} +$$

$$-10\log_{10}\left[1 + \left(\frac{\left| \frac{f_c^2}{f} - f \right|}{(f_{hi}-f_{lo})} \right)^{2n} \right] \tag{6.35}$$

6.3.2 Model Measured Saturated Power Output Data Files

Measured saturated power output data can be modeled by performing a curve fit to the data file using the methods described above in Section 6.2.3 to model measured small signal gain. The resulting equation has the form

$$P_{sat}(f) = P_{sat}(f_c) + \Delta S_1 a_1 \log\left[\frac{10^{\left(\frac{f-f_1}{a_1}\right)}}{1+10^{\left(\frac{f-f_1}{a_1}\right)}} \right] + \Delta S_2 a_2 \log\left[\frac{10^{\left(\frac{f-f_2}{a_2}\right)}}{1+10^{\left(\frac{f-f_2}{a_2}\right)}} \right]$$

$$+ \ldots\ldots\ldots + \Delta S_3 a_3 \log\left[1+10^{\left(\frac{f-f_3}{a_3}\right)}\right] + \Delta S_4 a_4 \log\left[1+10^{\left(\frac{f-f_4}{a_4}\right)}\right] + \ldots\ldots\ldots \tag{6.36}$$

Either simulated or measured $P_{sat}(f)$ models can then be used in (5.1) and (5.2) to model class AB amplifier power output and phase as a function of power input over frequency and in (4.4) and (4.14) for class A amplifier models.

6.4 Bias Coefficient as a Function of Frequency

Quiescent bias coefficient b_q is defined in Section 5.2.1.2 to be the ratio of average bias current i_q to i_{dss} or i_{max} with no RF (no frequency). This definition provides a quantification of the designer's selection of quiescent bias current and a relative measure of how far into class AB operation the amplifier is with respect to class A. Bias current i_q is the variable in quiescent bias coefficient.

A dynamic bias coefficient b_f is now defined to be the ratio of quiescent bias current i_q to peak current along the load line which occurs when the RF current saturates (at any frequency). Quiescent bias current i_q (no signal present) is fixed; peak current i_{peak} at saturated power output varies as a function of load line impedance value which varies with frequency. Quiescent bias current i_q does not change; the saturation peak current changes. This gives rise to different bias coefficient values b_f as a function of frequency. An approximate value for $i_{peak}(f)$, which occurs at saturated power output, is obtained by assuming load line impedance variation causes change in P_{sat} as a function of frequency. Given either measured or simulated values for $P_{sat}(f)$, peak saturation current $i_{peak}(f)$ is then determined to be

$$i_{peak}(f) = \sqrt{\left(\frac{4 * 10^{\frac{P_{sat}(f)-30}{10}}}{(v_{ds} - v_k)}\right)^2 + i_q^2} \tag{6.37}$$

Using the value expressed by (6.37) for i_{peak}, the definition of dynamic bias coefficient is then determined to be

$$b_f = \frac{i_q}{\sqrt{\left(\frac{4 * 10^{\frac{P_{sat}(f)-30}{10}}}{(v_{ds} - v_k)}\right)^2 + i_q^2}} \tag{6.38}$$

This dynamic (6.38) for bias coefficient as a function of $P_{sat}(f)$ provides a relationship between compression coefficient $K(b)$ and frequency of operation. Recall that compression coefficient K is a function of bias coefficient.

6.5 Compression Coefficient as a Function of Frequency

Compression coefficient for MESFETs is defined by (5.11) and for bipolar transistors by (5.12). Both of these equations relate compression coefficient $K(b)$ to quiescent bias coefficient. Dynamic value b_f given by (6.38) can be substituted into (5.11) and (5.12) in place of the static value b_q to calculate a dynamic compression coefficient $K(f)$. This dynamic value for compression coefficient as a function of frequency completes the parameters needed to extend the class A and class AB amplifier behavioral models for power output and phase shift over the frequency domain.

An alternate, more involved method of determining compression coefficient $K(f)$ from measured data is described in Chapter 4, Section 4.1.1.2. Small signal gain $G_{ss}(f)$ and saturated power output $P_{sat}(f)$ as a function of frequency have to be measured and known. Amplifier power input equal to $[P_{sat}(f) - G_{ss}(f)]$ has to be set at each measured frequency to obtain measured amplifier power output $P_{outK}(f)$. A data file containing measured P_{sat} and $P_{outK}(f)$ is then developed and compression coefficient as a function of frequency is calculated using (4.6).

6.6 Average Bias Current as a Function of Frequency and Power Input

Bias current with no RF input, i_q, is constant across the frequency domain. Its value relative to the peak current obtained by the saturated RF current waveform i_{peak} determines the amplifier's class of operation (class A or class AB). Bias current changes as power input increases and the amplifier approaches saturation. The change is minimal for class A amplifiers biased at 50 percent I_{dss}. Change in average bias current as a function of power input for class AB amplifiers at a single frequency is illustrated in Examples 3.1, 3.2, and 3.3. As input signal frequency changes with a large signal input, the load line impedance value changes causing i_{peak} to change. Change in load line impedance value as a function of frequency is difficult to determine unless details of the load impedance match circuit are known. Average bias current as a function of power input and frequency needs to be measured in order to accurately model power-added efficiency and heat dissipation as a function of frequency. An approximate value for $i_{peak}(f)$, which occurs at saturated power output, is given in (6.37) above and repeated here

$$i_{peak}(f) = \sqrt{\left(\frac{4 * 10^{\frac{P_{sat}(f)-30}{10}}}{(v_{ds} - v_k)}\right)^2 + i_q^2}$$

Using (5.15) from Chapter 5, Section 5.2.2 as the basis, a model of bias current as a function of frequency is developed.

$$i_{dc}(f) = i_{peak}(f)\left\{b_q + \frac{(0.5 - b_q)}{2}\log_{10}\left[\frac{1 + 10^{\frac{2(P_{in}-P_{sat}(f)-P_1(b))}{(P_2(b)-P_1(b))}}}{1 + 10^{\frac{2(P_{in}-P_{sat}(f)-P_2(b))}{(P_2(b)-P_1(b))}}}\right]\right\} \quad (6.39)$$

where $i_{peak}(f)$ is from (6.37) and normalized power levels $P_1(b)$ and $P_2(b)$ come from (5.19) and (5.18). Values for $P_1(b)$ and $P_2(b)$ are specific to MESFETs and bipolar transistors and are determined as shown in Example 5.1 for the square law MESFET. Value $P_2(b)$ is the same for square law and step-doped MESFETS, and the difference $[P_2(b) - P_1(b)]$ has a constant but different value for each transistor type.

6.7 Noise Figure as a Function of Frequency

A rule of thumb for estimating amplifier noise figure at a single frequency is proposed by (4.17), in Section 4.1.6, Chapter 4. Noise figure in dB as a function of frequency can be estimated by noting noise figure value at a reference frequency, for example the frequency in the center of a bandpass f_c, then adding to that value a value of gain or loss equal to $(G_c - G(f))$, (small signal gain at center frequency minus small signal gain at frequency points across the band). The rule of thumb for estimating noise figure in dB over frequency then becomes

$$NF \approx 10\sqrt{i_{dc}(f)} + 0.5 + (G_c - G(f)) \tag{6.40}$$

The term $i_{dc}(f)$ accounts for bias current variation over frequency with power input as occurs in class AB amplifiers. If small signal gain over frequency is simulated to be an n pole Butterworth response, noise figure estimate as a function of frequency is then

$$NF(f) = 10\sqrt{i_{ds}(f)} + 0.5 + 10\log_{10}\left[1 + \left(\frac{\left|\frac{f_c^2}{f} - f\right|}{(f_{hi} - f_{lo})}\right)^{2n}\right] \tag{6.41}$$

If small signal gain over frequency is simulated to be a two-pole Chebishev response with R dB of ripple, noise figure estimate as a function of frequency is then

$$NF(f) = 10\sqrt{i_{ds}(f)} + 0.5(1 + R) - 2QR\left\{\frac{3(f - f_{lo})}{(f_{hi} - f_{lo})}\right.$$

$$+\log_{10}\left[\frac{\left(1+10^{\frac{(6f-3f_{hi}-3f_{lo})}{(f_{hi}-f_{lo})}}\right)}{\left(1+10^{\frac{(6f-f_{hi}-5f_{lo})}{(f_{hi}-f_{lo})}}\right)\left(1+10^{\frac{(6f-5f_{hi}-f_{lo})}{(f_{hi}-f_{lo})}}\right)}\right]\Bigg\}$$

$$+10\log_{10}\left[1+\left(\frac{\left|\frac{f_c^2}{f}-f\right|}{(f_{hi}-f_{lo})}\right)^{2n}\right] \quad (6.42)$$

Equations for higher order networks can be developed easily following the procedures outlined in Section 6.2.2.3.

6.8 Power-Added Efficiency as a Function of Frequency

Power-added efficiency as a function of power input over frequency is now calculable using the dynamic equations developed in this chapter for power output and average DC current. Power output $P_{out}(f)$ as a function of $P_{in}(f)$, $G(f)$, $K(f)$, and $P_{sat}(f)$ is now defined. Average DC current $i_{ds}(f)$ as a function of $P_{in}(f)$, $P_{sat}(f)$, $P_1(b)$, $P_2(b)$, b_q, and $i_{max}(f)$ is also defined. All of these parameters dynamically affect power-added efficiency in the equation

$$PAE=\frac{\left(10^{\frac{P_{out}(f)-30}{10}}-10^{\frac{P_{in}(f)-30}{10}}\right)}{v_{ds}i_{ds}(f)}*100 \quad (6.43)$$

Example 6.1

Given : A class AB MESFET amplifier exhibits flat saturated power output of +30 dBm and operates over a 3 dB bandwidth 2.5 GHz to 3.2 GHz. Maximum small signal gain within the operating bandwidth is 14 dB. Small signal gain as a function of frequency exhibits Chebishev 1 dB peak-to-peak ripple, two poles. Power supply voltage applied is 8 volts. Quiescent bias current with no RF is 0.114 amps.

Determine:

1. Average DC current i_{ds} as a function of power input over the frequency band.

2. Power output as a function of P_{in} over the frequency band.
3. Compression depth as a function of P_{in} over the frequency band.
4. Phase shift as a function of P_{in} over the frequency band.
5. Power-added efficiency as a function of P_{in} over the frequency band.
6. The compression depth at which maximum *PAE* occurs over the frequency band.

Procedure:
Create a spreadsheet and enter all of the given parameters including those parameters directly calculable from given values such as bandpass mean frequency f_c, and i_{max} at f_c based on P_{sat} and voltages given (assume $v_k = 1$ volt). Establish a column for frequency values that cover at least the band of interest. Using Equation (6.18) calculate P_{sat} as a function of frequency and place values in the column next to frequency. Similarly, using appropriate equations, establish columns for i_{max}, i_q, $P_1(b)$, $P_2(b)$, i_{ds}, $b(f)$, $K(f)$, $G(f)$, $P_{out}(f)$, compression, $\varphi(f)$, and $PAE(f)$, and calculate all of the parameters. Average DC current as a function of power input over frequency is shown in Figure 6.10.

Notice that average current remains at the quiescent bias level until power input begins to drive the transistor into the nonlinear region of operation. Power output as a function of power input over the frequency band is plotted in Figure 6.11. Notice that at low power levels, power output follows the small signal gain Chebishev 1 dB ripple and that as power input increases driving the transistor into the nonlinear region power output flattens across the band reproducing flat saturated power output. Saturated power output in band is +30 dBm.

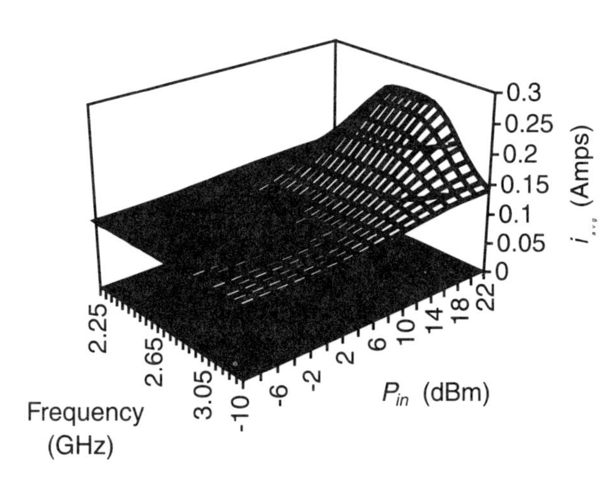

Figure 6.10 Example 6.1 average DC current as a function of power input over frequency.

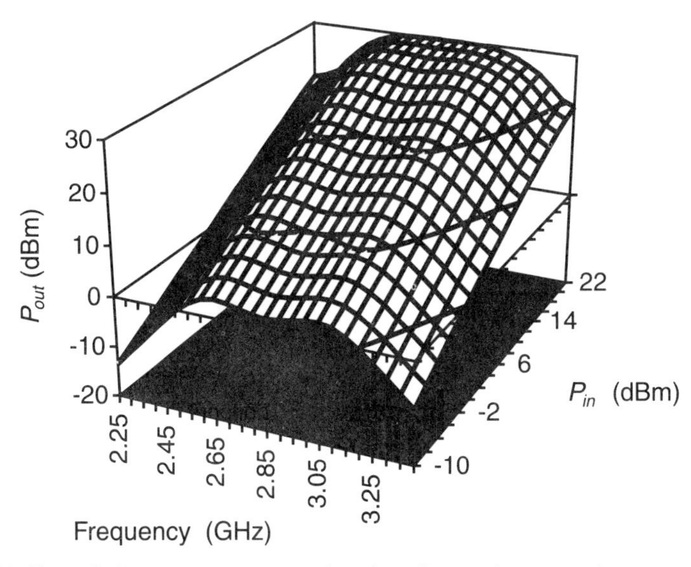

Figure 6.11 Example 6.1 power output as a function of power input over frequency.

Compression depth as a function of power input is plotted in Figure 6.12. Notice how gain ripple has little effect on compression where power input is low. As power input drives the transistor into the nonlinear region, gain ripple causes ripple in compression depth. Compression coefficient $K(f)$ is also changing with frequency, causing compression to peak at band edges.

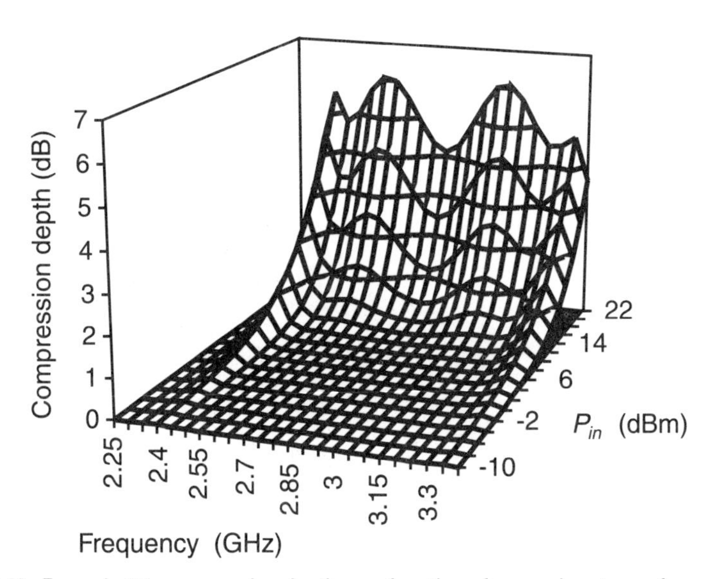

Figure 6.12 Example 6.1 compression depth as a function of power input over frequency.

Phase shift as a function of frequency follows compression depth since phase shift is determined by multiplying the compression term by phase sensitivity of five degrees per dB of compression. Use the compression data shown in Figure 6.12 and multiply the vertical scale by a factor of five to understand how phase shift values change as a function of power input over frequency.

Power-added efficiency as a function of power input over frequency plotted in Figure 6.13 shows the advantage of operating at a specific power input level. *PAE* peaks then falls off as the transistor is driven harder into compression. The effect of gain ripple on *PAE* also shows clearly in the plotted surface. *PAE* is greatest where gain peaks occur.

Power-added efficiency is studied more closely at 2.65 GHz, 2.8 GHz, and 3.05 GHz in Figure 6.14 to show that peak *PAE* occurs when compression depth is 2.0 dB.

6.9 Summary

Expansion of behavioral modeling into the frequency domain has been accomplished in this chapter. A technique for simulating maximally flat or Butterworth frequency response of n pole networks has been developed resulting in a model equation

$$G_{bp}(f) = G_c(f_c, b, i_{ds}) - 10\log_{10}\left[1 + \left(\frac{\left|\frac{f_c^2}{f} - f\right|}{(f_{hi} - f_{lo})}\right)^{2n}\right] \tag{6.44}$$

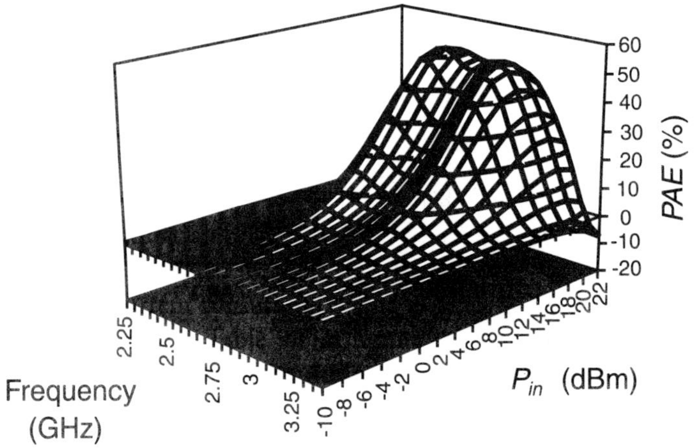

Figure 6.13 Example 6.1 power-added efficiency as a function of power input over frequency.

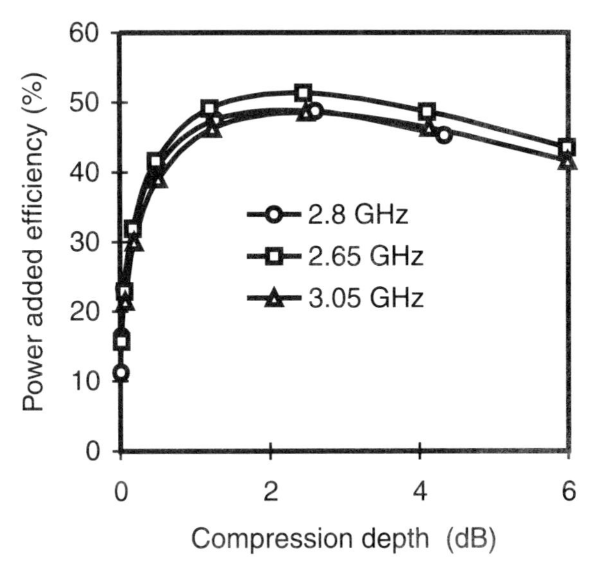

Figure 6.14 Peak power-added efficiency occurs at 2 dB compression.

This technique was expanded to include simulation of n pole Chebishev ripple of amplitude R in the bandpass resulting in equations for two-pole and three-pole ripple and a technique for developing higher order networks with in-band ripple. The equations for Chebishev ripple (see (6.12) and (6.20)) replace the term $G_c(f_c, b, i_{ds})$ in (6.44) above.

Equations for saturated power output as a function of frequency similar to those for small signal gain were developed. Saturated power output of an amplifier with an n pole, output impedance matching network, that is maximally flat across the operating band, can be represented by

$$P_{sat}(f) = P_{sat}(f_c) - 10 \log_{10}\left[1 + \left(\frac{\left|\frac{f_c^2}{f} - f\right|}{(f_{hi} - f_{lo})}\right)^{2n}\right] \tag{6.45}$$

Techniques for adding Chebishev ripple to saturated power output were also developed.

A new bias coefficient b_f whose value is dependent on load line impedance and saturated power output was defined. Since load line impedance value and saturated power output vary with frequency, this new bias coefficient b_f varies with frequency. Compression coefficient K is dependent on bias coefficient, hence $K(f)$ is defined and equations for the value of $K(f)$ are developed for the different transistor types.

Average bias current is a function of saturated power output P_{sat}, power input P_{in}, and quiescent bias coefficient b_q. Since saturated power output is a function of frequency, average bias current also varies with frequency. The relationship is developed into the model equation

$$i_{dc}(f) = i_{peak}(f)\left\{ b_q + \frac{(0.5 - b_q)}{2} \log_{10}\left[\frac{1 + 10^{\frac{2(P_{in} - P_{sat}(f) - P_1(b))}{(P_2(b) - P_1(b))}}}{1 + 10^{\frac{2(P_{in} - P_{sat}(f) - P_2(b))}{(P_2(b) - P_1(b))}}} \right] \right\} \quad (6.46)$$

The rule of thumb for estimating amplifier noise figure, (4.17), is expanded into the frequency domain for circuit simulations using Butterworth and Chebishev responses. Noise figure as a function of frequency for the n pole Butterworth response is modeled as

$$NF(f) = 10\sqrt{i_{ds}} + 0.5 + 10 \log_{10}\left[1 + \left(\frac{\left| \frac{f_c^2}{f} - f \right|}{(f_{hi} - f_{lo})} \right)^{2n} \right] \quad (6.47)$$

The dynamic relationship of *PAE* over frequency and input power level is explained and a model equation is developed.

References

[1] Dishal, M., "Filters, Coupling Networks, and Attenuators," In *Electronics Engineer's Handbook*, Section 12, D. Fink (editor in chief), New York, NY, McGraw-Hill, 1975.

[2] Dishal, M., "Filters, Image Parameter Design, Modern Network Theory Design, Simple Band-pass Design," In *Reference Data For Radio Engineers, Fifth Edition*, Chapters 7-9, H. P. Westman (ed), Indianapolis, IN, Howard W. Sams & Co., Inc., 1969.

7

Waxing Hot and Cold

7.1 Adding Temperature as a Variable

Temperature affects just about every parameter associated with electronic circuitry. Small signal gain, saturated power output, bias current, noise figure, S-parameters, to name a few, are all affected by temperature. Sensitivity of a given parameter to temperature may not be constant as a function of other variables. For instance, gain sensitivity and saturated power output sensitivity to temperature often vary as a function of frequency. Methods are developed in this chapter to include the effect of temperature sensitivities in the behavioral models developed in earlier chapters.

Temperature sensitivities referred to in this chapter are for a single amplifier stage. Gain sensitivity of multiple cascaded stages is determined by multiplying the number of stages by the sensitivity of a single stage. Typical value for amplifier stage gain sensitivity to temperature and saturated power output sensitivity value to temperature is −0.015 dB/degrees C. Typical value of amplifier stage noise figure sensitivity to temperature is +0.015 dB/degrees C. Typical saturated power output sensitivity to temperature for a single amplifier stage is independent of all other stages. If a multistage amplifier is designed wisely, the saturated power output temperature sensitivity of the final stage is the only P_{sat} temperature sensitivity that need be of concern. Noise figure sensitivity to temperature of the first stage of a multistage amplifier is the predominant noise figure sensitivity, unless first stage amplifier gain is low. Room temperature of 25 degrees C is used as a reference temperature.

7.2 Small Signal Gain Sensitivity to Temperature

Amplifier small signal gain as used in behavioral models is expressed in decibels (dB). Small signal gain sensitivity to temperature is therefore expressed as change in gain ΔdB per change in temperature Δ degrees C. It is recognized that parameter temperature sensitivity is often not constant with frequency and is sometimes a nonlinear function of frequency, and that the definition needs to include that variability. Parameter temperature sensitivity is defined therefore as

$$U(f) = \frac{\Delta dB(f)}{\Delta °\text{C}} \tag{7.1}$$

This definition is universally applied to small signal gain, saturated power output, noise figure, and any other parameter that has a temperature sensitivity that can be expressed as change in gain ΔdB per change in temperature Δ degrees C with variability over temperature. Change in G_{ss}, P_{sat}, or NF is always referenced to ambient temperature of 25 degrees C. Small signal gain of a single amplifier stage as a function of frequency and temperature is then determined to be

$$G_{ss}(f, T) = G_{ss}(f, 25) + U_G(f)(T - 25) \tag{7.2}$$

The term $G_{ss}(f, 25)$ is small signal gain as a function of frequency measured or simulated at temperature 25 degrees C. If the behavioral model of small signal gain at 25 degrees C is created from curve fit to measured data, $G_{ss}(f, 25)$ has the general form

$$G_{ss}(f, 25) = A + Bf + \sum_{1}^{i} C_i * \log_{10}\left[\frac{10^{\frac{f-f_i}{c_i}}}{1 + 10^{\frac{f-f_i}{c_i}}}\right] + \sum_{1}^{k} D_k * \log_{10}\left[1 + 10^{\frac{f-f_k}{d_k}}\right] \tag{7.3}$$

If the model equation for small signal gain is developed as a simulation to a Butterworth or Chebishev response, $G_{ss}(f, 25)$ has the general form

$$G_{ss}(f, 25) = G_c(f_c, b, i_{ds}) - 0.5R - 10 \log_{10}\left[1 + \left(\frac{\left|\frac{f_c^2}{f} - f\right|}{(f_{hi} - f_{lo})}\right)^{2n}\right]$$

$$+ 2QR\left\{\frac{3(f - f_{lo})}{(f_{hi} - f_{lo})} + \log_{10}\left[\frac{\left(1 + 10^{\frac{(6f-3f_{hi}-3f_{lo})}{(f_{hi}-f_{lo})}}\right)}{\left(1 + 10^{\frac{(6f-f_{hi}-5f_{lo})}{(f_{hi}-f_{lo})}}\right)\left(1 + 10^{\frac{(6f-5f_{hi}-f_{lo})}{(f_{hi}-f_{lo})}}\right)}\right]\right\} \tag{7.4}$$

Small signal gain temperature sensitivity $U_G(f)$ can range from a constant over frequency, to a linear function of frequency, and even to a more complex function of frequency describable by a polynomial expression. Figure 7.1 shows small signal gain as a function of temperature over the operating frequency band 2 GHz to 28 GHz.

Temperature sensitivity of gain $U_G(f)$ shown in Figure 7.1 is -0.015 dB/degrees C at the low end of the band, at mid-band -0.022 dB/degrees C, and at the upper end of the band -0.03 dB/degrees C. Gain temperature sensitivity of this MMIC device is not linear with frequency. There are three data points that can be used to develop a second-order polynomial equation to describe $U_G(f)$. Expand a polynomial equation about frequency at 10 GHz. Table 7.1 lists the data used to determine coefficients u_i in the second-order polynomial equation

$$U_G(f) = u_0 + u_1(f - 10) + u_2(f - 10)^2 \tag{7.5}$$

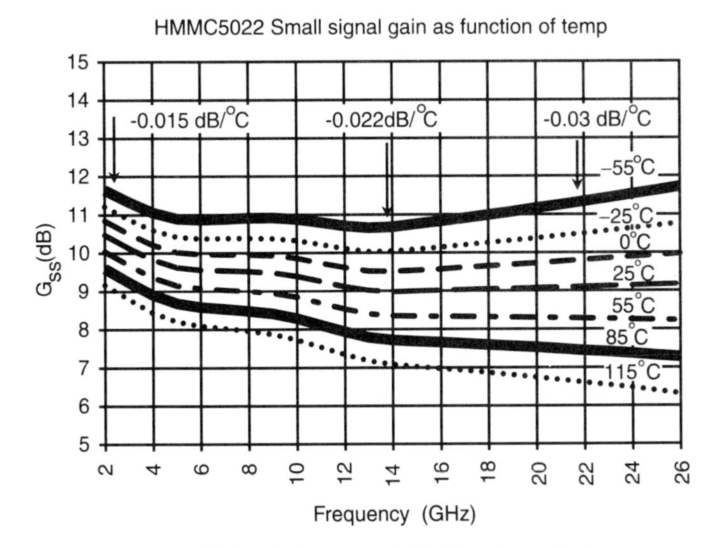

Figure 7.1 Temperature sensitivity of a broadband MMIC varies with frequency.

Table 7.1
Data From Figure 7.1 Used To Develop MMIC Temperature Sensitivity Equation

f	*U*(*f*)
2.5 GHz	−0.015 dB/°C
15 Ghz	−0.022 dB/°C
25 Ghz	−0.03 dB/°C

Matrix methods are used to solve for the second-order polynomial equation coefficients.

$$\begin{bmatrix} u_0 \\ u_1 \\ u_2 \end{bmatrix} = \begin{bmatrix} -0.015 \\ -0.022 \\ -0.3 \end{bmatrix} * \begin{bmatrix} 1 & (2.5 - 10) & (2.5 - 10)^2 \\ 1 & (15 - 10) & (15 - 10)^2 \\ 1 & (25 - 10) & (25 - 10)^2 \end{bmatrix}^{-1} \tag{7.6}$$

The resulting equation

$$U_G(f) = -0.0188 - 0.0005866(f - 10) - 0.00001066(f - 10)^2 \tag{7.7}$$

is plotted in Figure 7.2 showing the modeled value of MMIC temperature sensitivity over frequency.

The use of gain as a function of frequency and temperature as expressed by (7.2) in class A amplifier behavioral model (4.4) and (4.14) and in class AB amplifier behavioral model (5.1) and (5.2) adds to the utility of the amplifier models. However, the amplifier behavioral model's response to temperature change is not complete until sensitivity of saturated power output to temperature is added.

7.3 Saturated Power Output Sensitivity to Temperature

Amplifier stage saturated power output sensitivity to temperature $U_{sat}(f)$ has typical value of −0.015 dB/degrees C. Depending on the amplifier design, its value can be constant as a function of frequency, linear, or nonlinear as a function of frequency. A model for saturated power output as a function of frequency and temperature

$$P_{sat}(f, T) = P_{sat}(f, 25) + U_{sat}(f)(T - 25) \tag{7.8}$$

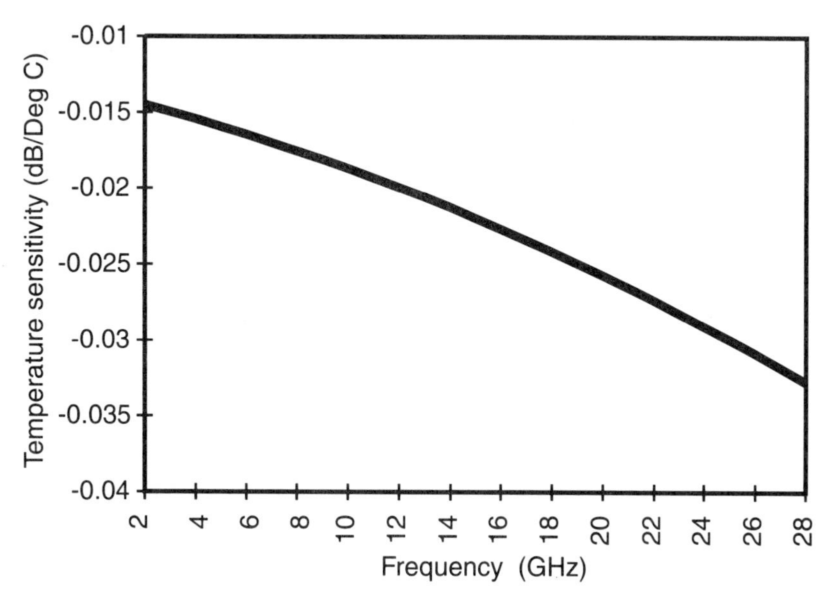

Figure 7.2 Modeled MMIC small signal gain sensitivity to temperature as a function of frequency.

uses either measured data or simulated data for $P_{sat}(f, 25)$, saturated power output over frequency at 25 degrees C. If measured data is used $P_{sat}(f, 25)$ has the form

$$P_{sat}(f, 25) = A + Bf + \sum_{1}^{i} C_i * \log_{10}\left[\frac{10^{\frac{f-f_i}{c_i}}}{1 + 10^{\frac{f-f_i}{c_i}}}\right] + \sum_{1}^{k} D_k * \log_{10}\left[1 + 10^{\frac{f-f_k}{d_k}}\right] \quad (7.9)$$

Saturated power output temperature sensitivity $U_{sat}(f)$ is easily determined from measured data taken over frequency at different temperatures. If $U_{sat}(f)$ is not a constant and does not change linearly as a function of temperature, refer to the procedure described above using measured data to fill in Table 7.1 to develop a polynomial equation by expanding data about an in-band frequency (see (7.6) and (7.7)).

If simulated data is used to formulate a model of saturated power output as a function of temperature, the term $P_{sat}(f, 25)$ in (7.8) will have the form

$$P_{sat}(f, 25) = P_{sat}(f_c, b, i_{ds}) - 0.5R - 10\log_{10}\left[1 + \left(\frac{\left|\frac{f_c^2}{f} - f\right|}{(f_{hi} - f_{lo})}\right)^{2n}\right]$$

$$+ 2QR\left\{\frac{3(f - f_{lo})}{(f_{hi} - f_{lo})} + \log_{10}\left[\frac{\left(1 + 10^{\frac{(6f - 3f_{hi} - 3f_{lo})}{(f_{hi} - f_{lo})}}\right)}{\left(1 + 10^{\frac{(6f - f_{hi} - 5f_{lo})}{(f_{hi} - f_{lo})}}\right)\left(1 + 10^{\frac{(6f - 5f_{hi} - f_{lo})}{(f_{hi} - f_{lo})}}\right)}\right]\right\} \quad (7.10)$$

Equation (7.10) describes a P_{sat} response for an amplifier having a Chebishev two-pole matching network with in-band ripple amplitude R, and saturated power output at band center of $P_{sat}(f_c, b, i_{ds})$. Simulated P_{sat} temperature sensitivity $U_{sat}(f)$ can be modeled to be constant, linear, or nonlinear over frequency.

The model for saturated power output as a function of frequency and temperature (7.8) is used in unison with the model for small signal gain as a function of frequency and temperature (7.2) to give the behavioral models for class A amplifiers (4.4) and (4.14) and for class AB amplifiers (5.1) and (5.2) full sensitivity to power input, frequency, and temperature.

7.4 Noise Figure Sensitivity to Temperature

Typical noise figure sensitivity to temperature value has the same magnitude as gain and saturated power output sensitivity but has the opposite sense. Amplifier stage noise figure increases with temperature by approximately +0.015 dB/degrees C. Noise figure sensitivity to temperature can be constant over frequency, or it too can be linear or nonlinear over frequency. An equation for noise figure as a function of frequency and temperature is of the same form as the equations for gain and saturated power output

$$NF(f, T) = NF(f, 25) + U_{NF}(f)(T - 25) \quad (7.11)$$

A model equation for noise figure over frequency at 25 degrees C, $NF(f, 25)$ can be developed from a curve fit to measured data, estimated data, or curve fit to data generated by RF circuit computer aided design software. In either case the curve-fit equation gives the term for $NF(f, 25)$ the form

$$NF(f, 25) = A + Bf + \sum_{1}^{i} C_i * \log_{10}\left[\frac{10^{\frac{f-f_i}{c_i}}}{1 + 10^{\frac{f-f_i}{c_i}}}\right] + \sum_{1}^{k} D_k * \log_{10}\left[1 + 10^{\frac{f-f_k}{d_k}}\right] \quad (7.12)$$

A technique for estimating amplifier noise figure over frequency is developed in Section 6.7 resulting in a model equation for n pole Butterworth response noise figure at 25 degrees C.

$$NF(f, 25) = 10\sqrt{i_{ds}} + 0.5 + 10\log_{10}\left[1 + \left(\frac{\left|\frac{f_c^2}{f} - f\right|}{(f_{hi} - f_{lo})}\right)^{2n}\right] \quad (7.13)$$

A second model developed in Section 6.7 adds Chebishev ripple for a two-pole network to noise figure in the bandpass. That equation is repeated here to give noise figure over frequency at 25 degrees C.

$$NF(f, 25) = 10\sqrt{i_{ds}} + 0.5(1 + R) - 2QR\left\{\frac{3(f - f_{lo})}{(f_{hi} - f_{lo})} + \log_{10}\left[\frac{\left(1 + 10^{\frac{(6f - 3f_{hi} - 3f_{lo})}{(f_{hi} - f_{lo})}}\right)}{\left(1 + 10^{\frac{(6f - f_{hi} - 5f_{lo})}{(f_{hi} - f_{lo})}}\right)\left(1 + 10^{\frac{(6f - 5f_{hi} - f_{lo})}{(f_{hi} - f_{lo})}}\right)}\right]\right\} + 10\log_{10}\left[1 + \left(\frac{\left|\frac{f_c^2}{f} - f\right|}{(f_{hi} - f_{lo})}\right)^{2n}\right] \quad (7.14)$$

Equations (7.12), (7.13), or (7.14) can be entered into (7.11) as a model for amplifier noise figure at 25 degrees C. Section 6.2.2.3 describes the procedure for developing equations for higher order Chebishev ripple. Noise figure sensitivity to temperature $U_{NF}(f)$ can be a constant, linear, or nonlinear as a function of frequency. If $U_{NF}(f)$ is nonlinear, a polynomial equation can be found using the methods outlined above by building a table like Table 7.1, and by referring to (7.5), (7.6), and (7.7).

Example 7.1

Simulated gain variation over frequency as a function of temperature.

Given:
A class A MESFET amplifier stage at 25 degrees C has $G_c = 12$ dB gain and $P_{sat} - 18.75$ dBm at center frequency $f_c = 2.5$ GHz. Saturated power output over the amplifier's 40 percent bandwidth has 0.5 dB Chebishev ripple from a matching network with three poles. Small signal gain exhibits the same 0.5 dB ripple. The amplifier is biased at 6 volts and 0.03 amperes. Small signal gain temperature coefficient $U_G(f) = -0.015$ dB/degrees C is constant over frequency, saturated power output temperature coefficient varies linearly over frequency obeying the equation $U_{sat}(f) = 0.01 - 0.01f$. Noise figure temperature coefficient is constant over frequency and has value $U_{NF}(f) = +0.015$ dB/degrees C.

Required:
Ambient temperature is 55 degrees C. 1) Simulate P_{1dB} as a function of frequency. 2) Simulate P_{3rdOIP} as a function of frequency. 3) Simulate noise figure as a function of frequency.

Procedure:
Determine that the operating bandwidth $BW = 0.4f_c = 1$ GHz. Knowing that $(f_{hi} - f_{lo}) = 1.0$ and that $\sqrt{f_{hi} * f_{lo}} = 2.5$, determine that $f_{lo} = 2.05$ and $f_{hi} = 3.05$. Set up a spreadsheet with a column containing frequency from 1.5 to 3.5 GHz in increments of 0.05 GHz. Use (6.35) to compute P_{sat} over frequency at 25 degrees C for a three-pole Chebishev bandpass realizing from Table 6.2 that $Q = 1.27$

$$P_{sat}(f, 25) = (P_{sat}(f_c, 25) - 0.5R) + 2QR\left\{\frac{5(f - f_{lo})}{(f_{hi} - f_{lo})}\right.$$

$$\left. + \log_{10}\left[\frac{\left(1 + 10^{\frac{(6f-3f_{hi}-7f_{lo})}{(f_{hi}-f_{lo})}}\right)\left(1 + 10^{\frac{(10f-7f_{hi}-3f_{lo})}{(f_{hi}-f_{lo})}}\right)}{\left(1 + 10^{\frac{(10f-f_{hi}-9f_{lo})}{(f_{hi}-f_{lo})}}\right)\left(1 + 10^{\frac{(10f-5f_{hi}-5f_{lo})}{(f_{hi}-f_{lo})}}\right)\left(1 + 10^{\frac{(10f-9f_{hi}-f_{lo})}{(f_{hi}-f_{lo})}}\right)}\right]\right\}$$

$$- 10\log_{10}\left[1 + \left(\frac{\left|\frac{f_c^2}{f} - f\right|}{(f_{hi} - f_{lo})}\right)^{2n}\right]$$

and plot the results as shown in Figure 7.3. Add correction to the above equation for temperature $P_{sat}(f, T) = P_{sat}(f, 25) + U_{sat}(f)(T - 25)$, calculate $P_{sat}(f, T)$ at $T = 55$ degrees C in another column in the spreadsheet and plot results as shown in Figure 7.3.

This is a class A MESFET amplifier; bias current is assumed to remain constant with power input, $b_q = 0.5i_{dss}$ and $K = 6$ (see Figure 5.4). However, bias coefficient b_f will vary with frequency and temperature as $P_{sat}(f, T)$ varies (see Section 6.4, and (6.37)). Compute b_f as a function of frequency at 55 degrees C in a new column using (6.37). Compression coefficient $K(f)$ is then determined using (5.11) and is entered in a new column on the spreadsheet. One dB compression point power as a function of frequency and temperature $P_{1dB}(f, T)$ can now be simulated using the data $P_{sat}(f, T)$ and $K(f)$ and (4.7). The result of this calculation for $P_{1dB}(f, T)$ is displayed in Figure 7.4.

Amplifier output third-order intercept P_{3rdOIP} can also be simulated as a function of frequency and temperature using (4.12) and the data generated for $P_{sat}(f, T)$ and $K(f)$. Results are displayed in Figure 7.5.

Noise figure as a function of frequency and temperature is estimated by

$$NF(f, T) = 10\sqrt{i_{ds}(f)} + 0.5 + (G_c - G(f)) + U_{NF}(f)(T - 25)$$

Gain at band center $G_c = 12$ dB, and quiescent bias current $i_q = 0.03$A. Since this is a class A amplifier, bias current $i_{ds}(f)$ is assumed to be constant with small signal input over frequency. Noise figure temperature sensitivity

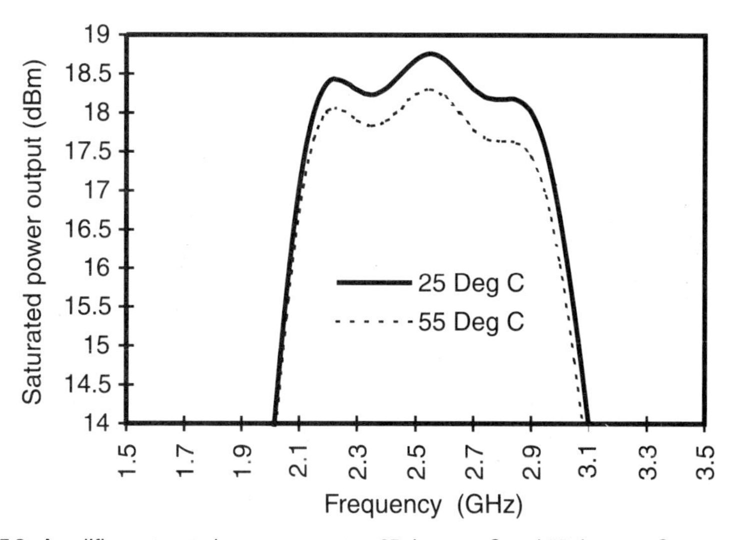

Figure 7.3 Amplifier saturated power output at 25 degrees C and 55 degrees C.

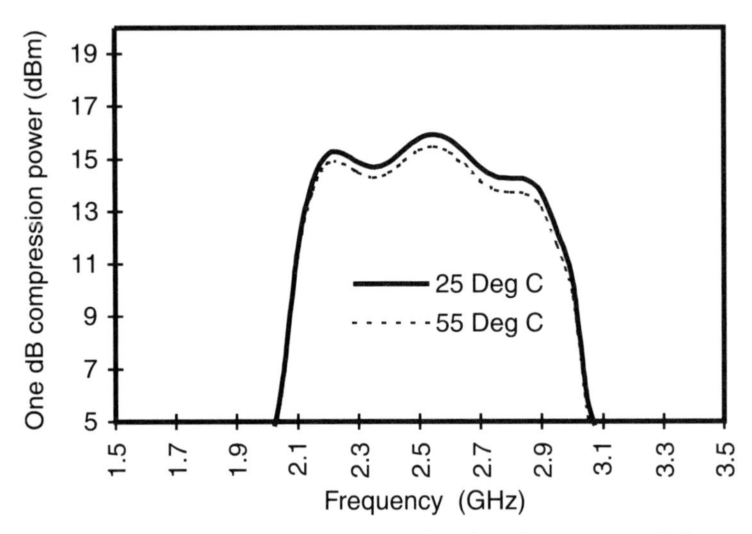

Figure 7.4 One dB compression power output as a function of frequency at 25 degrees C and 55 degrees C.

$U_{NF}(f) = +0.015$ dB/degrees C is given to be constant over frequency. Small signal gain as a function of frequency is a three-pole, 0.5 dB ripple, Chebishev response which is modeled by (6.6) where the term $G_c(f, b, i_{ds})$ is (6.20), which adds the Chebishev ripple. All of this comes together in the equation for noise figure as a function of frequency and temperature

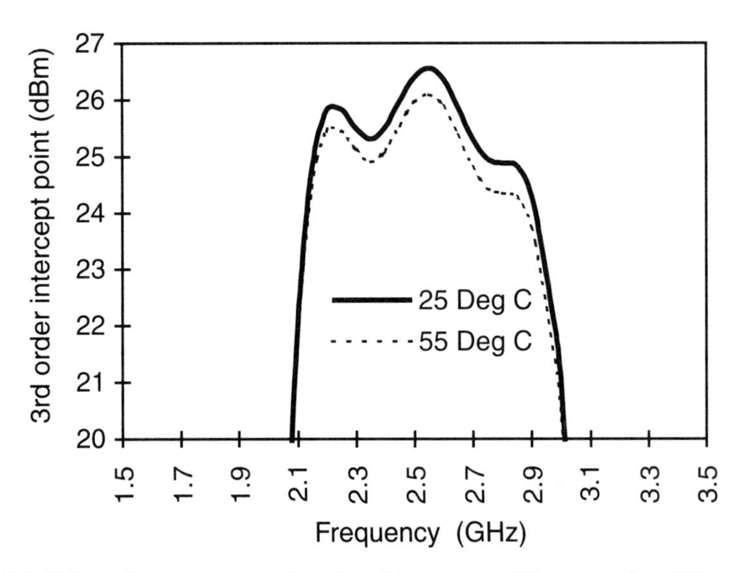

Figure 7.5 Third-order intercept as a function of frequency at 25 degrees C and 55 degrees C.

$$NF(f, T) = 10\sqrt{i_{ds}(f)} + 0.5(1 + R) - 2QR\left\{\frac{5(f - f_{lo})}{(f_{hi} - f_{lo})}\right.$$

$$\left. + \log_{10}\left[\frac{\left(1 + 10^{\frac{(6f - 3f_{hi} - 7f_{lo})}{(f_{hi} - f_{lo})}}\right)\left(1 + 10^{\frac{(10f - 7f_{hi} - 3f_{lo})}{(f_{hi} - f_{lo})}}\right)}{\left(1 + 10^{\frac{(10f - f_{hi} - 9f_{lo})}{(f_{hi} - f_{lo})}}\right)\left(1 + 10^{\frac{(10f - 5f_{hi} - 5f_{lo})}{(f_{hi} - f_{lo})}}\right)\left(1 + 10^{\frac{(10f - 9f_{hi} - f_{lo})}{(f_{hi} - f_{lo})}}\right)}\right]\right\}$$

$$+ 10\log_{10}\left[1 + \left(\frac{\left|\frac{f_c^2}{f} - f\right|}{(f_{hi} - f_{lo})}\right)^{2n}\right] + U_{NF}(f)(T - 25)$$

Calculated values for estimated noise figure as a function of frequency at 25 degrees C and 55 degrees C are plotted in Figure 7.6.

7.5 Summary

Methods have been developed for the inclusion of temperature sensitivity to amplifier parameters of small signal gain, saturated power output, and noise

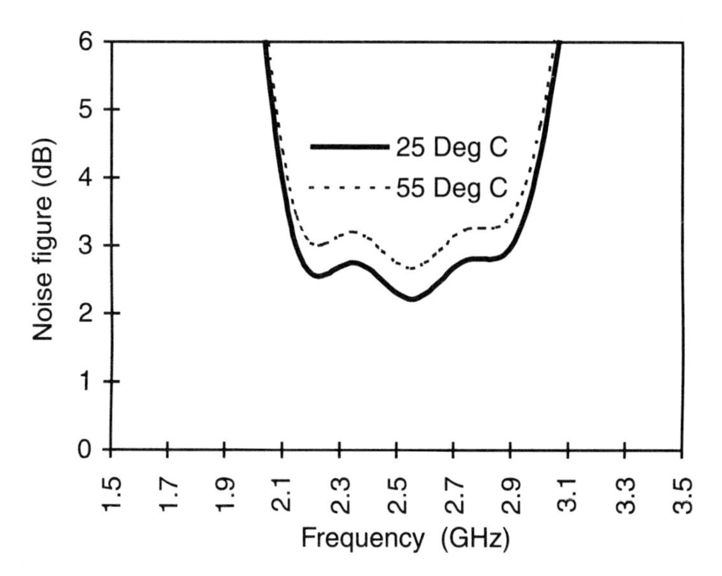

Figure 7.6 Estimated noise figure as a function of frequency at 25 degrees C and 55 degrees C.

figure. By example it is shown how these three sensitivities affect and give value to temperature sensitivity of b_f bias coefficient, $K(f)$ compression coefficient, P_{1dB}, one dB compression point and P_{3rdOIP}, amplifier third-order intercept point.

The utility of behavioral modeling has been expanded into multidimensional relationships that explain single amplifier stage properties. The following chapters add to this utility by expanding into the dimension of probability distributions of parameters over large populations and by expanding into the dimension of cascading multiple amplifier stages. New and useful design trade spaces will be revealed and methods of assigning specifications and limits to parameters will be developed.

7.6 Problems

Problem 7.1

Using the data in Example 7.1 simulate and compare small signal gain over frequency at 25 degrees C and 55 degrees C.

Problem 7.2

Using the data in Example 7.1 simulate and compare amplifier compression depth at 25 degrees C and 55 degrees C over frequency with power input of +7 dBm.

8

Probably Not as Expected

8.1 Accounting for Parameter Variability

Manufacturing technology has advanced to the point where microcircuits can be fabricated thousands at a time on a single wafer, and factory manufacturing capability is measured in terms of thousands of wafers processed per month. The net result is the capability to produce millions of microcircuits per month, hundreds of thousands per day at a single facility. The processes that fabricate these quantities of microcircuits are often not as stable as one would like. Process instability results in variability of specific parameters that affect the overall performance of subsystems and systems that the microcircuits are installed in. The simple task of placing limits of acceptability on performance parameters becomes one of profound significance. Product yield is directly affected, cost is directly impacted, and average product performance can degrade when improper limits are used.

Probability density functions are introduced in this chapter as a tool for modeling parameter variability. Specific probability density functions are suggested as models for amplifier parameters, and values relating to parameter variability are offered from practical experience. Methods of applying stochastic variables to models for multistage amplifiers are described. The use of unique risk analysis spreadsheet add-in software such as CRYSTAL BALL [1], or AT RISK [2], for the modeling of parameter value variability, and for modeling circuit and subsystem performance variability is discussed and demonstrated.

8.2 Risk Analysis Spreadsheet Add-In Software

Several risk analysis spreadsheet add-in programs exist at the time of this writing. Two of the most popular are CRYSTAL BALL created by Decisioneering [1], and AT RISK created by Palisade Corp. [2]. Both add-in programs are designed for assessing business risk and decisionmaking. CRYSTAL BALL is favored because it has the flexibility of allowing access to and formating of raw data files that it generates for engineering analysis and modeling. These risk analysis programs allow the user to select a probability density function from a library of functions and assign that density function to a value in a spreadsheet cell. A cell that is assigned a probability density function receives a new random number each time the spreadsheet is calculated. The values assigned are weighted in frequency of occurrence according to the probability density function assigned to that cell. Any number of cells in the spreadsheet can be assigned specific probability density functions. These cells, called assumptions, can be reference cells that affect calculated values in the spreadsheet. Spreadsheet cells that contain final calculated results can be designated as forecast values. The risk analysis program saves forecast values and uses them as data when it automatically assembles and formats data into reports. The number of trials to be run is defined in a risk analysis software dialog box and the add-in program is commanded to run. The risk analysis add-in program then takes control of the spreadsheet. It assigns a random value to each of the assumption cells according to the probability density defined for that cell. All calculations are performed by the spreadsheet generating values in the cells designated as forecast cells. The forecast values are recorded by the risk analysis program and a new set of random variables are placed in the assumption cells which generate new forecast values that are again recorded. The process is repeated over and over according to the number of trials commanded. Thousands of trials can be requested, each generating new and different assumption and forecast values. Forecast values are automatically presented in chart form as histograms or probability distributions when all trials are completed. CRYSTAL BALL allows the user access to the raw data files, which gives great flexibility in the way the data can be formated and used.

8.3 Useful Probability Density Functions

Two probability density functions are most useful in modeling statistical behavior of microwave device and circuit parameters. The Gaussian or normal probability density function [3, 4] is applicable to parameters that are found to vary about a mean value with equal certainty of having values higher than the mean as well as lower than the mean. The certainty of finding values significantly higher than or lower than the mean decreases exponentially. The population of parameter values characterized in units of volts, amperes, watts, voltage gain, current gain, transconductance, and so forth is typically characterized by a Gaussian probability density function.

The Weibull probability density function [5–7] is adaptable and can be configured to apply to a skewed data population. Skewed data populations are found in practice when a linear valued dataset such as a population of voltage or current ratios is converted to a decibel-valued population. The same dataset when converted to units of decibels transforms into a new probability density that, depending on the magnitude of data variance, can be modeled by either a Gaussian or Weibull probability density distribution. Empirical transformation relationships between the Gaussian and Weibull functions are developed in this chapter and examples are given showing typical microcircuit parameter variability due to production process variances.

Nonlinear processes create greatly skewed datasets. Consider the nonlinear amplification that occurs in a power amplifier as described by (4.4). If amplifiers of a population are each tested by driving them with a fixed power input level P_{in}, if the small signal gain G_{ss} of each amplifier is different due to process variation, and the saturated power output P_{sat} and compression coefficient K of each amplifier are also different due to process variations, test results for power output, phase shift, and amplifier compression depth will vary significantly. The population of test result data in decibel values will be significantly skewed. A Weibull function is needed to model the test results.

8.3.1 The Normal (Gaussian) Probability Density Function

The Gaussian or normal probability density function is defined by the equation

$$F(x) = \frac{1}{\sigma_g \sqrt{2\pi}} e^{-\frac{1}{2}\left(\frac{x - m_g}{\sigma}\right)^2} \tag{8.1}$$

where m_g is the population's mean value and σ_g is the standard deviation from the mean. Mean value of a Gaussian population density function is defined

$$m_g = \frac{\sum x_i}{n} \tag{8.2}$$

and standard deviation from the mean (or root mean square, RMS, value) of a Gaussian probability density function is defined

$$\sigma_g = \sqrt{\frac{\sum(x_i - m_g)^2}{n}} \tag{8.3}$$

where n is the number of objects in the data set, each having randomly variable value x_i.

The probability that any one value x_i will deviate from the mean by more than $\pm 3\sigma_g$ is less than 0.2 percent, which is to say that 99.8 percent of the population's values will fall within $\pm 3\sigma_g$ of the mean. Another valuable observation is the fact that 68.2 percent of the population will have values x_i that fall within $\pm 1\sigma_g$ of the mean. The normal probability density function (8.1) is plotted in Figure 8.1.

If the area under the curve in Figure 8.1 is set equal to the number of objects n in a population, and the total range of values x_i is divided into bins of equal width, the curve traced by (8.1) becomes an envelope predicting the frequency of occurrence or number of objects that have values that fall into the range of each of the bins.

The Gaussian or normal probability density function is used to model expected density distributions for population parameter values of voltage in units of volts, current in units of amperes, transconductance in units of mhos, small signal gain in units of voltage ratio or current ratio, and S-parameter values. It

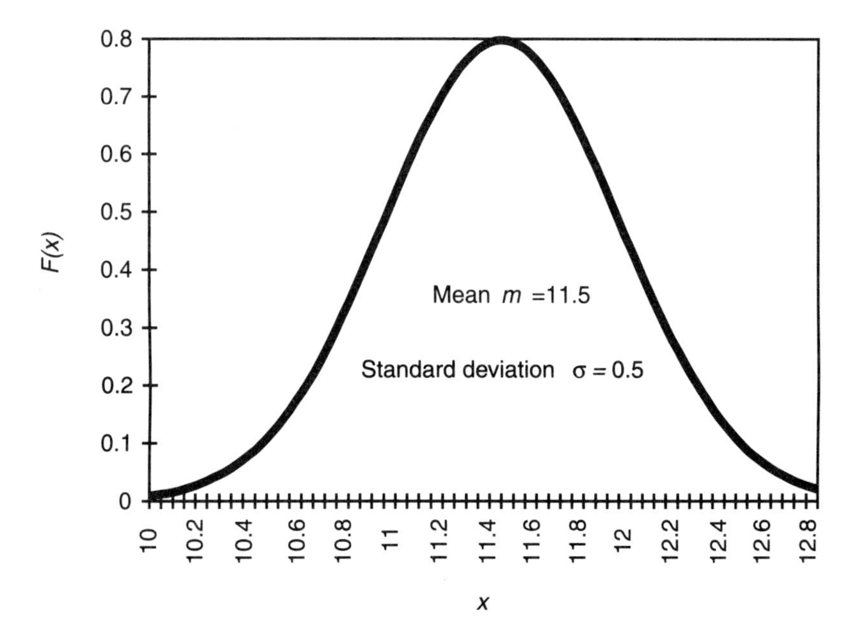

Figure 8.1 The normal or Gaussian probability density function.

can also be used to model a dataset with units of watts, values of resistance, and impedance, and in some cases datasets with units of decibels.

8.3.2 The Weibull Probability Density Function

Several variations of the Weibull probability density function equation exist. The Weibull probability density function equation

$$F(x) = \left(\frac{x - x_0}{\alpha}\right)^{(\beta-1)} e^{-\left(\frac{x-x_0}{\alpha}\right)^{\beta}} \tag{8.4}$$

is defined for all values of $x > x_0$. This Weibull probability density function is plotted in Figure 8.2 where shape factor $\beta = 8$, scale factor $\alpha = 2$, and location $x_0 = 10$.

The Weibull probability density function is adaptable in that its location factor x_0, shape factor β, and scale factor α are adjustable. Location factor x_0 is a Weibull probability density function scale reference point and has no physical significance other than providing a common scale reference. Scale factor α relates the Weibull probability density function range to standard deviation of the dataset values. Shape factor β determines the extent of skew and places skew to the high or low side of the mean. Shape factor values of $\beta < 3.2$ skew the

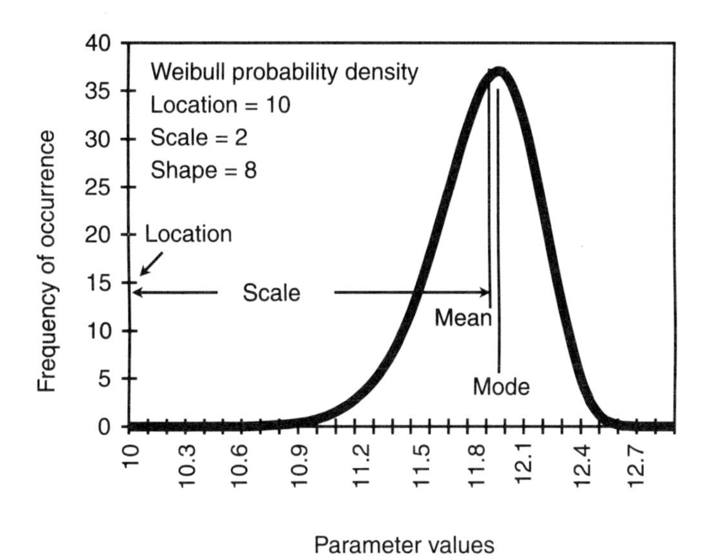

Figure 8.2 Weibull probability density where $\beta = 8$, $\alpha = 2$ and $x_0 = 10$.

population's mode value to the left of the Weibull distribution mean, while shape factor values of $\beta > 3.2$ skew the population's mode value to the right of the Weibull distribution mean.

8.3.2.1 Weibull Shape Factor $\beta = 8$ Is Most Useful for Behavioral Modeling

The most useful Weibull shape factor for behavioral modeling purposes is β – 8. This shape best characterizes the probability density of decibel values of parameters such as voltage and current ratios, S-parameters, small signal gain, and saturated power output. Probability density of linear-valued sets of these parameters is usually Gaussian. When a linear set of values is transformed into a set of decibel values, the minimum expected Weibull distribution value in decibels corresponds to the $-3\sigma_g$ or smallest expected value of the linear Gaussian distribution. The maximum expected Weibull distribution value in decibels corresponds to the $+3\sigma_g$ or largest expected linear Gaussian distribution value. Standard deviation of the decibel value data set in decibel units is closely related to standard deviation of linear values of the Gaussian distribution. If the linear Gaussian distributed dataset represents values of voltage or current ratio, or transconductance, standard deviation in decibel values is

$$\sigma_w = -20 \log_{10}\left[\frac{m_g - \sigma_g}{m_g}\right] \tag{8.5}$$

where m_g is the linear Gaussian distribution mean value and σ_g is the Gaussian standard deviation value. The logarithm argument of (8.5) is the normalized standard deviation σ_n of the Gaussian distribution. If the Gaussian data set is in units of watts or a power ratio, standard deviation in decibel units is computed by 10 log$[\sigma_n]$. The random variable value that occurs most frequently (probability density is greatest) in a Weibull population, the mode value, x_{mode}, is a decibel value determined by the empirical equation

$$x_{mode} = M_g + 0.19\sigma_w \tag{8.6}$$

where M_g is the decibel value of the linear data set (or Gaussian) mean value and σ_w is the Weibull standard deviation from (8.5) in decibel units.

The Gaussian mean value bin is expected to contain more units (probability density is greater) than any other bin, and the mode value bin in the Weibull distribution will contain a proportionately greater probability density than any other cell in the Weibull distribution. Mean value and standard deviation of the decibel-valued dataset are easily calculated once the Gaussian distributed linear values are converted to decibel values.

Numeric data is not always available when modeling the density of values known to follow a Weibull distribution. Empirical equations need to be developed for Weibull location, and scale values are needed in order to relate Weibull probability density of decibel values to Gaussian probability density of linear values. CRYSTAL BALL is used to generate sample populations and provide statistical data for the development of the empirical equations.

Multiple runs of CRYSTAL BALL are used to generate Gaussian distributions. Two trials are designed, one in which the decibel value is created for conversion of voltage and current ratios to decibels by using 20 log[x_i] and a second independent trial data set is created for conversion of power units into decibel units using 10 log[x_i]. Runs in each trial are made with Gaussian standard deviation as the variable ranging from 5 percent of the mean value to 30 percent of the mean. Each run generates 2000 random values. Each randomly valued data point generated to fill the Gaussian distributed data set is also converted to a decibel value and stored as a data point for the Weibull distribution data set. Data generated for both conversion trials, voltage to decibels and power to decibels, verifies empirical (8.5), which relates Weibull probability density function standard deviation σ_w to Gaussian standard deviation σ_g. Figure 8.3 illustrates the correlation between standard deviation of decibel values about the Weibull distribution mean determined by CRYSTAL BALL trials, and the standard deviation value computed using (8.5).

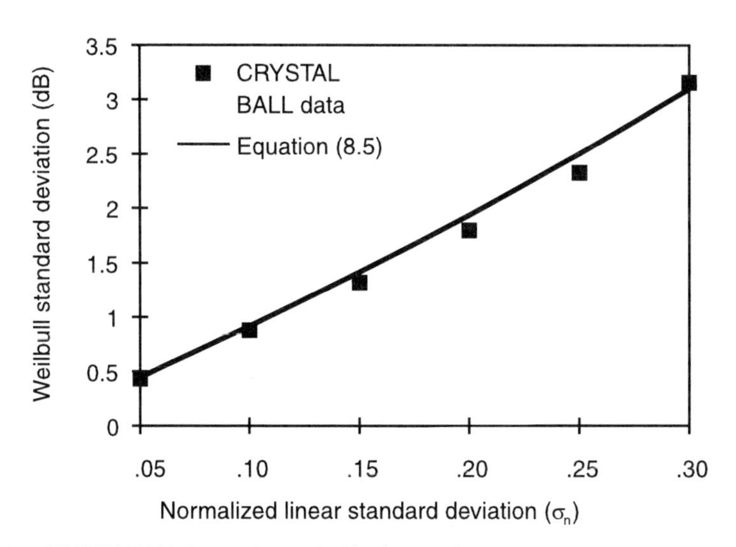

Figure 8.3 CRYSTAL BALL data and empirical (8.5) for Weibull standard deviation are compared.

The CRYSTAL BALL trials also yield data showing the Weibull probability density distribution mean decibel value is a function of Gaussian distribution mean value m_g and Weibull distribution standard deviation value σ_w from (8.5). The equation for Weibull mean value depends on whether the conversion is from voltage or power data. The empirical equation for Weibull mean value M_w where voltage and current ratios are converted to decibel values is

$$M_w = 20 \log_{10}[m_g] - 0.00076\sigma_w - 0.04978\sigma_w^2 \tag{8.7}$$

When power ratios are converted to decibels the first term of (8.7) is 10 $\log[m_g]$. The significant point to be made about (8.7) is that the Weibull mean value in decibels is less than the decibel value of the Gaussian mean. The difference between the two mean values in units of decibels is

$$\Delta M_{w-g} = -0.00076\sigma_w - 0.04978\sigma_w^2 \tag{8.8}$$

CRYSTAL BALL data for linear Gaussian to decibel Weibull conversion with normalized Gaussian standard deviation ranging from 5 percent to 30 percent (0.05 to 0.3) confirms the difference in population mean values predicted by (8.8) as shown in Figure 8.4.

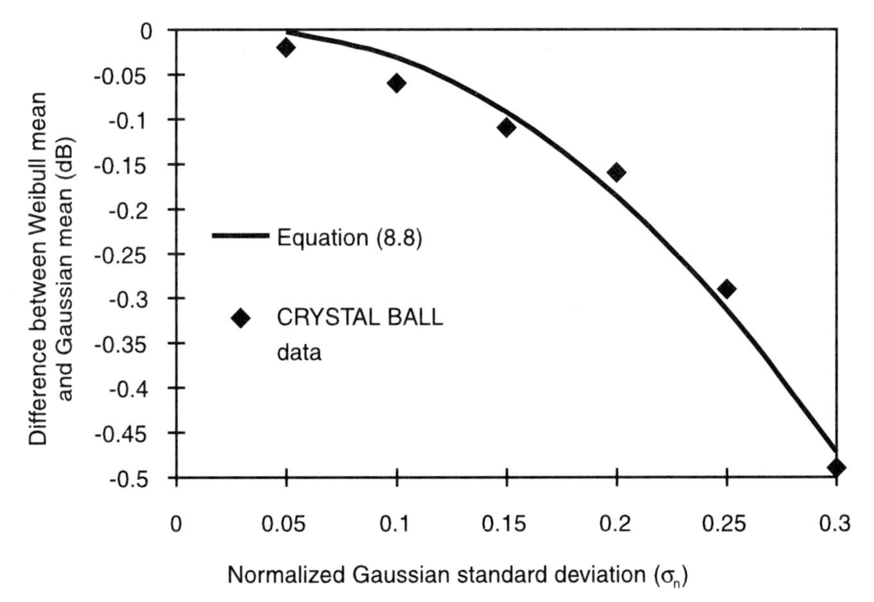

Figure 8.4 CRYSTAL BALL data confirms empirical (8.8) for difference in Weibull and Gaussian population mean values.

The CRYSTAL BALL experiments performed above confirmed equations that transform Gaussian dataset mean value and standard deviation into Weibull mean value, mode value, and standard deviation. Relationships between Weibull probability density function parameters x_0, α, and β and Weibull distribution mean value and standard deviation are now developed.

The CRYSTAL BALL decibel data generated by the above trials are plotted as histograms for each of the normalized Gaussian standard deviation values ranging from 0.05 to 0.3. Weibull shape factor $\beta = 8$ gives best fit to the distributions of decibel data. For this special case, where Weibull shape parameter $\beta = 8$, the scale parameter α is directly related to the Weibull distribution standard deviation σ_w (in decibel units) by the empirical equation

$$\alpha = 0.5423 + 5.6351\sigma_w \tag{8.9}$$

and the location parameter x_0 is related to Weibull distribution mean value M_w (in decibel units) and Weibull standard deviation σ_w in decibel units by the empirical equation

$$x_0 = M_w - 0.54 - 5.3\sigma_w \tag{8.10}$$

Once the scale factor, the location, and shape factor are determined, the Weibull probability density function (8.4) can be used to create an envelope that fits measured or simulated data. Figure 8.5 illustrates the function fit to data where location $x_0 = -7.78$ dB, scale $\alpha = 18.3$ dB, shape $\beta = 8$, Weibull mean value = 9.51 dB, mode value = 10.24 dB, and standard deviation value = 3.16 dB.

8.3.2.2 A Special Case: $\beta = 3.2$

If the Weibull shape factor $\beta = 3.2$, and the scale factor $\alpha = 3.125\sigma_g$ where σ_g is the Gaussian standard deviation value, the probability density values forecast by the two functions are essentially the same. A small adjustment of the Weibull location parameter is needed to align value scales. The near equivalence is illustrated in Figure 8.6.

Weibull function with shape factor $\beta \approx 1$ can be made to equate to the Raleigh probability density function with proper adjustment of scale and location factors.

8.3.2.3 Which to Use, Gaussian or Weibull?

A study of Weibull probability density plot overlays on CRYSTAL BALL generated data leads to the conclusion that when decibel values of standard deviation are 1.0 dB or less, there is not much difference between the density of decibel

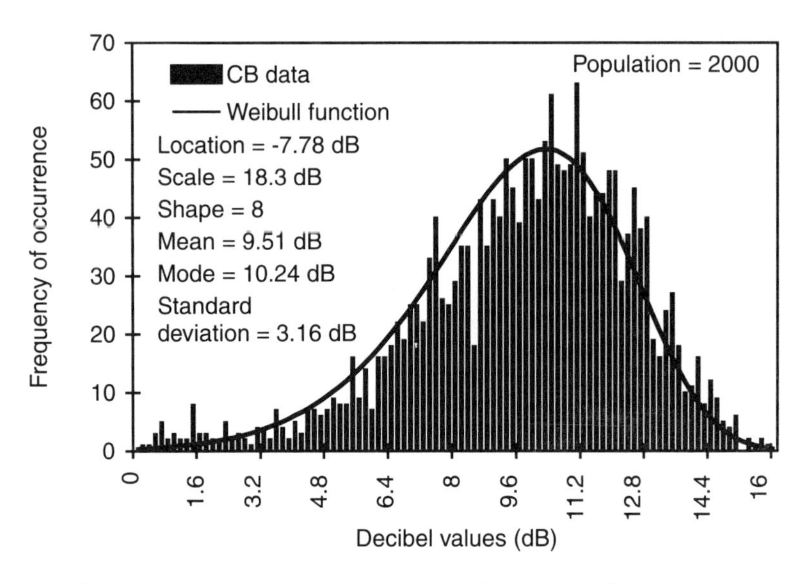

Figure 8.5 Empirical equations provide envelope functions for modeling measured and simulated Weibull distribution data.

data generated, the decibel-valued Weibull probability density function, and a decibel-valued Gaussian probability density function having the same decibel value of standard deviation. Figure 8.7 illustrates the point with a comparison of plot overlays for Weibull standard deviations of 0.92 dB. Figure 8.8 shows a comparison of plot overlays where standard deviation is 2.50 dB and the Weibull probability density function is the correct model to use. The conclusion reached through this study is that a decibel-valued Gaussian probability density function is an acceptable model for where standard deviation is 1.0 dB or less. This convention of application simplifies the formulation of complex amplifier models and has little impact on the accuracy of predicted statistical distributions.

8.3.2.4 A Software Utility for Determining Weibull Probability Density Parameters

A software utility "Probability Density Calculator" for determining Weibull probability density parameter decibel values that relate to a given Gaussian probability density distribution of linear values is included in the disk that is supplied with this book. The utility also computes Gaussian and Weibull probability densities for a population of any size if the mean value and standard deviation of the population are known.

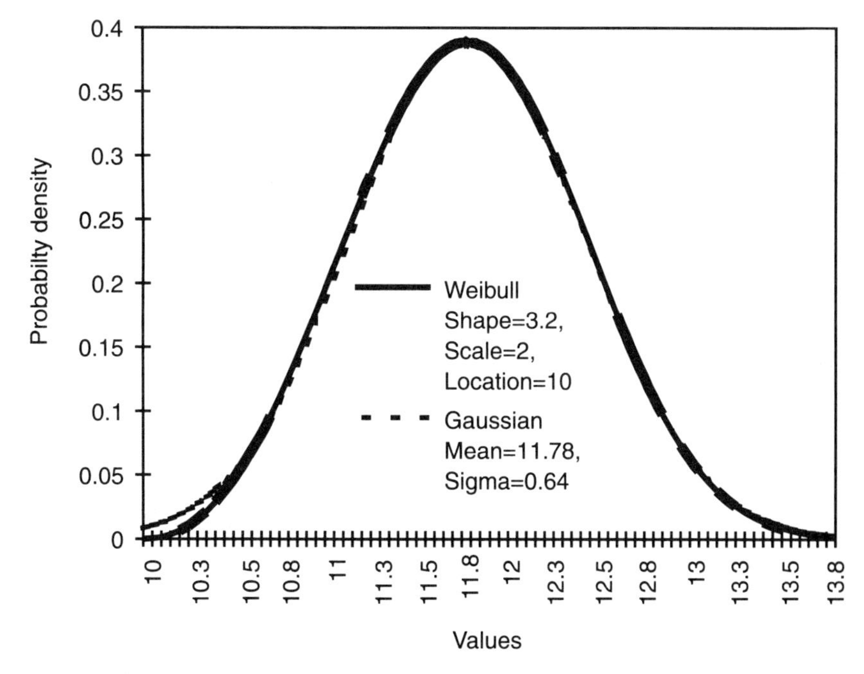

Figure 8.6 Weibull probability density function can be made essentially equal to Gaussian.

8.4 Develop Conventions for Applying Probability Density Distributions in Device and Circuit Models

The goal of this section is to gain understanding of the use of the Gaussian probability density function and the relationship that exists between the Gaussian probability density function, the Weibull probability density function, and amplifier parameters. Amplifier parameter statistical models will be postulated and stochastic variables typical of solid state foundry practices will be generated to demonstrate the correlation those variables have with specific models.

The manufacturing process that creates solid state microwave transistors and monolithic circuits consists of no less than 14 serial steps beginning with the raw Silicon or Gallium Arsenide wafer. The first steps involve doping that wafer with specific doping profiles either by ion implantation or by epitaxial growth, to create the desired semiconductor material characteristics. The wafer is then subjected to repeated etching and dielectric and metal deposition steps to create specific layer thickness, to add metal contacts, to provide thermal vias for heat removal, and to create interconnect wiring and metal pads for electrical connection. One of the final steps is deposition of a passivation layer to protect the transistor or circuit from corrosive effects of the atmosphere. Tolerances placed on each step of the complicated process are balanced in level of severity

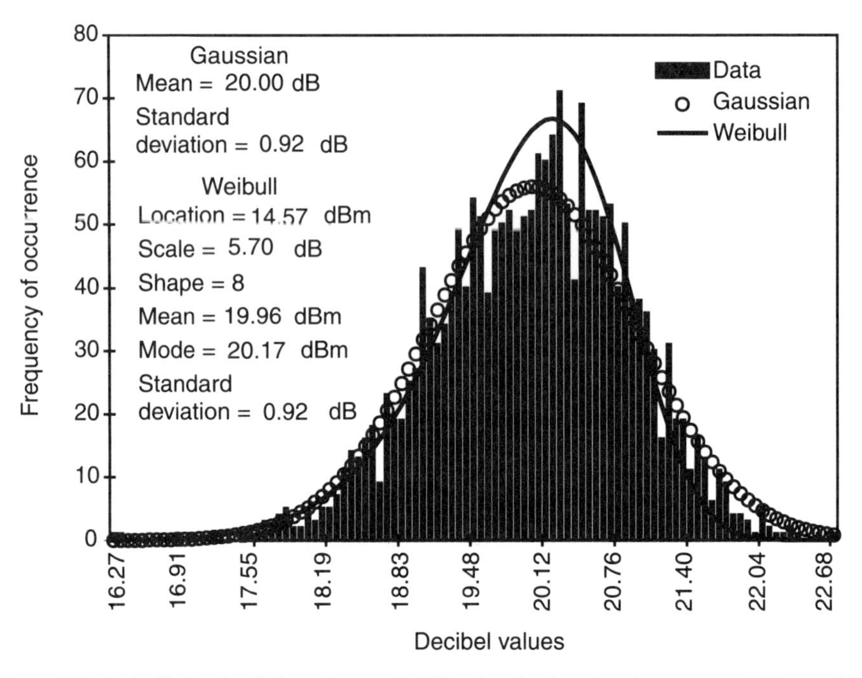

Figure 8.7 A decibel-valued Gaussian probability density function is an acceptable model where standard deviation is 1.0 dB or less.

such that the total process produces a device that is usable by the industry and returns an acceptable financial profit to the foundry. The net result is a process that turns out transistors that exhibit transconductance with standard deviation of 5 percent to 6 percent of population mean value. This value of standard deviation is observed over a period of many months and includes data from many production lots of the same transistor or MMIC chip. The greatest variance in parameter values occurs from lot to lot as the process line is cleaned and recharged for the next lot. Wafers common to a single lot (usually ten to twenty wafers) will have smaller standard deviation of values with respect to that lot's mean value. Parameter values observed within the population of chips on a single wafer will have even smaller standard deviation. The values of standard deviation of concern to the designer are the lot to lot values that can affect the performance of systems manufactured over several year periods.

Bias current drawn by a transistor with specific bias voltages applied will also exhibit standard deviation of current equal to 5 percent to 6 percent of the population mean. Transistors placed in close proximity on a wafer or MMIC will experience common variations in processing, and their parameters will vary in unison by the same magnitudes. Single amplifier stages exhibit standard deviation of voltage gain 5 percent to 6 percent of the mean voltage gain value.

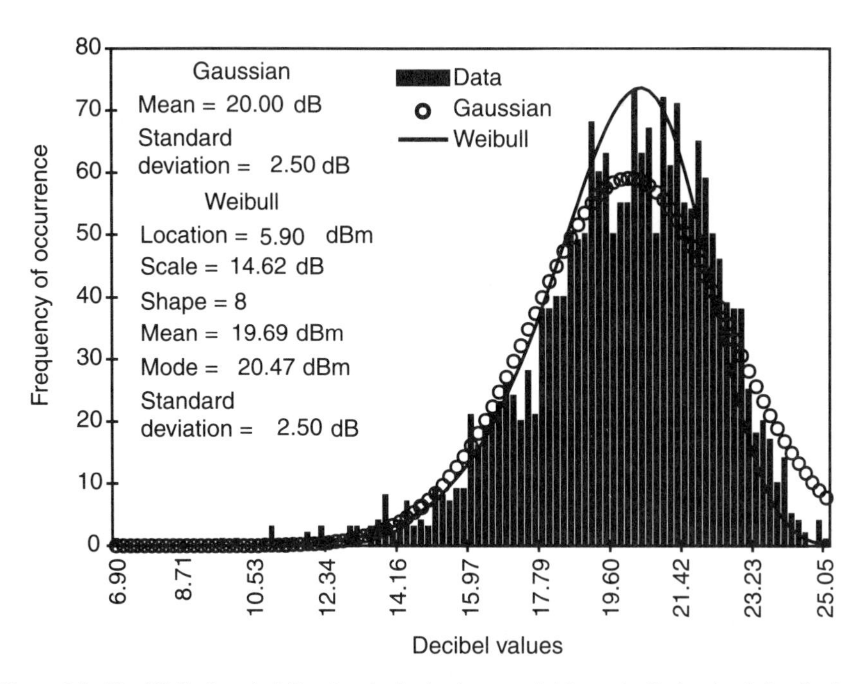

Figure 8.8 The Weibull probability density is the best model for a decibel-valued distribution where standard deviation is greater than 1.0 dB.

This translates into standard deviation of power gain of approximately 0.5 dB for a single amplifier stage. Cascaded stages in a single MMIC chip exhibit greater variation of voltage gain because all stages experience the same process variations, and gain variation occurs in unison. The result is an amplification of variability particularly where the adjacent transistors are connected in cascade on a MMIC. Methods of forecasting the variability of transistors and MMIC amplifiers are discussed in this section.

8.4.1 Modeling Single Stage Amplifier Small Signal Gain Statistics

Create a CRYSTAL BALL spreadsheet that generates voltage gain values with Gaussian density distribution about a mean of 3.16 with standard deviation of 0.158 (5 percent of the mean). Compute power gain $G_i = 20\log_{10}[g_i]$ for each random voltage gain value generated and save both voltage gain and power gain values for a run of 2000 transistors. The Gaussian probability density distribution of voltage gain values will have mean value $m_g = 3.16$, standard deviation $\sigma_g = 0.158$, minimum expected voltage gain $-3\sigma_g = 2.68$, and maximum expected gain value $+3\sigma_g = 3.63$. The decibel values of gain will have minimum expected value corresponding to the Gaussian $-3\sigma_g$ of 8.56 dB and maximum expected value of 11.2 dB.

Parameters for a Weibull probability density simulation are determined. Standard deviation of the decibel-valued Weibull population is determined by (8.5) to be $\sigma_w = 0.45$ dB. Mean value of the decibel-valued Weibull population is determined by (8.7) to be 9.98 dB. The decibel-valued Weibull population's mode value is determined by (8.6) to be 10.07 dB. Also determine parameters for a decibel-valued Gaussian probability density distribution. Mean value for a simulated Gaussian decibel-valued distribution is 20 log[3.16] = 9.99 dB. Let standard deviation for a simulated Gaussian decibel-valued population be the same as the Weibull standard deviation 0.45 dB. Minimum expected value $-3\sigma = 8.64$ dB, and maximum expected value $+3\sigma = 11.34$ dB. All of these values are tabulated in Table 8.1.

The issue here is whether to simulate single-stage amplifier gain populations with Gaussian or Weibull probability density distributions. Notice that the difference between the decimal value of the Gaussian mean 20 log[3.16] = 9.99 dB and the decibel value Weibull population mode value is only 0.08 dB. Also notice that the expected maximum and minimum decibel values are not severely shifted with respect to the decibel value of the Gaussian mean. When standard deviation of the voltage or current ratio is small, decibel-valued populations related to that voltage or current-valued population can be approximated by the Gaussian probability density function with very little error.

Determine the Weibull function location, scale, and shape factors. With (8.9) to obtain Weibull scale factor $\alpha = 3.08$ and (8.10) to obtain Weibull location factor $x_0 = 7.05$, Weibull probability density function values are calculated to overlay on the decibel-valued CRYSTAL BALL data histogram shown in Figure 8.9. A Gaussian probability density function is also calculated using the decibel values shown in the Gaussian simulation data in Table 8.1. This Gaussian function is also plotted as an overlay on the decibel value histogram

Table 8.1
Single Stage Amplifier Gain Value Statistics

	Minimum Expected	Mean	Mode	Maximum Expected	Standard Deviation
Linear	2.68	3.16	3.16	3.63	0.158
Weibull Simulation	8.56 dB	9.98 dB	10.07 dB	11.2 dB	0.45 dB
Gaussian Simulation	8.64 dB	9.99 dB	9.99 dB	11.34 dB	0.45 dB

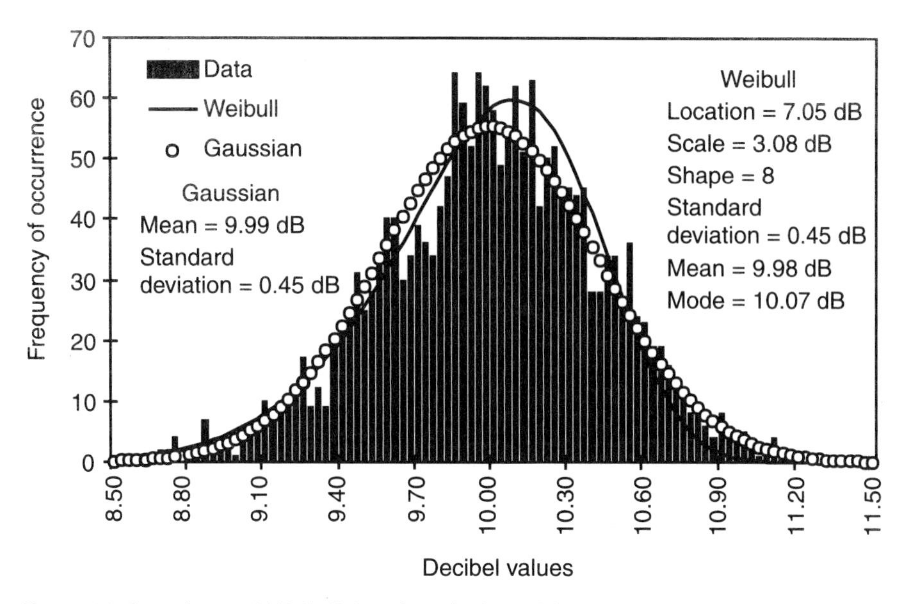

Figure 8.9 Gaussian and Weibull functions both model populations with small standard deviations.

to demonstrate the close correlation of the Gaussian distribution to data when the standard deviation is small.

This near equivalence between Gaussian density function representation of decibel data populations and Weibull density function representation of the same decibel-valued dataset is useful for voltage and current ratios having linear value normalized standard deviations as large as 10 percent.

8.4.1.1 Single Stage Amplifier Statistical Model Convention

The convention used for modeling statistical variations of small signal gain is that of a Gaussian probability density function having mean value equal to the decibel value of the linear distribution mean value and standard deviation of 0.5 dB.

8.4.2 Modeling Multiple Cascaded Stages on the Same MMIC

Amplifier stages in close proximity on the same MMIC chip all experience the same process variations and all exhibit the same gain deviation from the population mean. This behavior is modeled by assigning a voltage gain value to a single spreadsheet cell and creating an assumption of a Gaussian probability density function having mean value 3.16 and standard deviation 5 percent of the mean. A second cell is created which calculates total MMIC voltage gain by raising the assumption cell gain to a power n equal to the number of cascaded stages on the MMIC. A decibel value is calculated, $20n\log[x_i]$, to represent mean value of the MMIC population's power gain. The decibel value of standard deviation is expected to be $20n\log[\sigma_n]$ where σ_n is the normalized linear valued standard deviation value ($\sigma_n = 0.95$) for a single amplifier stage having standard deviation equal to 5 percent of the mean value.

Two multiple staged MMICs are modeled, one having three stages, the other having six stages. CRYSTAL BALL data is generated for both models and decibel-valued Gaussian and Weibull probability density function parameters are computed. Figure 8.10 shows the results obtained for the three-staged MMIC. The Gaussian function overlay is a more acceptable match to the CRYSTAL BALL data than the Weibull function overlay. The decibel value of standard deviation is 1.33 dB, three times that of a single amplifier stage. This distribution is simply a magnification of the Gaussian distribution shown in Figure 8.9 which represents a single-stage amplifier.

Study the results of a similar CRYSTAL BALL test for a six stage MMIC. The single-stage voltage gain is raised to the sixth power and converted to a decibel value. The linear normalized standard deviation is also raised to the

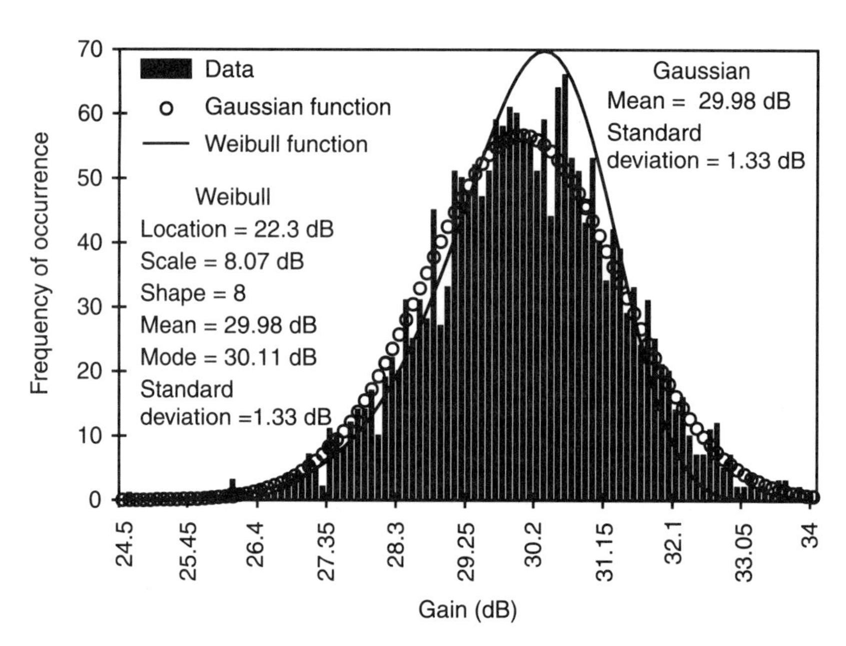

Figure 8.10 Small signal gain statistics for a three stage MMIC are best modeled with a Gaussian probability density function.

sixth power and converted to a decibel value. CRYSTAL BALL data is presented as a histogram and Gaussian and Weibull probability density function overlays are plotted. The results of the test are shown in Figure 8.11. The Gaussian probability density function is again the best model for representing the MMICs small signal gain statistics. The Gaussian distribution that best represents a single-stage amplifier is simply multiplied in decibel scale factor by the number of correlated cascaded stages on the MMIC.

8.4.2.1 MMIC Small Signal Gain Statistical Model Convention

The Gaussian probability density function is used to model populations of small signal gain decibel values for MMIC chips. Standard deviation of small signal gain for a single amplifier stage is found from experience to vary from one foundry to the next. The most common value found throughout the industry for standard deviation of small signal gain for a single stage is 0.45 dB (standard deviation of transconductance or voltage gain 5.0 percent).

When cascaded amplifiers are designed into a single monolithic circuit, they still exhibit variation in small signal gain with a standard deviation over a large population of 0.49 to 0.5 dB per amplifier stage. Because the cascaded stages on a single microcircuit chip are so close in proximity, whatever happens to one stage in the process of fabrication happens to all stages on that one

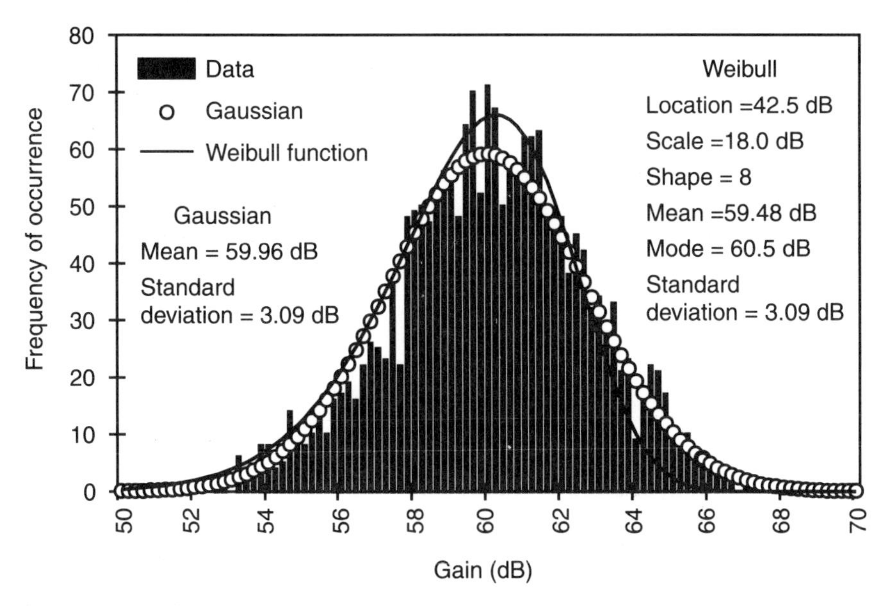

Figure 8.11 The Gaussian probability density function best models small signal gain statistics of a six-stage MMIC.

monolithic chip. If one stage experiences an increase in gain, all of the stages on that chip experience the same increase in gain. Standard deviation of MMIC gain in decibels is simply the number of stages multiplied by the decibel value of standard deviation of a single stage

$$\sigma_{MMIC} = N\sigma_i \tag{8.11}$$

Mean decibel value of small signal gain of a population of monolithic circuits is simply the sum of mean decibel values of the stages cascaded on the monolithic circuit

$$G_{MMIC} = G_1 + G_2 + G_3 + \cdots + G_i \tag{8.12}$$

Figure 8.12 illustrates the expected small signal gain and standard deviation of gain for various combinations of multistage MMICs.

8.4.3 Modeling Cascaded Independent Amplifiers

A population of single amplifier stages will exhibit standard deviation of small signal gain on the order of 0.5 dB. When several independent single amplifier stages, selected for gain and saturated power output, are cascaded together to create an amplifier assembly, mean gains of each of the amplifier stage populations expressed in decibels are added to obtain total amplifier assembly mean value of gain. Each cascaded amplifier stage has a unique small signal gain value which is independent from gain values of the other stages. Each cascaded amplifier stage is selected to have a P_{1dB} compression point greater than the stage that drives it so that no one stage saturates before the final stage compresses. All of the stages are therefore different in design and independent in statistical variation of parameters. These independent amplifier stages are drawn from differ-

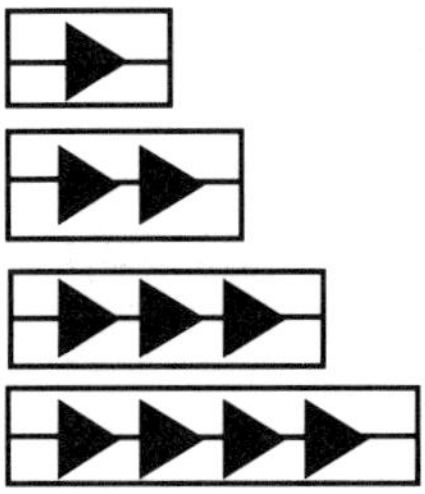

	Mean Gain	Standard deviation	3 σ
1 stage	G (dB)	0.5 dB	1.5 dB
2 stages	2G (dB)	1.0 dB	3.0 dB
3 stages	3G (dB)	1.5 dB	4.5 dB
4 stages	4G (dB)	2.0 dB	6.0 dB

Figure 8.12 Standard deviation of gain expected in MMICs having multiple cascaded amplifier stages.

ent populations and created on different process lines. Each population of independent amplifier stages will have its unique value of standard deviation σ_i of small signal gain.

When a population of amplifier assemblies is created by cascading independent amplifier stages drawn from different populations, the amplifier assembly population will exhibit a decibel-valued standard deviation of small signal gain that is the root-sum-square of the decibel-valued standard deviations of the independent amplifier populations

$$\sigma_{cas} = \sqrt{\sigma_1^{\,2} + \sigma_2^{\,2} + \sigma_3^{\,2} + \circ\circ\circ\circ + \sigma_i^{\,2}} \tag{8.13}$$

where each σ_i is in units of decibels. The population of cascaded amplifier stages will have mean value of small signal gain equal to the sum of mean values of gain of the individual stage populations

$$G_{cas} = G_1 + G_2 + G_3 + \circ\circ\circ\circ + G_i \tag{8.14}$$

where each G_i is in units of decibels. Create a CRYSTAL BALL model of independent amplifier stages to understand the probability density function that best characterizes cascaded independent amplifier stages.

Amplifier stages are cascaded to obtain greater small signal gain. There is a practical limit to the amount of gain that can be designed into a single package using cascaded independent MMIC chips as gain blocks. That practical limit is approximately 60 dB. It becomes extremely difficult to isolate package output from package input to greater than 60 dB thus avoiding unity feedback and preventing unwanted oscillation. Sixty decibels of gain can be developed using six to nine cascaded amplifier stages depending on frequency of operation and transistor technology used. Study the design where nine cascaded stages of gain are used. Several combinations of MMIC designs are possible. Case 1) a combination of a four-stage and a five-stage MMIC; Case 2) a combination of 3 three-stage MMICs; and Case 3) three two-stage and a three-stage MMIC.

Configure a spreadsheet that uses separate cells to represent voltage gain of each MMIC chip used in the design. Assign a Gaussian density function assumption to each of the cells and assign voltage gain to each cell that is commensurate with the number of amplifier stages in the MMIC represented by that cell. Assign standard deviation of voltage gain to each assumption equal to 5 percent of the mean gain of that cell. Establish a cell that calculates the product of voltage gains of each of the MMICs cascaded in the design, compute the decibel value of total cascaded gain and declare that cell to be the forecast value. Use the conventions defined above to develop parameters for a Gaussian probability density model and a Weibull probability density model of the cascaded set of MMICs, run the CRYSTAL BALL models, and plot the Gaussian and Weibull functions as overlays to the CRYSTAL BALL generated data.

8.4.3.1 Case 1: Design Using a Four-Stage MMIC and a Five-Stage MMIC

Mean cascaded gain for the population is expected to be 89.9 dB. Standard deviation of gain in decibel units is expected to be

$$\sigma = \sqrt{(4 * 0.4455)^2 + (5 * 0.4455)^2}$$

$$\sigma = 2.85 \text{ dB}$$

These are the values that determine a Gaussian probability density distribution overlay. A Weibull probability density distribution will have mean value

$$M_w = 20 \log_{10}[(3.16)^9] - 0.00076\sigma - 0.04978\sigma^2$$

$$M_w = 89.54 \text{ dB},$$

location value $x_0 = M_w - 0.54 - 5.3\sigma = 73.89$ dB, and a scale value $\alpha = 0.5423 + 5.6351\sigma$, which is 16.6 dB. Figure 8.13 shows the CRYSTAL BALL data generated for small signal gain distribution of the cascaded four-stage MMIC and five-stage MMIC amplifier. The data is overlaid with Gaussian and Weibull probability density functions based on the parameters derived above. The conclusion reached is that the Gaussian distribution best characterizes the small signal gain statistics of a nine-stage amplifier made by cascading a four-stage and a five-stage MMIC.

8.4.3.2 Case 2: Design Using Three Cascaded Three-Stage MMIC Amplifiers

Mean cascaded gain for the population is expected to be 89.9 dB. Standard deviation of gain in decibel units is expected to be

$$\sigma = \sqrt{2 * (3 * 0.4455)^2 + (3 * 0.4455)^2}$$

$$\sigma = 2.31 \text{ dB}.$$

These are the values that determine a Gaussian probability density distribution overlay. A Weibull probability density distribution will have mean value

$$M_w = 20 \log_{10}[(3.16)^9] - 0.00076\sigma - 0.04978\sigma^2$$

$$M_w = 89.67 \text{ dB},$$

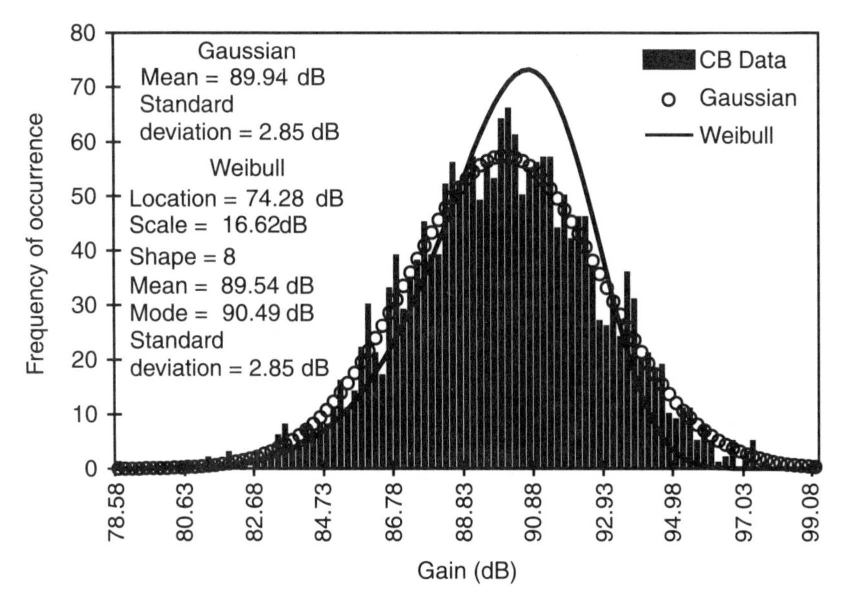

Figure 8.13 A Gaussian probability function best characterizes small signal gain statistics of cascaded four-stage and five-stage MMIC amplifiers.

location value $x_0 = M_w - 0.54 - 5.3\sigma = 76.89$ dB, and a scale value $\alpha = 0.5423 + 5.6351\sigma$, which is 13.6 dB. Figure 8.14 shows the CRYSTAL BALL data generated for small signal gain distribution of the three cascaded three-stage MMICs. The data is overlaid with Gaussian and Weibull probability density functions based on the parameters derived above. The conclusion reached is that the Gaussian distribution best characterizes the small signal gain statistics of a nine-stage amplifier made by cascading 3 three-stage MMICs.

8.4.3.3 Case 3: Design Using 3 Two-Stage MMICs and a Three-Stage MMIC

Mean cascaded gain for the population is expected to be 89.9 dB. Standard deviation of gain in decibel units is expected to be

$$\sigma = \sqrt{3 * (2 * 0.4455)^2 + (3 * 0.4455)^2}$$

$$\sigma = 2.04 \text{ dB}.$$

These are the values that determine a Gaussian probability density distribution overlay. A Weibull probability density distribution will have mean value

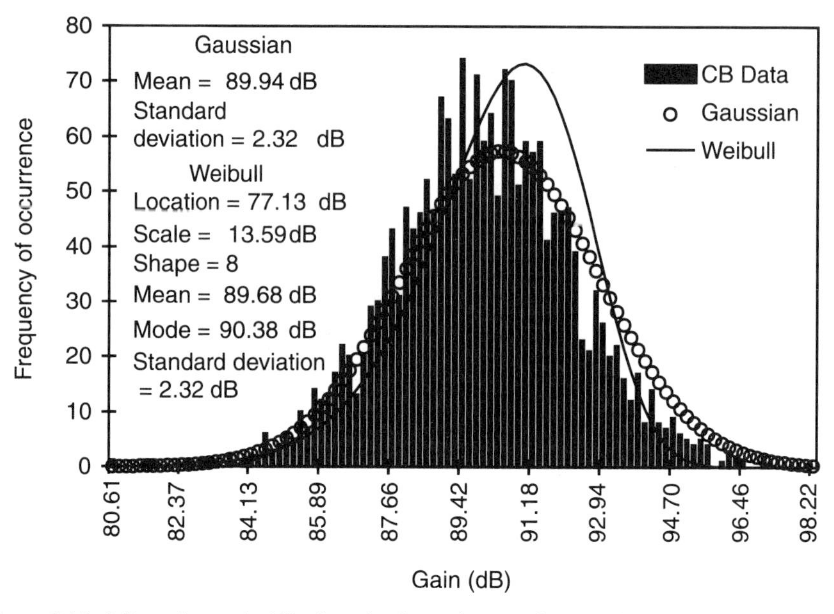

Figure 8.14 A Gaussian probability function best characterizes small signal gain statistics of three cascaded three-stage MMIC amplifiers.

$$M_w = 20 \log_{10}[(3.16)^9] - 0.00076\sigma - 0.04978\sigma^2$$

$$M_w = 89.73 \text{ dB},$$

location value $x_0 = M_w - 0.54 - 5.3\sigma = 78.38$ dB, and a scale value $\alpha = 0.5423 + 5.6351\sigma$, which is 12.04 dB. Figure 8.15 shows the CRYSTAL BALL data generated for small signal gain distribution of the three cascaded three-stage MMICs. The data is overlaid with Gaussian and Weibull probability density functions based on the parameters derived above. The conclusion reached is that the Gaussian distribution best characterizes the small signal gain statistics of a nine-stage amplifier made by cascading 3 two-stage MMICs and a three-stage MMIC.

8.4.4 Amplifier Saturated Power Output Statistics

Amplifier saturated power output in units of watts is calculated using (4.2). The value of saturated power output is the product of two independent linear values, current i_{dss} and voltage v_{ds}. Both of these parameters vary from amplifier to amplifier and each can be represented by a Gaussian probability density function. Standard deviation of current value i_{dss} is expected to be 5 percent of the

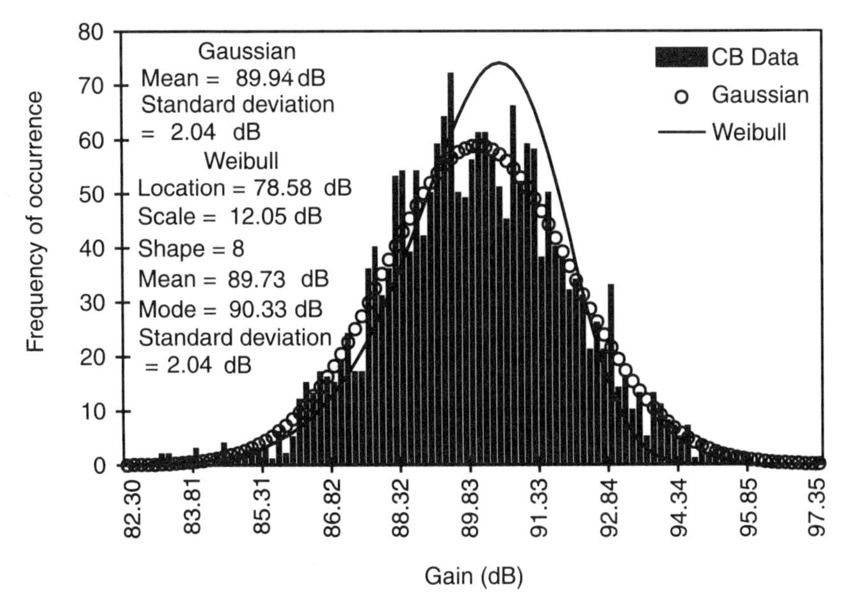

Figure 8.15 A Gaussian probability function best characterizes small signal gain statistics of three cascaded two-stage MMIC amplifiers and a three-stage MMIC.

mean value. Standard deviation of voltage v_{ds} is more a function of power supply regulation and is typically found to have a value on the order of 2 percent of the mean from load to load.

A spreadsheet model is constructed using CRYSTAL BALL to provide stochastic variability to values for current and voltage. A cell is created for i_{dss} and a mean value of 0.50 amperes is arbitrarily assigned with standard deviation of 0.025 A. A cell for voltage v_{ds} is created, and a mean value of 9 volts is arbitrarily assigned with standard deviation of 0.18 V. Knee voltage in the transistor of 1.0 volt is assumed. Mean value of saturated power output from (4.2) is 1.0 watt or +30.0 dBm. A spreadsheet cell is established to compute the decibel value of each CRYSTAL BALL trial and that cell is the forecast value. The forecast cell contains the logarithm of a value of power (watts) obtained by multiplying current and voltage. The decibel value of the forecast quantity is obtained by multiplying the logarithm by ten. A value of 30 is added to the logarithm to reference the decibel value to 1 milliwatt. Expected decibel value of standard deviation of saturated power output is calculated by

$$\sigma = \frac{\sqrt{[20\log_{10}(\sigma_1)]^2 + [20\log_{10}(\sigma_2)]^2}}{2} \tag{8.15}$$

$$\sigma = \sqrt{[10\log_{10}(0.95)]^2 + [10\log_{10}(0.98)]^2} = 0.24 \text{ dB}$$

This mean value and standard deviation are used to compute Weibull probability density function mean value $M_w = 30 - 0.00076\sigma - 0.04978\sigma^2 = 29.99$ dBm, location value $x_0 = 28.19$, and scale value $\alpha = 1.89$. Plot the Weibull and Gaussian functions as overlays on the CRYSTAL BALL data as shown in Figure 8.16. The conclusion is that saturated power output data populations can be simulated by either Gaussian or Weibull probability density functions. Experience has taught that the Weibull model for saturated power output is more likely to be the correct model.

8.4.4.1 Saturated Power Output Statistical Model Convention

Saturated power output of an amplifier stage in decibel values is modeled by convention to be a Weibull probability density function. Standard deviation of saturated power output is typically 0.24 dB. Weibull scale value is typically $\alpha = 1.89$, and Weibull location value is typically 1.81 dB less than the decibel value of the Gaussian mean.

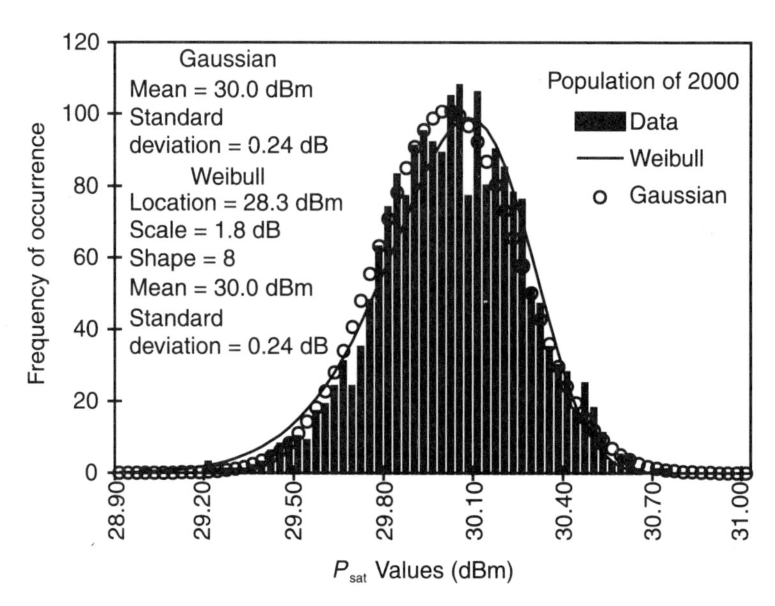

Figure 8.16 Saturated power output in decibel values can be modeled as either a Gaussian or Weibull probability density.

8.5 Modeling Population Variations of Small Signal Gain as a Function of Frequency

Gain as a function of frequency is modeled in Chapter 6. The Butterworth response is modeled by (6.6), Chebishev response is modeled by (6.12) and (6.20), and measured data files are modeled by curve fitting resulting in an equation of the form of (6.30). Each of these model equations has a constant term that represents small signal gain at a reference frequency somewhere in the amplifier's bandpass. That gain term can be assigned a CRYSTAL BALL assumption probability density parameter. The frequency response equation is then used to calculate gain as a function of frequency by establishing a range of frequencies on the spreadsheet and using the frequency response equation with its current random gain constant term value to determine gain at each frequency. Each of the calculated gains for frequencies of interest can be designated as a CRYSTAL BALL forecast. New forecast cell values will be calculated for each trial, and a new frequency response will be generated when the risk analysis program is run. Figure 8.17 shows the result of running 100 trials of an

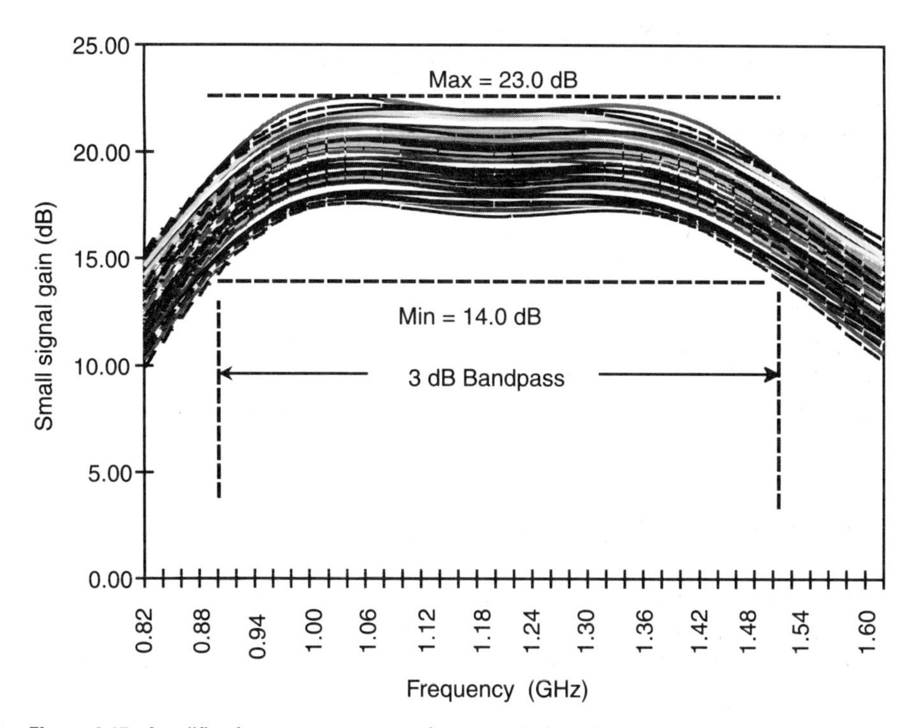

Figure 8.17 Amplifier frequency response for a population of assemblies can be easily modeled.

amplifier model that has mean gain of 20 dB with a standard deviation $1\sigma = 1$ dB, a Chebishev 2 pole, 0.5 dB ripple, with 3 dB bandpass from 0.9 to 1.5 GHz. The ripple factor R is given a standard deviation $1\sigma = 0.2$ dB.

8.6 Using Gain and Saturated Power Output Statistics in the Nonlinear Amplifier Behavioral Model

The nonlinear amplifier behavioral model, (5.1), calculates power output P_{out}, as a function of power input P_{in}, when given values for small signal gain, G_{ss}, saturated power output, P_{sat}, and compression coefficient K. Statistical models defined in this chapter can be applied to the parameters in the nonlinear amplifier behavioral model equation through the use of risk analysis spreadsheet add-in software. Risk analysis results give understanding of parameter variables characteristic of large populations.

For example, amplifiers in a production line population are typically tested for maximum power output by establishing an acceptable input drive level for testing, inserting the amplifier into the test fixture, and measuring output power. Amplifier small signal gain varies with Gaussian probability density, saturated power output varies with Weibull probability density, and compression coefficient varies as a function of bias current and i_{max}. Bias current varies with Gaussian probability density. If the drive power is not high enough, and gain happens to be low for a given amplifier, power output might not obtain compression, and the measured data will not represent maximum power output of that amplifier. The amplifier might not pass a minimum power output limit. On the other hand, if drive level is deliberately set high to assure that all amplifiers are compressed, amplifiers with high gain will be driven hard into compression during the test with a risk of damage to the amplifier. Variable bias current also affects noise figure as determined by the rule of thumb (4.17), giving noise figure a Gaussian probability density distribution.

Example 8.1 Testing a Single Stage Amplifier

Assume a single stage class A amplifier is to be tested. Amplifier design center parameters and expected standard deviations are listed in Table 8.2. A risk analysis spreadsheet is developed based on Table 8.2 parameters, and 1000 trials are run to obtain a representative statistical sample. Power output at $P_{in} = +7$ dBm is determined for each amplifier in the population using the amplifier behavioral model equation

Table 8.2
Amplifier Design Parameters Estimated by Behavioral Model

Parameter	Value	Standard Deviation
Operating Voltage	8 volts	
Bias Current	0.0285 amperes	0.004 amperes
Small Signal Gain G_{ss}	14 dB	0.5 dB
Saturated Power Output P_{sat}	+20 dBm	0.24 dB
Noise Figure NF	2.19 dB	
Compression Coefficient K	6.08	
Compression Depth @ P_{in} = +7 dBm	2.38 dB	
P_{out} @ P_{in} = +7 dBm	18.62 dBm	

$$P_{out} = P_{in} + G_{ss} - K\log\left[1 + 10^{\left(\frac{P_{in} + G_{ss} - P_{sat}}{K}\right)}\right]$$

A cell is established on the spreadsheet to generate statistical values for small signal gain. The Gaussian probability density function is used. Another cell is established to generate statistical values for current; the Gaussian probability density function is used. Noise figure is calculated for each amplifier in another cell using the rule of thumb (4.17). Another cell is established to generate statistical values for saturated power output; the Weibull probability density function is used, location at P_{sat} − 1.81 dB, scale α = 1.89 dB and shape $\beta = 8$. Compression coefficient K is assumed to be nonvariant and is given the constant value $K = 6.08$. This statistical data is used to calculate power output and compression depth for each of the 1000 amplifiers where power input is +7 dBm.

Another cell is established to calculate the input power needed to obtain 2.38 dB compression for each amplifier. This value is calculated by setting the amplifier behavioral model compression term equal to 2.38 dB and solving for power input $P_{in}(2.38)$. The result obtained is the equation

$$P_{in}(2.38) = P_{sat} - G_{ss} + K\log\left[10^{\frac{2.38}{K}} - 1\right] \tag{8.16}$$

The power input needed to obtain 2.38 dB compression for each amplifier is then used as input power to that specific amplifier, and power output at 2.38 dB compression is calculated. The results of a 1000 trial run of the spreadsheet where input power is fixed at +7 dBm are tabulated in Table 8.3. The

Table 8.3
Test Results from 1000 Trials with P_{in} = +7 dBm

	Mean	Standard Deviation	Range Minimum	Range Maximum	Range Width
G_{ss}	14.01 dB	0.51 dB	12.23 dB	15.5 dB	3.27 dB
P_{sat}	19.9 dBm	0.26 dB	18.71 dBm	20.52 dBm	1.81 dB
P_{out} (+7)	18.47 dBm	0.41 dB	16.6 dBm	19.67 dBm	3.07 dB
Compression	2.54 dB	0.48 dB	1.31 dB	4.16 dB	2.85 dB
Current	0.03 A	0.004 A	0.01 A	0.04 A	0.03 A
Noise Figure	2.18 dB	0.12 dB	1.67 db	2.61 dB	0.94 dB

results of a 1000 trial run where input power is adjusted to obtain 2.38 dB compression for each amplifier are tabulated in Table 8.4.

Notice that with power input for all amplifiers fixed at +7 dBm the population's mean power output is 0.15 dB less than the expected power output at P_{in} = +7 dBm listed in Table 8.2. Standard deviation of power output is 0.41 dB with a range width of 3.07 dB with power input fixed at +7 dBm. Range minimum of power output is 16.6 dBm. Of 1000 amplifiers, at least one was driven 4.16 dB into compression. Mean compression depth for the population was 2.54 dB, 0.16 dB greater than expected, and at least one amplifier did not have sufficient gain to obtain more than 1.31 dB compression.

Notice that with input power adjusted for 2.38 dB compression for every amplifier, the population mean power output is still 0.28 dB less than the 18.62 dBm that was expected from Table 8.2. Also notice that the mean power input required to establish 2.38 dB compression in each amplifier is 6.71 dB, 0.29 dB less than the calculated, expected, value of +7 dBm shown in Table 8.2.

The conclusion drawn from this example is that acceptance test limits have to be carefully configured in order that yields not be impacted and costs increased. Another important conclusion to be drawn from the example is that test

Table 8.4
Test Results from 1000 Trials where Compression = 2.38 dB Each Amplifier

	Mean	Standard Deviation	Range Minimum	Range Maximum	Range Width
P_{in}	6.71 dBm	0.85 dB	2.99 dBm	8.98 dBm	6.0 dB
P_{out}	18.34 dBm	0.66 dB	14.35 dBm	19.55 dBm	5.2 dB

procedures have to be carefully configured. The wrong approach to testing an amplifier can cause poor yields. For instance, if minimum test limits are set at +18.0 dBm (2 dB less than P_{sat}) for power output with +7 dBm power input, only 73.3 percent of the population would pass as illustrated in Figure 8.18.

The statistical parameters of standard deviations and probability density distributions used in this example are typical for a single class A amplifier having compression coefficient $K = 6$. Compression characteristic is soft, causing power output at 2.38 dB compression to be 1.38 dB less than P_{sat}. A class AB amplifier having compression coefficient $K = 2$ would exhibit compressed power output 0.1 dB less than P_{sat} at 2.0 dB compression.

8.7 Summary

Probability density functions have been introduced as models for population statistics that apply to small signal gain, current consumption, saturated power output, and other amplifier parameters. The Gaussian or normal probability density function is applied to small signal gain and current. The Weibull probability density function is applied to saturated power output. Typical values are given for standard deviations based on observation over many years of experience. Typical values are also given for the Weibull scale and location parameters.

Equations are developed that assist in assigning values to the Weibull location parameter and scale factor based on measured standard deviation values. An example is given showing how the equations are used.

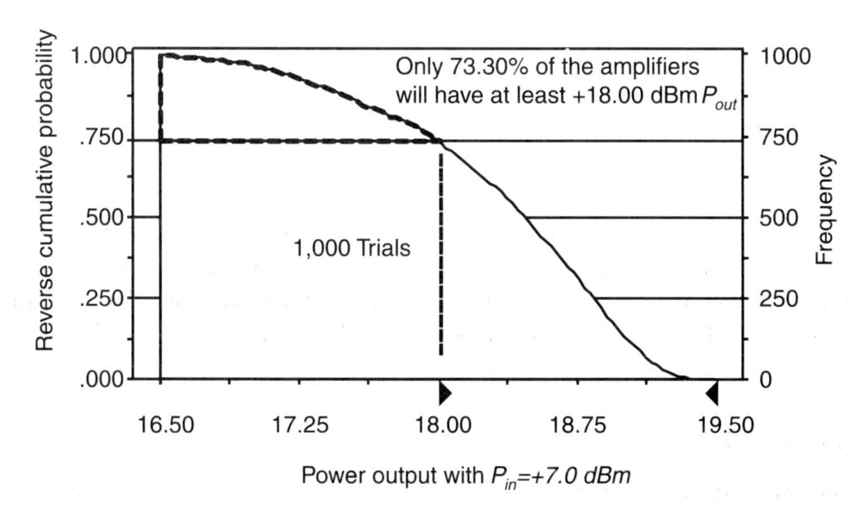

Figure 8.18 Only 73.3 percent of the population has power output greater than +18.0 dBm with P_{in} = +7 dBm.

Risk analysis spreadsheet add-in software is introduced, and two existing programs CRYSTAL BALL and AT RISK are discussed. CRYSTAL BALL is preferred because it is more adaptable to engineering applications. It allows direct access to raw data files so data can be arranged as desired and plotted in graphs different than those provided by the add-in software. Risk analysis software is designed primarily for business applications.

An example of the use of CRYSTAL BALL in the analysis of a population of amplifiers is given. The class A amplifier population does not obtain mean value of compressed power output expected by simple calculation using the nonlinear amplifier behavioral model. Yield to what might be considered a modest test limit is only 73.3 percent. This type of analysis is not possible with existing CAD software because none of the existing software considers a compression coefficient, much less variability of compression coefficient, in an amplifier population.

8.8 Problems

Problem 8.1

A two-stage monolithic MESFET class A amplifier population has mean value of small signal gain of 20 dB. Each stage has 10 dB nominal small signal gain. The chip operates at $v_{dc} = 8$ volts. The first stage is designed to draw 0.02 amperes; the second stage is designed to draw 0.09 amperes. Compute expected amplifier noise figure. Compute expected amplifier saturated power output. Compute standard deviation of total current, small signal gain, and of noise figure.

Problem 8.2

A class AB MESFET amplifier is designed to operate at $v_{dc} = 8$ volts, has $i_{dss} =$ 0.18 amperes, and is biased at 0.2 i_{dss}. Small signal gain is designed to be 12 dB. Compute saturated power output. Determine the power input that causes nominal 2.0 dB compression. Use risk analysis software to compute power output statistics for a population tested with +15 dBm power input. Determine what percentage of the population is compressed more than 2.0 dB. Determine what percentage of the population has more than +24 dBm power output when driven with +15 dBm.

Problem 8.3

Determine standard deviation of 1 dB compression value and standard deviation of the third-order intercept value of the amplifier population described in Problem 8.2.

References

[1] "CRYSTAL BALL," Risk analysis software spreadsheet add-in created by Decisioneering Inc., 1515 Arapahoe St., Suite 1311, Denver, Colorado, 80202. http//www.decisioneering.com.

[2] "AT RISK," Risk analysis software spreadsheet add-in created by Palisades Corp., 31 Decker Road, Newfield, N.Y., 14867. http//www.palisade.com.

[3] Munroe, M. E., *Theory of Probability*, New York, NY: McGraw-Hill, 1951, pp. 91–94.

[4] "Probability and Statistics," In *Reference Data for Radio Engineers*, H. P. Westman (editor), Indianapolis, IN: Howard W. Sams & Co. Inc., 1969, pp. 39-4 and 39-5.

[5] Weibull, W., "Ingenors Vetenskaps Aladiens Handlingar," The Royal Swedish Institute Of Engineering Research, Proc. No. 151, Stockholm, 1939.

[6] Brown, W. K., and K. H. Wohletz, "Derivation of the Weibull Distribution Based on Physical Principles and Its Connections to the Rosin-Rammler and Lognormal Distributions," Earth and Environmental Science Division, Los Alamos National Laboratory, 1998. http://geont1.lanl.gov/Wohletz/SFT2-dco.htm.

[7] "Distribution Functions Used in Reliability," In *Reference Data For Radio Engineers*, H. P. Westman (editor), Indianapolis, IN: Howard W. Sams & Co. Inc., 1969, pp. 40-16 and 40-17.

9

Making More Better

9.1 Obtaining Optimum Performance From Cascaded Amplifier Stages

The modeling techniques developed in the first eight chapters of this book are applied to single amplifier stages. Models include equations for determining one dB compression point P_{1dB}, third-order intercept point P_{3rdOIP}, saturated power output P_{sat}, compression coefficient *K*, power-added efficiency *PAE*, and noise figure *NF*. These modeling techniques are expanded in this chapter by applying them to multiple-stage amplifiers. Methods are developed for the optimization of multiple-stage low-noise amplifier design, multiple-stage power amplifier design, and for the selection of a number of amplifier stages where multiple stages are needed to generate a large amount of gain.

System applications for low-noise amplifiers often require low noise, low power consumption and high dynamic range all in a single design. Battery operated receivers found in wireless cellular handsets clearly need to exhibit low power consumption. Their dynamic range, specified as high-input third-order intercept point value, has to be able to handle large in-band and out-of-band interference from competing cellular services while processing own service signals with minimal cross modulation and intermodulation distortion. The worst case scenario occurs when receiving a message while driving past a competing service cellular tower. The same receiver needs lowest possible noise figure in order to obtain acceptable signal to noise ratio of its own service signals when far from the nearest cellular tower. These three-receiver parameters, low-power consumption, high third-order intercept, and low-noise figure are mutually exclusive. A trade space exists that defines the relationship between the three parameters. This trade space is developed in this chapter. The low-noise amplifier trade space principles developed in this chapter are applied in an Excel spreadsheet workbook found in the software programs that accompany this book. The spreadsheet workbook calculates optimized amplifier stage saturated power output, input and output third-order intercept, noise figure, and DC power consumption parameters that give a required input third-order intercept point value for a cascaded set of amplifier stages having required total gain and operating from a specified power supply voltage. Cascaded amplifier noise figure and power consumption are also estimated.

Multiple amplifier stages are used to develop high gain in receivers and power amplifiers. Different transistor technologies offer different gain values per amplifier stage for the same frequency band. Trades between using many low-gain-per-stage amplifiers versus fewer high-gain-per-stage amplifiers are developed. Advantages and disadvantages are studied. Comparisons are made of noise figure, third-order intercept, and power consumption for multistage amplifiers having the same overall gain but different numbers of amplifier stages. These trades reveal the advantages of PHEMT technology versus MESFET technology.

Power amplifiers used in transmitter applications often require multiple gain stages in order to develop sufficient gain to drive the final amplifier stage into compression where maximum power-added efficiency is realized. The design of each successive stage must be carefully thought out. If the first stages of a cascaded amplifier set compress before the latter amplifier stages compress, the final power stage may never be driven into the level of compression needed for the desired power output and *PAE*. Small signal gain G_{ss}, saturated power output P_{sat}, and compression coefficient K of each amplifier stage needs to be carefully chosen to achieve the desired performance. Models developed for single amplifier stages are applied to multiple cascaded stages and cascaded amplifier equations are developed giving insight into the design options. A new trade space for power amplifier design is developed.

9.2 The Noise Figure, Third-Order Intercept, Power Consumption Trade Space

The trade space developed in this section applies to low-noise amplifiers that must have high-input third-order intercept point values and at the same time must have the lowest possible noise figure and lowest possible power consumption. Extremely low power consumption low-noise amplifiers are sometimes designed as class AB amplifiers for spacecraft applications where large dynamic range is not needed, and power consumption is at a premium. The methods discussed here do not apply to such extreme design parameters. The high dynamic range, low-noise, optimized power consumption amplifier modeled here is typically designed to be a class A amplifier, biased at 50 percent i_{dss}. A single amplifier stage is characterized by the parameters of quiescent bias current i_q, operating voltage v_{dc}, small signal gain G_{ss}, saturated power output P_{sat}, compression coefficient K, one dB compression point P_{1dB}, output third-order intercept point P_{OIP}, input third-order intercept point P_{IIP}, and noise figure NF, all defined and developed for the class A amplifier stage in Chapter 4. These parameter definitions are applied here to each stage of a two-stage amplifier where parameters of the first stage are identified with subscript 1 and the second stage with subscript 2. Figure 9.1 illustrates the parameters assigned to each of the two stages.

The object is to develop a relationship between total current consumption $i_t = (i_1 + i_2)$, the cascaded noise figure NF_c, and the cascaded input intercept point P_{IIPc}.

9.2.1 Joint Intercept Point (*JIP*) Is Defined

The concept of a joint intercept point value, *JIP*, is introduced. When the output port of circuit element 1 is joined to the input port of circuit element 2, each having a definable intercept point value looking into the port that is to be joined with the other, the intercept point resulting from the union is defined to be

$$jip = \frac{1}{\dfrac{1}{oip_1} + \dfrac{1}{iip_2}} = \frac{oip_1 * iip_2}{oip_1 + iip_2} = oip_1\left(\frac{1}{1 + \dfrac{oip_1}{iip_2}}\right) \tag{9.1}$$

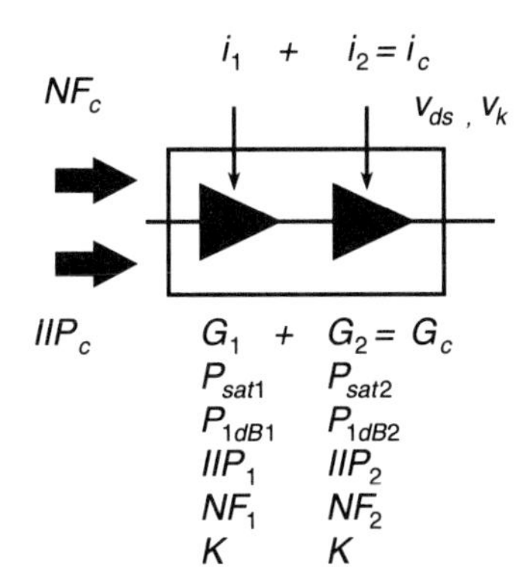

Figure 9.1 A two-stage amplifier is studied to develop a NF, P_{IIP}, i_{dc} trade space.

where oip_1 is the output intercept point value of circuit element 1, and iip_2 is the input intercept point of circuit element 2, and jip is joint intercept point, and all values expressed in units of watts. The third equality in (9.1) reveals interesting aspects of the joint intercept point. If the ratio $\frac{oip_1}{iip_2} \gg 1$, the jip value approaches that of iip_2. If the ratio $\frac{oip_1}{iip_2} \ll 1$, the jip value approaches that of oip_1. If the ratio $\frac{oip_1}{iip_2} = 1$, oip_1 and iip_2 are equal, and the jip value is half that of oip_1. The joint intercept value range is bounded by the values of iip_2 and oip_1.

The joint intercept point value is expressed as a decibel quantity relative to a milliwatt by the equation

$$JIP = 10 \log_{10}[jip] + 30 \text{ dBm} \tag{9.2}$$

The relationship between input intercept point of the cascaded circuit elements P_{IIPc} and the joint intercept point JIP is

$$P_{IIPc} = JIP - G_1 \text{ dBm} \tag{9.3}$$

Similarly, the relationship between the cascaded circuit element output intercept point P_{OIPc} and the joint intercept point is

$$P_{OIPc} = JIP + G_2 \text{ dBm} \tag{9.4}$$

The simple relationship $P_{OIPc} = P_{IIPc} + G_1 + G_2$ follows from (9.3) and (9.4).

Figure 9.2 illustrates the newly defined parameters of the two circuit elements being joined. Simplify third-order intercept point notation by letting OIP represent P_{OIP} and IIP represent P_{IIP}.

9.2.2 A Given *JIP* Can Be Satisfied by an Infinite Set of IIP_2 and OIP_1 Value Combinations

If a specific input third-order intercept point IIP_c is required of cascaded circuit elements, and first stage gain value G_1 is defined, then the cascaded element joint intercept point JIP value is clearly defined by (9.3). The values of OIP_1

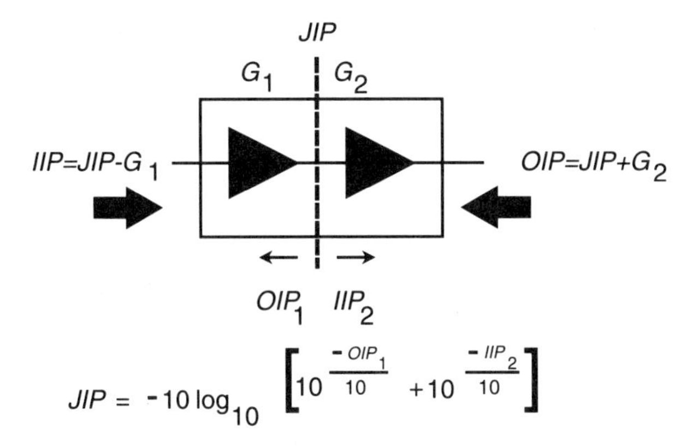

Figure 9.2 Joint intercept of two joined circuit elements is defined.

and IIP_2, however, can take on an infinite number of combinations as long as neither value is less than the defined *JIP*. If circuit element gain G_1 is not defined but is allowed to vary over a range, joint intercept point *JIP* value can take on a range of values also defined by (9.3). This is the case most useful to the development of a trade space. It provides the ability to optimize gain distribution between two cascaded amplifier stages. Let circuit element gain G_1 vary over a fixed range commensurate with typical transistor small signal gain parameters. Let G_1 vary over the range 6 dB $\leq G_1 \leq$ 12 dB. Then *JIP* will vary over a 6 dB range of values. There is a clearly definable relationship between G_1 values and *JIP* values through (9.3).

9.2.3 Use a Spreadsheet to Develop the Trade Space

Begin development of the trade space by first identifying values for required total cascaded amplifier gain $G_c = (G_1 + G_2)$, required cascaded amplifier input third-order intercept IIP_c, total allowable current consumption $i_t = (i_1 + i_2)$, maximum allowable cascaded amplifier noise figure NF_c, circuit operating voltage v_{dc}, transistor knee voltage v_k, amplifier class of operation (class A), and compression coefficient $K = 6$. Open a spreadsheet and establish reference cells into which values for G_1, G_2, IIP_c, v_{dc}, v_k, and K will be placed. Design the spreadsheet such that compression coefficient value K can easily be varied in all of the equations that use it by changing the value in a single spreadsheet reference cell. Also establish a row of values identified as JIP_{min} to JIP_{max} with a total range of 6 dB, incrementing values 0.2 dB per cell, with center of the range $JIP = IIP_c + 0.5G_c$. This range of *JIP* values is directly related to cascaded element input intercept point and G_1 values by the relationship $JIP = IIP_c - G_1$. A specific G_1 value corresponds to each *JIP* value. Develop an array of circuit element 2 input intercept point values as a function of circuit element 1 output intercept point values using the equation

$$IIP_2 = 10\log_{10}\left[\frac{oip_1 * jip}{oip_1 - jip}\right] + 30 = -10 * \log\left[10^{\frac{-JIP}{10}} - 10^{\frac{-OIP_1}{10}}\right] \quad (9.5)$$

Let circuit element 1 output intercept point values OIP_1 range from 0.5 dB greater than JIP_{min} to 15 dB greater than JIP_{min}. Figure 9.3 illustrates the format of the spreadsheet array. See Appendix B for an expansion of (9.5) to compute IIP of multiple stages.

Develop a second spreadsheet array using the same JIP values across and the same circuit element 1 output intercept values OIP_1 down. Fill this array with values of circuit element 2 output intercept values determined by the relationship

$$OIP_2 = IIP_2 + G_c - G_1 \quad (9.6)$$

where G_c is the total required cascaded amplifier gain and IIP_2 values are values taken from corresponding cells in the first array. Figure 9.4 illustrates the form and content of the second array.

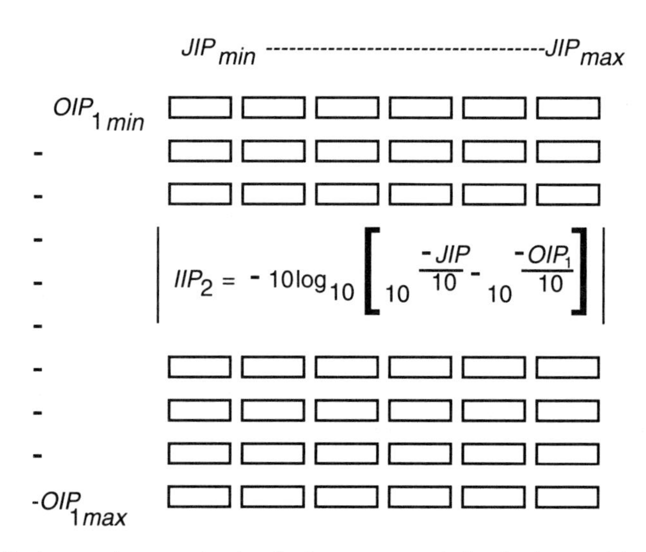

Figure 9.3 Begin a trade space by developing an array of circuit element 2 input intercept point values.

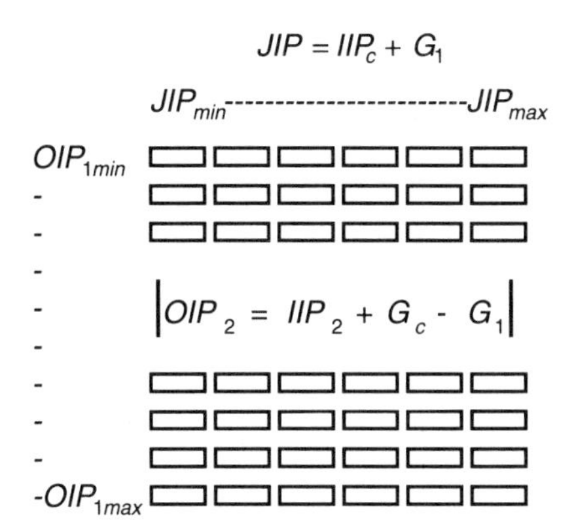

Figure 9.4 Develop a second array of circuit element two output intercept OIP_2 values.

Convert the second array, Figure 9.4, from output intercept point values to saturated power output values using the relationship developed in (4.12) and repeated here

$$P_{sat} = P_{OIP} - 9.63 - K\log_{10}\left[10^{\frac{1}{K}} - 1\right] \tag{9.7}$$

The amplifier stages are assumed to be class A; therefore compression coefficient value $K = 6$ is a valid approximation and a good starting value. This new array has a left-hand column containing saturated power output values for circuit element 1, and an array of circuit element 2 saturated power output values corresponding to *JIP* values. Figure 9.5 illustrates the new array form and content.

Convert the saturated power output array shown in Figure 9.5 into an array of current consumption for each amplifier stage using the relationship

$$i_{dc} = \frac{2 * 10^{\left(\frac{P_{sat} - 30}{10}\right)}}{(v_{ds} - v_k)} \tag{9.8}$$

where v_k is the transistor knee voltage. The assumption that $v_k = 1$ volt is a good approximation. This new array contains current consumed by circuit element 1 in the left column, and current consumption values for an array of different circuit element 2 stages in the remaining columns. Figure 9.6 illustrates the form and content of the two circuit element DC current consumption array. Create another new array that contains the sum of currents drawn by the two circuit elements as a function of *JIP* value with the same range of *JIP*

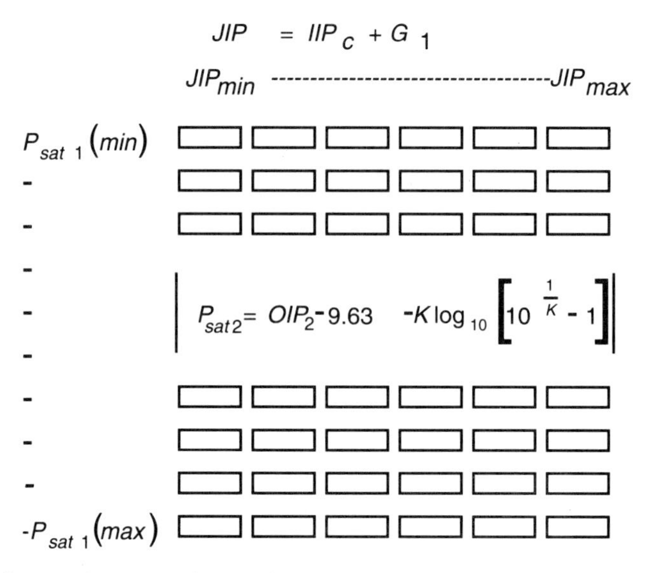

Figure 9.5 Convert data array of output intercept point to saturated power output values.

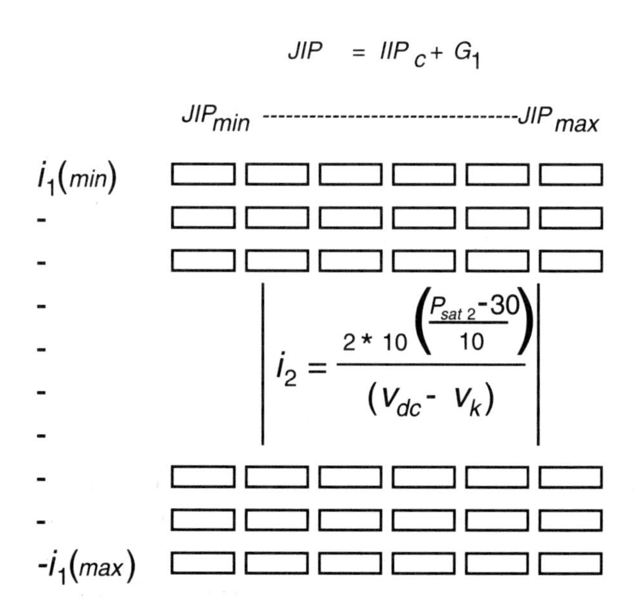

Figure 9.6 Convert saturated power output array into an array containing circuit element DC current.

values as row headings. Figure 9.7 illustrates the form and content of the total DC current array.

Use the previously mentioned noise figure rule of thumb (4.17) to convert the circuit element current consumption array illustrated in Figure 9.6 into a circuit element noise figure array. Noise figure of circuit element 1 for different current consumption values fills the left-hand column of the new array. Noise figure of circuit element 2 for different *JIP* values fills the remaining columns in the array. Cascaded amplifier total gain $G_c = (G_1 + G_2)$ is constant for each of these data arrays and cascaded amplifier input third-order intercept point IIP_c is fixed by requirement. Array column headings are easily converted from *JIP* values to amplifier stage gain values using (9.3) and (9.4) above. Change *JIP* values to corresponding circuit element 1 gain values G_1 on the circuit element noise figure array. Figure 9.8 illustrates the form and content of the circuit element noise figure array.

Develop a final cascaded amplifier noise figure array from the values in the circuit element noise figure array shown in Figure 9.8. Noise factor of each stage

$$F_n = 10^{\left(\frac{NF_n}{10}\right)} \tag{9.9}$$

and small signal gain of each stage

$$g_n = 10^{\left(\frac{G_n}{10}\right)} \tag{9.10}$$

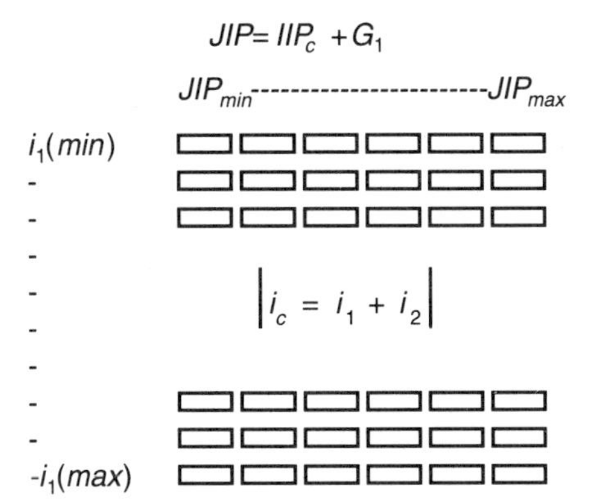

Figure 9.7 Sum the two-circuit element current consumption values to obtain a total current consumption array.

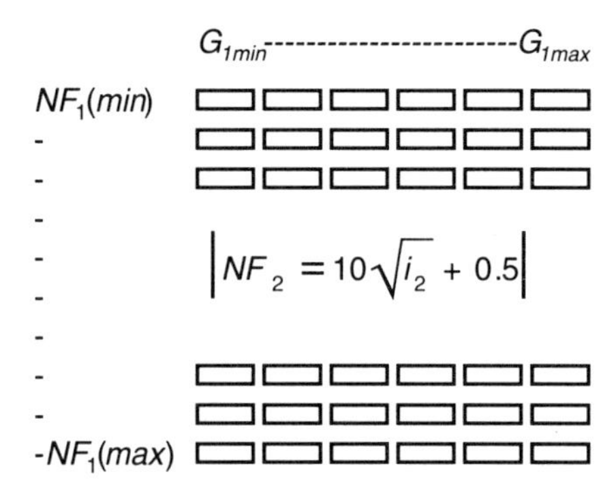

Figure 9.8 Convert circuit element current consumption into a circuit element noise figure array.

are used in the familiar cascaded amplifier stage noise figure equation

$$NF_c = 10\log_{10}\left[F_1 + \frac{F_2 - 1}{g_1} + \frac{F_3 - 1}{g_1 g_2} + \text{ooo} + \frac{F_{n-1} - 1}{g_1 g_2 \text{ ooo } g_{n-1}}\right] \quad (9.11)$$

The form and content of the cascaded circuit element noise figure array is illustrated in Figure 9.9.

Select corresponding columns of total current consumption from the total current array, Figure 9.7, and from the cascaded noise figure array, Figure 9.9, and plot total drain current as a function of cascaded noise figure for gain ratios where $G_1 > G_2$, $G_1 = G_2$, and $G_1 < G_2$. Match column data for cascaded noise figure and total current consumption by matching first-stage output intercept values. Use the spreadsheet's scatter plot utility to plot matched data.

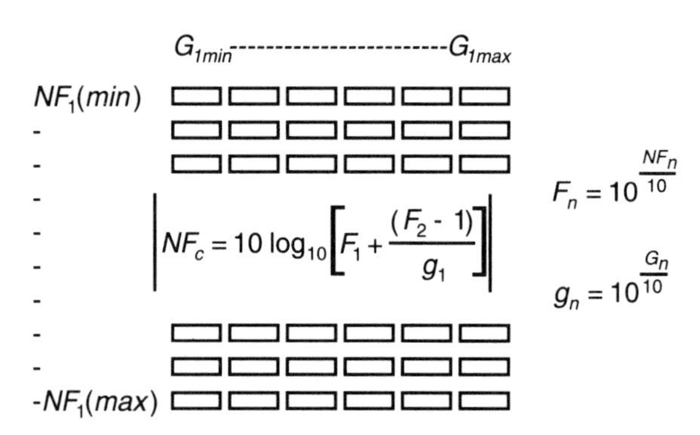

Figure 9.9 Cascaded circuit element noise figure is estimated for each combination of gain and stage 1 output intercept point.

The result is a plot that shows how optimum noise figure and optimum current consumption are obtained when $G_1 = G_2$ as illustrated in Figure 9.10. The plot also shows that there is an optimum combination of noise figure and current consumption for gain ratios $G_1 \neq G_2$. Each curve represents a continuum of combinations of circuit element 1 output third-order intercept point and circuit element 2 input third-order intercept point values that always satisfy the required cascaded element input third-order intercept point for a given gain distribution ratio. This continuum of circuit element input and output intercept combination values represents an infinite number of design possibilities that constitute the trade space.

The realization that there are optimum combinations of circuit element gain that give lowest noise figure and lowest current consumption is a profound discovery. Also, the realization that there is an optimum combination of circuit element input and output intercept points that results in low-noise figure and low current consumption, requires further development of the trade space. Expand the trade space to show the trades between noise figure and current consumption as a function of input third-order intercept values.

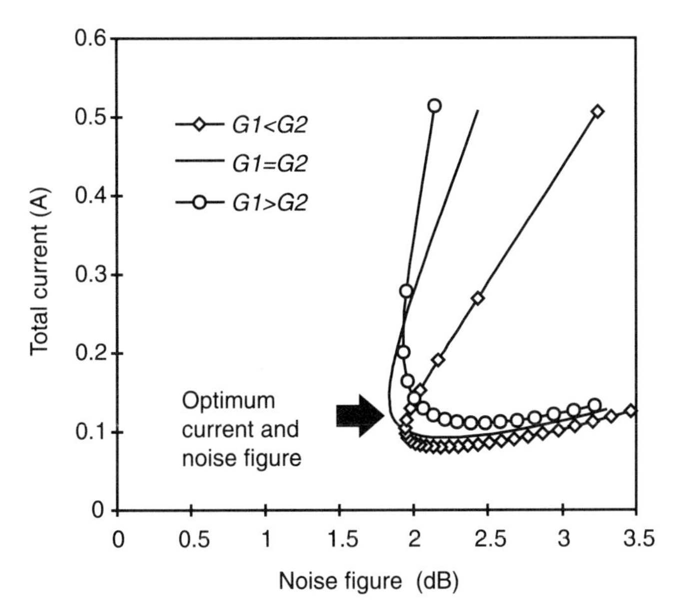

Figure 9.10 Optimum noise figure and current consumption occur when $G_1 = G_2$.

9.2.4 The Noise Figure, Current Consumption, Input Third-Order Intercept Trade Space

Given the conclusion that optimum gain distribution is achieved when $G_1 = G_2$, vary the cascaded third-order input intercept IIP_c spreadsheet cell value plus and minus several dB around the required value. Since G_1 is fixed by the relationship $G_1 = G_2$, variation of IIP_c causes required IIP values to change. Columns in the arrays developed in Figures 9.3 through 9.9 can be retitled to corresponding IIP_c values. The arrays previously developed now give optimum noise figure current consumption trade values for the several incremented cascaded circuit element input third-order intercept point IIP_c values. Use the spreadsheet scatter plot utility to plot total current versus noise figure for the several cascaded amplifier input third-order intercept point IIP_c values as illustrated in Figure 9.11. This trade space is the objective of this trade study development. Figure 9.11 clearly shows the relationship between cascaded amplifier current consumption, noise figure, and input third-order intercept point. Each cascaded circuit element input third-order intercept curve in Figure 9.11 represents a continuum of combinations of circuit element 1 output third-order intercept and circuit element 2 input third-order intercept values that satisfy the cascaded circuit element input third-order intercept IIP_c value. This continuum of combinations represents the design choices available to the circuit designer when selecting specific transistor sizes, bias voltages, and bias currents for the two circuit element stages. Circuit element design parameters of small signal gain, current consumption, noise figure, saturated power output,

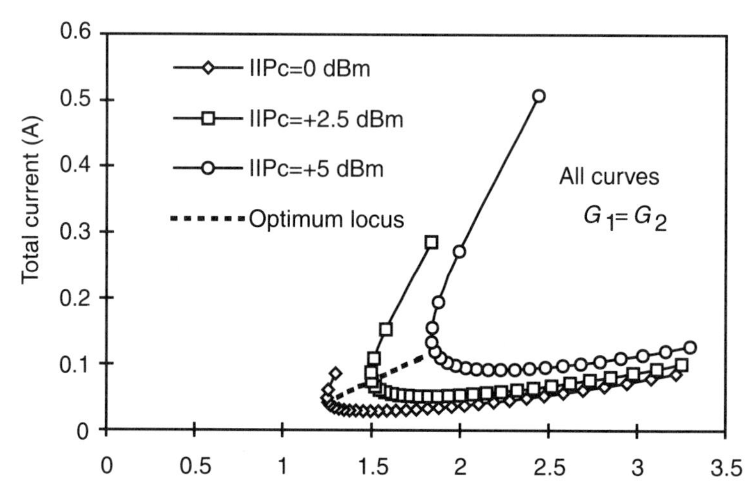

Figure 9.11 Trade space showing the relationship between current consumption, noise figure, and input third-order intercept for cascaded amplifier stages.

and input and output third-order intercept that achieve the optimum design are found in the various arrays illustrated in Figures 9.4, 9.5, 9.6, and 9.8. The optimum design for a required cascaded amplifier input third-order intercept is a combination that falls on the locus traced through the apices of the trade space curves (see Figure 9.11).

An EXCEL workbook software utility that solves for optimum two-stage, three-stage, and four-stage class A amplifier noise figure and current consumption parameters given a desired IIP_c and v_{dc} is included on the disk supplied with this book. Instructions for using the utility are included on sheet 1 of the EXCEL workbook.

9.2.5 Expanding the Trade Space to More Than Two Amplifier Stages

Low-noise amplifier gain required by some systems is often higher than can be provided by two cascaded circuit element stages. Three or four stages might be required to realize the required total small signal gain. The trade space developed above for two cascaded circuit element stages has to be expanded to apply to the increased number of gain stages. The more stages of gain required for a given application, the greater the current consumption will have to be to maintain a given level of input third-order intercept point. It is not unusual to arrive at design allocations specifying a 1 watt saturated power output in the final stage of a three-stage low-noise amplifier where cascaded circuit element input third-order intercept of +5 dBm is required.

9.2.5.1 Four-Stage Low-Noise Amplifier Optimization Approach

The driving function in every multistage low-noise amplifier design is input third-order intercept point. Noise figure and current consumption are optimized for a given IIP_c. If the number of gain stages is an even count of four, the first two stages and last two stages are partitioned together condensing the problem to two circuit elements of equal gain as illustrated in Figure 9.12. The two circuit elements are treated as if they were two amplifier stages having twice the normal gain, drawing individual currents and having individual noise figures. Optimum design parameters for the two circuit elements are developed using the procedures developed in Sections 9.2.2 and 9.2.3. Optimum design parameter values for current and noise figure for each of the two circuit elements containing two amplifier stages are preliminary at this point of the trade process. Input third-order intercept values for the two circuit elements derived in the optimization are valid and are used in the next optimization step. Each of the circuit elements are next broken down into two-stage amplifiers that exhibit total gain and cascaded input third-order intercept values derived from the circuit element trade results. Amplifier stage parameter values of noise

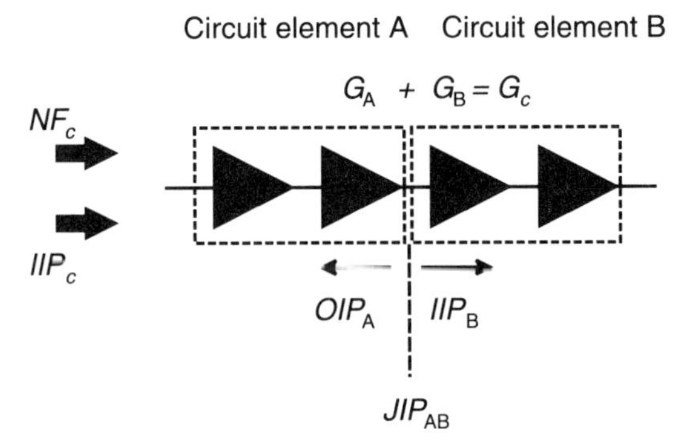

Figure 9.12 Four stages are partitioned into a pair of two-stage circuit elements.

figure, current consumption, saturated power output, *IIP*, and P_{1dB} derived from this step of the optimization process are valid.

An optimum design for the two cascaded amplifier stages in each of the two circuit elements is then developed using the procedures described in Sections 9.2.2 and 9.2.3. A well-designed spreadsheet will accept variables of specified gain, input third-order intercept, voltage, total current consumption, and noise figure and can be used to quickly arrive at optimum amplifier design parameters. An Excel workbook application for the analysis and optimization of four cascaded stages is included in the *Amplifier Analysis* software package in the back of the book.

9.2.5.2 Three-Stage Low-Noise Amplifier Optimization Approach

The approach to optimizing three cascaded amplifier stages is slightly different. The discovery of an optimum design for the required cascaded amplifier input third third-order intercept can only be satisfied by partitioning into two circuit elements of unequal gain. Small signal gain is not infinitely variable or conveniently divisible. The three stages have to be partitioned into a one-stage circuit element model and a two-stage circuit element model. The second circuit element will have roughly twice the gain of the first. Proceed through the optimization process as described in Sections 9.2.2 and 9.2.3 and derive parameters for the two unequal gain circuit element blocks. The data derived for the first circuit element block are the optimum values for saturated power output, current consumption, noise figure, input and output third-order intercept for the first amplifier stage. Only the optimum input intercept and the assigned cascaded gain value for circuit element B are useful at this point in the analysis; ignore circuit element B noise figure and current consumption values. Decompose circuit element B into its two amplifier stages and determine the amplifier stage parameters that satisfy the optimum IIP_B and small signal gain values derived for circuit element B. Optimum current consumption, noise figure, saturated power output, and input and output third-order intercept values for amplifier stages two and three are derived from this analysis of circuit element B. Figure 9.13 shows the circuit element approach for a three cascaded amplifier stage low-noise amplifier. Example 9.2 later in this chapter illustrates the optimization procedure. An Excel workbook for the analysis and optimization trade space of three-stage amplifiers is included in the software package in this book.

9.2.5.3 A Software Utility for Noise Figure, 3rd OIP, Current Consumption Optimization

A software utility "Amplifier Analysis" that solves for optimum two-stage, three-stage, and four-stage class A amplifier parameters given a desired IIP_c, G_c, v_k, and v_{dc}, is included on the disk in the back of this book. The utility is a Microsoft EXCEL workbook. Instructions for using the program appear on the first page of the workbook, and successive pages contain the utilities needed to optimize two-stage, three-stage, and four-stage amplifiers.

9.3 Trades Involving Number of Amplifier Stages and Amplifier Stage Small Signal Gain

A trade space is developed in Section 9.2 showing the relationship between noise figure, input third-order intercept, and current consumption for two cascaded amplifier stages and is extended in Section 9.2.4.2 to apply to three cascaded amplifier stages. The question arises in the process of designing a low-noise amplifier, "What advantage is there in using three low-gain MESFET transistor amplifier stages versus two higher gain PHEMT amplifier stages?"

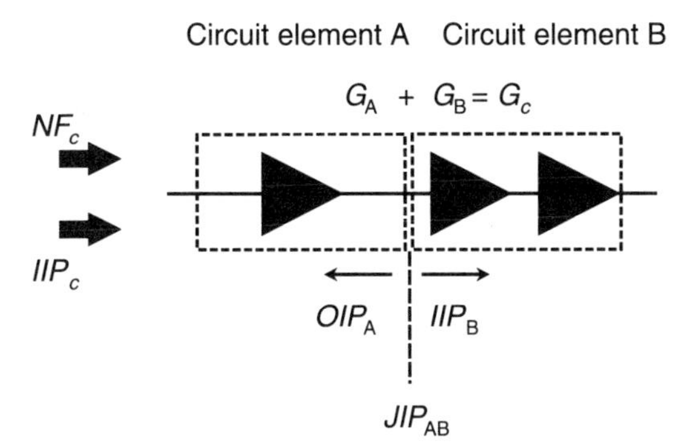

Figure 9.13 Three amplifier stages are partitioned into two unequal gain circuit elements.

Trade spaces developed earlier are applied here to answer this question. A design example is used to illustrate the trade methodology and trade space results.

Just three parameters are specified for a proposed X band amplifier design. Total cascaded amplifier small signal gain G_c has to be 21 dB, cascaded amplifier input third-order intercept IIP_c has to be +5 dBm and operating voltage is 6 volts (transistor knee voltage is assumed to be v_k = 1V and class A amplifier bias is assumed giving compression coefficient K = 6). The trade results to be compared are noise figure and total current consumption. A complete design parameter set is derived for a two-stage PHEMT amplifier and a three-stage MESFET amplifier based on just these three parameters. Figure 9.14 illustrates the two amplifier configurations under study.

The trade is developed in the two following examples. Example 9.1 considers the two-stage PHEMT amplifier where small signal gain is 10.5 dB per stage. Example 9.2 looks at the three-stage MESFET amplifier where amplifier stage small signal gain is 7 dB per stage.

Example 9.1 A Two-Stage Low-Noise Amplifier Based on PHEMT Technology.

The PHEMT transistor has higher gain than the MESFET and can easily be configured to produce 9.5 to 11.5 dB small signal gain per amplifier stage at frequencies in X band. Trade study results shown in Figure 9.10 indicate optimum noise figure and current consumption for a two-stage amplifier occur when $G_1 = G_2$. Let amplifier stage small signal gains be equal and 10.5 dB each. Since required cascaded amplifier input third-order intercept IIP_c = +5

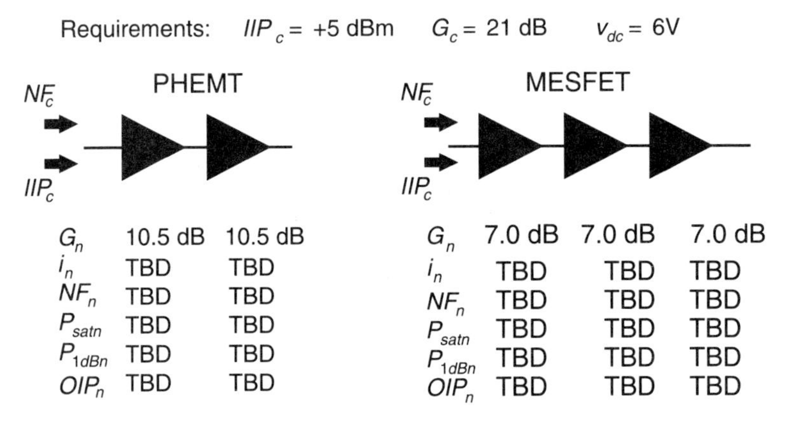

Figure 9.14 Study the trade between two high gain stages and three low gain stages.

dBm, and amplifier stage one gain $G_1 = 10.5$ dB, joint intercept point value $JIP = +15.5$ dBm. Develop a table showing stage two input third-order intercept point values that combine with stage 1 output third-order intercept values which result in $JIP = +15.5$ dBm. Let stage 1 output third-order intercept values range from $+16 \text{ dBm} \leq OIP_1 \leq +31$ dBm. Figure 9.15 illustrates the calculation being made and contains a table of the calculated values.

Convert input and output third-order intercept values shown in the table in Figure 9.15 into amplifier stage one and amplifier stage two saturated output values using (9.6) and (9.7) as shown in Table 9.1. The values of saturated power output for each amplifier stage are then converted into current consumption values for each amplifier stage using (9.8). Calculated current consumption for each of the two stages and the total current consumption are listed in Table 9.2. Registry of corresponding values is maintained from table to table.

Noise figure for each of the two amplifier stages is estimated using the rule of thumb (4.17) and individual amplifier stage current consumption data tabulated in Table 9.2. Cascaded amplifier stage noise figure is then calculated

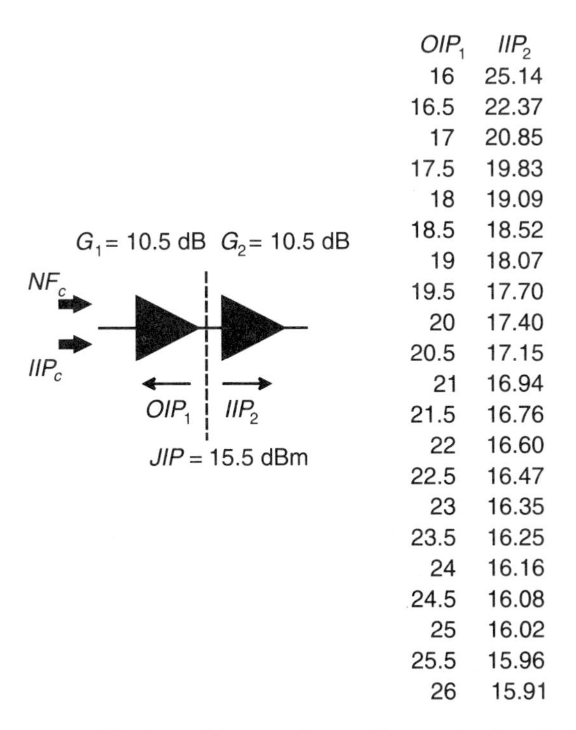

OIP_1	IIP_2
16	25.14
16.5	22.37
17	20.85
17.5	19.83
18	19.09
18.5	18.52
19	18.07
19.5	17.70
20	17.40
20.5	17.15
21	16.94
21.5	16.76
22	16.60
22.5	16.47
23	16.35
23.5	16.25
24	16.16
24.5	16.08
25	16.02
25.5	15.96
26	15.91

Figure 9.15 Compute combinations of input and output intercept values that result in $JIP = +15.5$ dBm.

Table 9.1
Amplifier Stage Output Third-Order Intercept Values (dBm)

OIP_1	OIP_2
16	35.64
16.5	32.87
17	31.35
17.5	30.33
18	29.59
18.5	29.02
19	28.57
19.5	28.20
20	27.90
20.5	27.65
21	27.44
21.5	27.26
22	27.10
22.5	26.97
23	26.85
23.5	26.75
24	26.66
24.5	26.58
25	26.52
25.5	26.46
26	26.41

using (9.11). Individual amplifier stage noise figure and cascaded amplifier stage noise figure is tabulated in Table 9.3.

Use the total current consumption data in Table 9.2 and the cascaded amplifier noise figure data in Table 9.3 as inputs to a spreadsheet scatter plot to create a noise figure-current consumption trade space plot for a +5 dBm input third-order intercept, 21 dB gain, two-stage amplifier as illustrated in Figure 9.16. Optimum two-stage amplifier parameters are discovered from the trade space and are listed in Figure 9.16.

Table 9.2
Amplifier Stage Current Consumption and Total Amplifier Current (A)

i_{idc1}	i_{dc2}	i_{total}
0.005	0.503	0.509
0.006	0.266	0.272
0.007	0.187	0.194
0.008	0.148	0.156
0.009	0.125	0.134
0.010	0.110	0.119
0.011	0.099	0.110
0.012	0.091	0.103
0.014	0.085	0.099
0.015	0.080	0.095
0.017	0.076	0.093
0.019	0.073	0.092
0.022	0.070	0.092
0.024	0.068	0.093
0.027	0.067	0.094
0.031	0.065	0.096
0.035	0.064	0.098
0.039	0.063	0.101
0.043	0.062	0.105
0.049	0.061	0.110
0.055	0.060	0.115

Example 9.2 A Three-Stage Low-Noise Amplifier Based on MESFET Technology

Consider a three-stage X band MESFET low-noise amplifier having 21 dB total cascaded amplifier gain and +5 dBm input third-order intercept point, operating from a 6-volt power source. Assume the transistor knee voltage $v_k = 1\text{V}$ and assume the low-noise amplifier is biased class A resulting in $K = 6$.

9.3.1 Assign 7 dB Small Signal Gain to All Three Stages

Partition 7 dB circuit element gain into circuit element A and 14 dB into circuit element B. Since required input third-order intercept point is +5 dBm and

Table 9.3
Individual Amplifier Stage Noise Figure and Cascaded Noise Figure (dB)

NF_1	NF_2	NF_{cas}
1.24	7.59	2.44
1.28	5.66	1.99
1.33	4.83	1.88
1.38	4.35	1.84
1.43	4.04	1.84
1.49	3.81	1.86
1.54	3.64	1.89
1.61	3.51	1.93
1.67	3.41	1.98
1.74	3.33	2.03
1.82	3.26	2.09
1.89	3.20	2.16
1.98	3.15	2.23
2.06	3.11	2.31
2.16	3.08	2.39
2.25	3.05	2.48
2.36	3.02	2.58
2.47	3.00	2.68
2.58	2.98	2.79
2.71	2.97	2.91
2.84	2.95	3.03

first element gain is 7 dB, joint intercept point between circuit elements A and B is forced to have value $JIP_1 = +12$ dBm. Develop a table of circuit element A third-order output intercept point OIP_A values that combine with circuit element B third-order input intercept point IIP_B values to give $JIP_1 = +12$ dBm. Let OIP_A values vary over the range $+12.5 \text{ dBm} \leq OIP_A \leq +27.5$ dBm. Convert these output and input third-order intercept values into saturated power output values for each of the circuit elements as listed in Table 9.4.

Convert saturated power output data into current consumption data and current consumption data into noise figure estimates for the two circuit elements as shown in Table 9.5. Total the current consumption for the two circuit elements and calculate cascaded element noise figure thus developing the trade space shown in Figure 9.17. Optimized parameter values for circuit element *A*

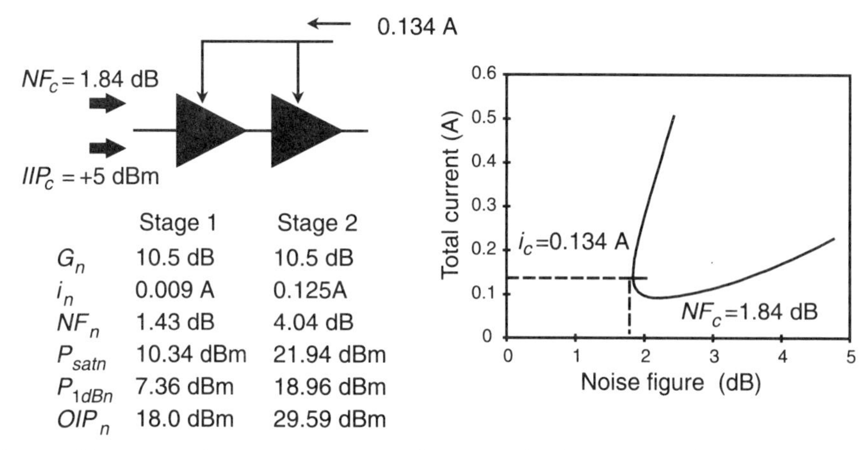

Figure 9.16 Optimum design parameters are discovered for the two-stage PHEMT low-noise amplifier.

at this point relate directly to the first amplifier stage and are listed in the table shown in Figure 9.17. Current values and noise figure values developed for circuit element *B* at this point of the trade study do not relate directly to amplifier stage two and three final optimized values, but the circuit element *B* value OIP_B = +28.2 dBm corresponding to optimum circuit element *A* OIP_A = +16.0 dBm from Table 9.4 is valid. Further decomposition of circuit element *B* is required.

Decompose circuit element *B*, which has 14.0 dB gain into a two-stage amplifier having cascaded input third-order intercept value IIP_B = +14.2 dBm as determined from the optimization process shown in Figure 9.17. This

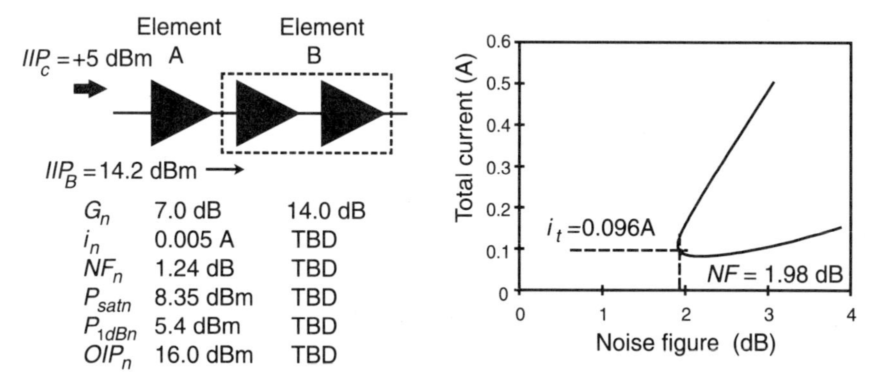

Figure 9.17 A trade space is developed for the two circuit elements to determine optimum amplifier stage one parameters.

Table 9.4
Circuit Element Output Third-Order Intercept Values Converted to Saturated Power Output Values

OIP_A	OIP_B	P_{satA}	P_{satB}
12.50	35.64	4.85	27.99
13.00	32.87	5.35	25.22
13.50	31.35	5.85	23.70
14.00	30.33	6.35	22.68
14.50	29.59	6.85	21.94
15.00	29.02	7.35	21.37
15.50	28.57	7.85	20.92
16.00	28.20	8.35	20.55
16.50	27.90	8.85	20.25
17.00	27.65	9.35	20.00
17.50	27.44	9.85	19.79
18.00	27.26	10.35	19.61
18.50	27.10	10.85	19.45
19.00	26.97	11.35	19.32
19.50	26.85	11.85	19.20
20.00	26.75	12.35	19.10
20.50	26.66	12.85	19.01
21.00	26.58	13.35	18.93
21.50	26.52	13.85	18.87

choice of input third-order intercept for circuit element *B* forces the joint intercept point between amplifier stages two and three to be $JIP_2 = +21.2$ dBm since small signal gain of amplifier stage two is assigned a value of 7.0 dB. Perform an optimization trade for amplifier stages two and three following the same procedure, arriving at a definitive amplifier stage two, stage three, current consumption, noise figure trade space shown in Figure 9.18.Final optimized parameter values for the three stage amplifier having the same gain in each stage are listed in Table 9.6.

9.3.2 Compare Results Obtained from the Two-Stage and Three-Stage Study

Compare results obtained from the two-stage PHEMT transistor low-noise amplifier optimization study from Example 9.1, Figure 9.16, to the results ob-

Table 9.5
Saturated Power Output Data Is Converted to Current Consumption and Current Consumption to Noise Figure Estimates

i_{dcA}	i_{dcB}	NF_A	NF_B
0.002	0.50	0.99	7.59
0.003	0.27	1.02	5.66
0.003	0.19	1.05	4.83
0.003	0.15	1.09	4.35
0.004	0.13	1.12	4.04
0.004	0.11	1.16	3.81
0.005	0.10	1.20	3.64
0.005	0.09	1.24	3.51
0.006	0.08	1.28	3.41
0.007	0.08	1.33	3.33
0.008	0.08	1.38	3.26
0.009	0.07	1.43	3.20
0.010	0.07	1.49	3.15
0.011	0.07	1.54	3.11
0.012	0.07	1.61	3.08
0.014	0.07	1.67	3.05
0.015	0.06	1.74	3.02
0.017	0.06	1.82	3.00
0.019	0.06	1.89	2.98

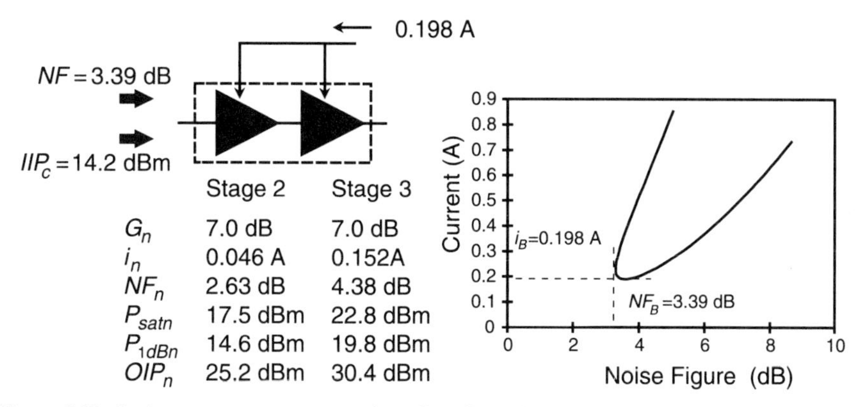

Figure 9.18 Optimum current consumption, P_{sat}, P_{1dB}, P_{OIP}, and NF for amplifier stages two and three of circuit element B are discovered.

Table 9.6
Three-Stage MESFET Low-Noise Amplifier Optimized Parameters

		Stage 1	Stage 2	Stage 3
Gain Per Stage (dB)		7.00	7.00	7.00
Current Per Stage (A)		0.005	0.046	0.151
NF Per Stage (dB)		1.24	2.63	4.39
P_{sat} Per Stage (dBm)		8.35	17.55	22.76
OIP Per Stage (dBm)		16.00	25.20	30.41
IIP Per Stage (dBm)		9.00	18.20	23.41
P_{1dB} Per Stage (dBm)		5.37	14.57	19.78
LNA Gain	21.00 dB		Total Current	0.202 Amps
LNA NF	1.95 dB		DC Power	1.212 Watts
LNA IIP	5.00 dBm			

tained from the three-stage MESFET low-noise amplifier optimization study from Example 9.2, Table 9.6. Small signal gain and input third-order intercept are identical by design requirement. The PHEMT amplifier combination of amplifier stage saturated power outputs that create the lowest noise figure (not the optimum balance of noise figure and power consumption) is discovered by finding the lowest cascaded amplifier noise figure value, NF_c = 1.84 dB, in Table 9.3. Total cascaded amplifier current consumption for that amplifier design, i_c = 0.134 A, is found in the corresponding row of Table 9.2. Output third-order intercept values for the cascaded amplifier stages that result in lowest possible noise figure are found in the corresponding row of Table 9.1. Two-stage PHEMT low-noise amplifier is 0.11 dB lower than the three-stage MESFET LNA. Current consumption of the PHEMT amplifier is 33 percent lower than that of the MESFET amplifier. It is clear from this trade study that there is benefit in using two high-gain amplifier stages instead of three low-gain stages when there is an option. Lower noise figure and lower current consumption result when the same input third-order intercept is designed into the amplifiers.

9.4 Developing a Power Amplifier Design Trade Space

The trade space developed here applies to multistage, class AB power amplifiers. Consider a two-stage amplifier designed to provide a specified small signal gain and to develop a given power output when driven to a specified compression depth by a specified input signal level. Power-added efficiency is expected to be as specified at the rated power output. The two-stage amplifier under consideration is depicted in Figure 9.19. Pertinent parameters for each amplifier stage are indicated in the figure. Cascaded amplifier gain $G_c = G_1 + G_2$.

Power output of each amplifier stage is modeled by the amplifier behavioral model equation

$$P_{out} = P_{in} + G_{ss} - K\log_{10}\left[1 + 10^{\frac{(P_{in}+G_{ss}-P_{sat})}{K}}\right] \quad (9.12)$$

It is assumed that each amplifier stage is biased at the same percent i_{dss} such that bias coefficient b_q is the same for both stages. That being the case, compression coefficient K is the same for both amplifier stages. The power output from amplifier stage one is the power input for amplifier stage two, $P_{out1} = P_{in2}$.

Rearrange Equation 9.12 to solve for power input as a function of power output. Subtract $(P_{in} + G_{ss})$ from both sides of Equation 9.12 and divide both sides by $-K$ to obtain

$$\frac{P_{in} + G_{ss} - P_{out}}{K} = \log_{10}\left[1 + 10^{\frac{(P_{in}+G_{ss}-P_{sat})}{K}}\right]$$

take the anti-logarithm of both sides

$$10^{\frac{(P_{in}+G_{ss}-P_{out})}{K}} = 1 + 10^{\frac{(P_{in}+G_{ss}-P_{sat})}{K}}$$

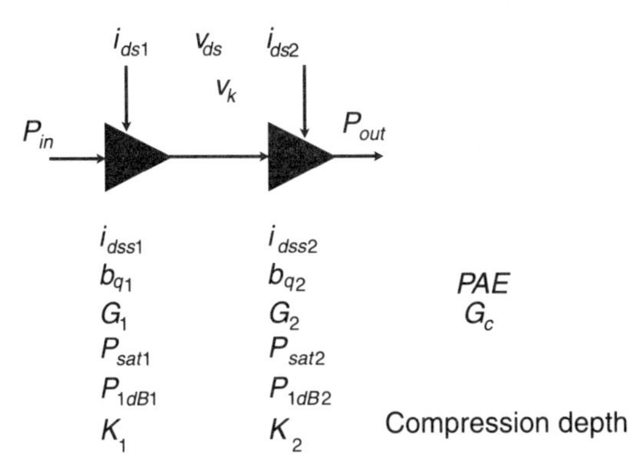

Figure 9.19 A two-stage class AB power amplifier and its parameters are considered.

combine terms of like exponents

$$10^{\frac{(P_{in}+G_{ss})}{K}}\left[10^{\frac{-P_{out}}{K}} - 10^{\frac{-P_{sat}}{K}}\right] = 1$$

divide both sides by the bracketed quantity and take the logarithm solving for P_{in} to obtain

$$P_{in} = K\log_{10}\left[\frac{1}{10^{\frac{-P_{out}}{K}} - 10^{\frac{-P_{sat}}{K}}}\right] - G_{ss} \tag{9.13}$$

Since power input of amplifier stage two is the same as power output of amplifier stage one, Equation 9.13 can be set equal to Equation 9.12. Place identifying subscripts on saturated power output P_{sat1} and small signal gain G_1 for the first stage and P_{sat2} and G_2 for the second stage.

$$P_{in} + G_1 - K\log_{10}\left[1 + 10^{\frac{(P_{in}+G_1-P_{sat1})}{K}}\right] = K\log_{10}\left[\frac{1}{10^{\frac{-P_{out}}{K}} - 10^{\frac{-P_{sat2}}{K}}}\right] - G_2$$

Rearrange this equality to solve for power output as a function of saturated power outputs, small signal gains, and compression coefficient K of the two amplifier stages. Add G_2 to both sides of the equality. Combine both logarithm terms into a single term. Divide both sides by compression coefficient K, and take the anti-logarithm to obtain

$$10^{\frac{(P_{in}+G_1+G_2)}{K}} = \frac{1 + 10^{\frac{(P_{in}+G_1-P_{sat1})}{K}}}{10^{\frac{-P_{out}}{K}} - 10^{\frac{-P_{sat2}}{K}}}$$

Rearrange this equality to read

$$10^{\frac{-P_{out}}{K}} = 10^{\frac{-P_{sat2}}{K}} + 10^{\frac{-(P_{in}+G_1+G_2)}{K}} + 10^{\frac{-(P_{sat1}+G_2)}{K}}$$

take the logarithm of both sides and multiply by $-K$ to finally obtain

$$P_{out} = -K\log_{10}\left[10^{\frac{-P_{sat2}}{K}} + 10^{\frac{-(P_{sat1}+G_2)}{K}} + 10^{\frac{-(P_{in}+G_1+G_2)}{K}}\right] \tag{9.14}$$

This equation gives profound insight into the operation and design of a two-stage power amplifier. It reveals the impact that the saturated power output of each stage has on final power output at any input power level. The term to the far right in the logarithm argument calculates linear power output, that is, the input power increased by the small signal gain of the two amplifier stages ($G_1 + G_2$). As long as the quantity ($P_{in} + G_1 + G_2$) is smaller than either P_{sat2} or ($P_{sat1} + G_2$), the right most term in the logarithm argument prevails, and (9.14) gives $P_{out} = P_{in} + G_1 + G_2$. When power input increases and the quantity ($P_{in} + G_1 + G_2$) becomes larger than either P_{sat2} or ($P_{sat1} + G_2$), power output saturates and becomes limited by the value of the terms P_{sat2} and ($P_{sat1} + G_2$). If the term ($P_{sat1} + G_2$) is larger than P_{sat2}, then power output is limited primarily by saturated power output of the final amplifier stage P_{sat2}. Equation (9.14) shows how much the final stage saturated power output is diminished by poor choice of saturated power output of amplifier stage one and small signal gain of amplifier stage two. The center term in the logarithm argument is the value of saturated power output of amplifier stage one, P_{sat1}, reflected to the amplifier's output through gain of the second amplifier stage, G_2. Setting amplifier stage two small signal gain G_2 as high as possible helps create the desired compression in the final stage. Figure 9.20 illustrates the impact on cascaded amplifier saturated power output for different ratios of ($P_{sat1} + G_2$) to P_{sat2}. In the limit where P_{in} is large, and the amplifier is fully saturated, maximum saturated power output of the two stages, (9.14), becomes

$$P_{sat(max)} = -K\log_{10}\left[10^{\frac{-P_{sat2}}{K}} + 10^{\frac{-(P_{sat1}+G_2)}{K}}\right]$$

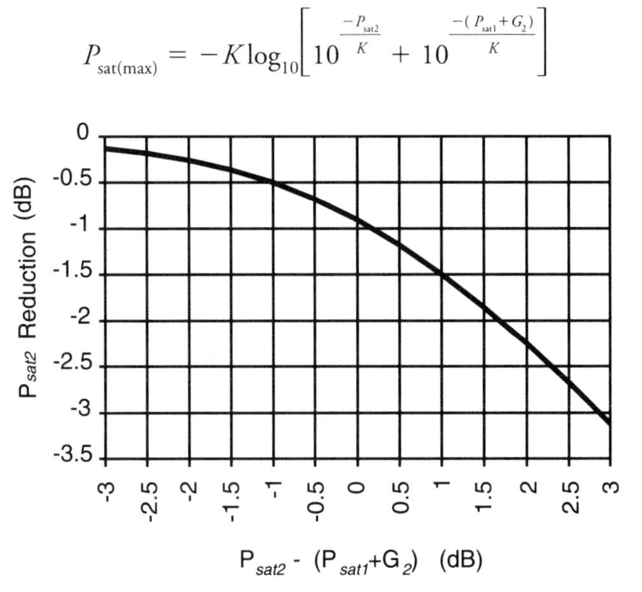

Figure 9.20 Final amplifier stage P_{sat} reduction as a function of ($P_{sat1} + G_2$).

Equation (9.14) is developed for a two-stage power amplifier. A similar equation can be developed for a three-stage power amplifier with the result

$$P_{out} = -K\log_{10}\left[10^{\frac{-P_{sat3}}{K}} + 10^{\frac{-(P_{sat2}+G_3)}{K}} + 10^{\frac{-(P_{sat1}+G_2+G_3)}{K}} + 10^{\frac{-(P_{in}+G_1+G_2+G_3)}{K}}\right] \tag{9.15}$$

The modification and addition of terms required for additional amplifier stages is obvious, and derivation is assigned to the reader.

9.4.1 Find an Optimum Ratio of Saturated Power Output Stage to Stage

Equation (9.14) shows the multidimensional relationship between a two-stage amplifier assembly saturated power output and parameters associated with two amplifier stages. The design trade space to be developed is one that gives an optimum ratio of saturated power output of amplifier stage one with respect to saturated power output of amplifier stage two. The goal is to find the P_{sat} ratio that maximizes power output from a two-stage amplifier assembly while realizing maximum power-added efficiency. A trade space of three dimensions is most easy to visualize. The power output relationship here deals with seven independent variables, P_{in}, G_1, G_2, P_{sat1}, P_{sat2}, quiescent bias coefficient b_q, current as a function of power input to each stage, and compression coefficient K. The power-added efficiency relationship (4.16) adds three more variables, applied voltage v_{dc}, knee voltage v_k, and average current consumption i_{dc}. Power amplifiers are usually class AB amplifiers which exhibit dynamic average current as a function of power input. Average current in class AB amplifier stages varies according to (5.15) or (5.17), and compression coefficient varies according to (5.11) or (5.12), depending on the transistor type used.

Example 9.3 Develop a Trade Space for Determining Optimum P_{sat} Ratio in a Two-Stage Power Amplifier

A method for developing the trade space is best taught by example, assigning values to a typical two-stage power amplifier and developing a spreadsheet analysis. With this in mind, begin trade space development by modeling the two-stage amplifier power output in a spreadsheet using (9.14). Set up the spreadsheet with data inputs G_1, G_2, P_{sat2}, v_{dc}, v_k, and K in rows 1 and 2 as shown in Figure 9.21.

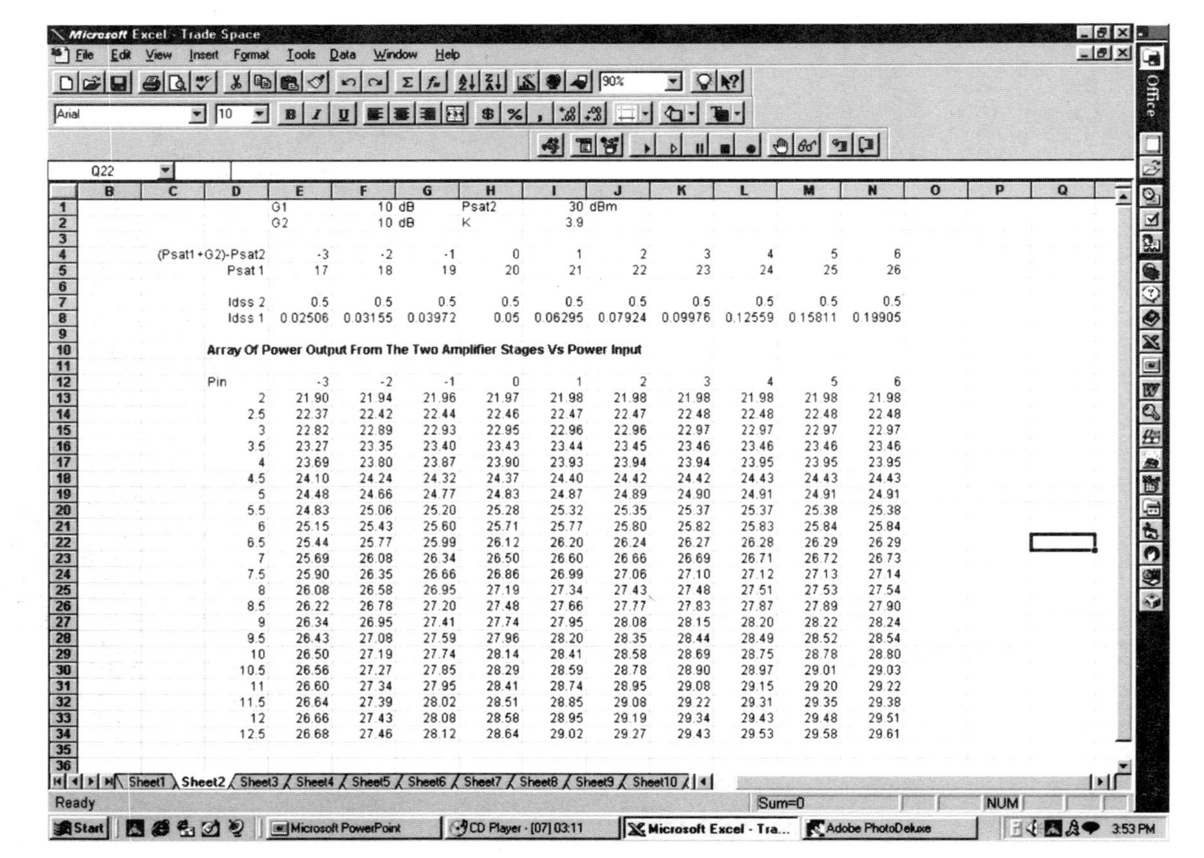

	D	E	F	G	H	I	J	K	L	M	N
1		G1	10 dB		Psat2	30 dBm					
2		G2	10 dB		K	3.9					
4	(Psat1+G2)-Psat2	-3	-2	-1	0	1	2	3	4	5	6
5	Psat 1	17	18	19	20	21	22	23	24	25	26
7	Idss 2	0.5	0.5	0.5	0.5	0.5	0.5	0.5	0.5	0.5	0.5
8	Idss 1	0.02506	0.03155	0.03972	0.05	0.06295	0.07924	0.09976	0.12559	0.15811	0.19905
10	**Array Of Power Output From The Two Amplifier Stages Vs Power Input**										
12	Pin	-3	-2	-1	0	1	2	3	4	5	6
13	2	21.90	21.94	21.96	21.97	21.98	21.98	21.98	21.98	21.98	21.98
14	2.5	22.37	22.42	22.44	22.46	22.47	22.47	22.48	22.48	22.48	22.48
15	3	22.82	22.89	22.93	22.95	22.96	22.96	22.97	22.97	22.97	22.97
16	3.5	23.27	23.35	23.40	23.43	23.44	23.45	23.46	23.46	23.46	23.46
17	4	23.69	23.80	23.87	23.90	23.93	23.94	23.94	23.95	23.95	23.95
18	4.5	24.10	24.24	24.32	24.37	24.40	24.42	24.42	24.43	24.43	24.43
19	5	24.48	24.66	24.77	24.83	24.87	24.89	24.90	24.91	24.91	24.91
20	5.5	24.83	25.06	25.20	25.28	25.32	25.35	25.37	25.37	25.38	25.38
21	6	25.15	25.43	25.60	25.71	25.77	25.80	25.82	25.83	25.84	25.84
22	6.5	25.44	25.77	25.99	26.12	26.20	26.24	26.27	26.28	26.29	26.29
23	7	25.69	26.08	26.34	26.50	26.60	26.66	26.69	26.71	26.72	26.73
24	7.5	25.90	26.35	26.66	26.86	26.99	27.06	27.10	27.12	27.13	27.14
25	8	26.08	26.58	26.95	27.19	27.34	27.43	27.48	27.51	27.53	27.54
26	8.5	26.22	26.78	27.20	27.48	27.66	27.77	27.83	27.87	27.89	27.90
27	9	26.34	26.95	27.41	27.74	27.95	28.08	28.15	28.20	28.22	28.24
28	9.5	26.43	27.08	27.59	27.96	28.20	28.35	28.44	28.49	28.52	28.54
29	10	26.50	27.19	27.74	28.14	28.41	28.58	28.69	28.75	28.78	28.80
30	10.5	26.56	27.27	27.85	28.29	28.59	28.78	28.90	28.97	29.01	29.03
31	11	26.60	27.34	27.95	28.41	28.74	28.95	29.08	29.15	29.20	29.22
32	11.5	26.64	27.39	28.02	28.51	28.85	29.08	29.22	29.31	29.35	29.38
33	12	26.66	27.43	28.08	28.58	28.95	29.19	29.34	29.43	29.48	29.51
34	12.5	26.68	27.46	28.12	28.64	29.02	29.27	29.43	29.53	29.58	29.61

Figure 9.21 Set up a spreadsheet for the development of a P_{sat} ratio trade space.

Row 4 contains a range of values −3 through +6 that give variability to the ratio of P_{sat2} to P_{sat1}. The number in row 4 is

$$N = (P_{sat1} + G_2) - P_{sat2} \tag{9.16}$$

from which values for P_{sat1} are determined. Row 5 contains values of P_{sat1} derived from the values of N in row 4 using (9.16). Amplifier stage i_{dss} current values in rows 7 and 8 are calculated using (5.3). The array is filled with cascaded amplifier power output, calculated using (9.14), as a function of power input values listed in column D. Other inputs for (9.14) are P_{sat2} value in row 4, P_{sat1} value tabulated in row 5, which varies from column to column, amplifier stage 2 gain G_2, and compression coefficient K. Power output data as a function of power input and value N is plotted as a surface as shown in Figure 9.22. Notice the decrease in saturated power output as P_{sat} ratio factor N becomes smaller and goes negative. This is due to the value $P_{sat1} + G_2$ becoming the predominant value in (9.14) as saturation occurs.

Several intermediate spreadsheet arrays are now needed to develop data specific to each amplifier stage. Create an array of amplifier stage one power output as a function of power input. Each column in this array represents a different amplifier stage one design having different values of P_{sat}, (values located in row 5), using transistors with different i_{dss} values (values located in row 8).

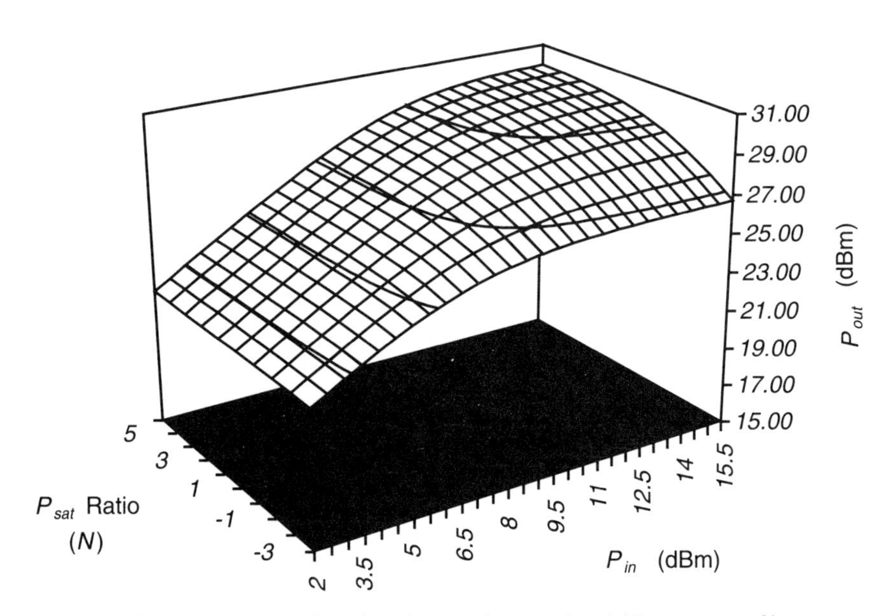

Figure 9.22 Power output as a function of power input and variable parameter N.

Power output of amplifier stage one appearing in this array is power input to amplifier stage two.

Two additional arrays needed are average current consumed by each of the amplifier stages as a function of power input. The class AB amplifier stages are biased at 20 percent i_{dss}. Quiescent bias coefficient $b_q = 0.2$. This gives compression coefficient K the value 3.9 (assuming MESFET transistors). Coefficients needed to calculate average current as a function of power input are a function of bias coefficient and for $b_q = 0.2$, $P_1(b) = -17.5$ and $P_2(b) = -5.8$. Equation (5.15) is then used to determine average current drawn by each amplifier stage. Total current drawn by the cascaded amplifier stages is the sum of the two individual current consumption arrays. Figure 9.23 shows the total current array.

A power-added efficiency data array is calculated using power output and power input from the first array shown in Figure 9.21, total current data from the array shown in Figure 9.23, and the applied DC voltage from row 1. The data in the power-added efficiency array is plotted as a surface as shown in Figure 9.24. Notice the peak in power-added efficiency as a function of amplifier stage P_{sat} ratio value N and power input.

The power-added efficiency peak is not sharp. It is difficult to determine which P_{sat} ratio optimizes power output and efficiency to the best advantage. Maximum power output is seen to occur at higher values of $N = 6$ when the amplifier assembly is driven into saturation in Figure 9.22. Maximum power-added efficiency is seen to occur in Figure 9.24 when $2 < N < 3$. The following procedure uses data in the power output array and in the power-added efficiency array to magnify the trade space and aid in arriving at an optimum ratio of saturated power output for the two amplifier stages. Determine the maximum power-added efficiency value in the *PAE* field of data used to generate Figure 9.24 and add 3 to the value. Divide each cell in the power output data field, element by element, by the maximum *PAE* value plus three, minus the corresponding cell value of the *PAE* data field to obtain the factor

$$M = \frac{P_{out}(P_{in}, \mathrm{P}_{sat1})}{(PAE_{max} + 3 - PAE(P_{in}, P_{sat1}))} \tag{9.17}$$

and build a new data array of M versus power input. Plot the optimization factor M as a surface as illustrated in Figure 9.25. The surface shown in Figure 9.25 peaks sharply at an optimum value of N and P_{in} from which the optimum value of P_{sat1} is determined.

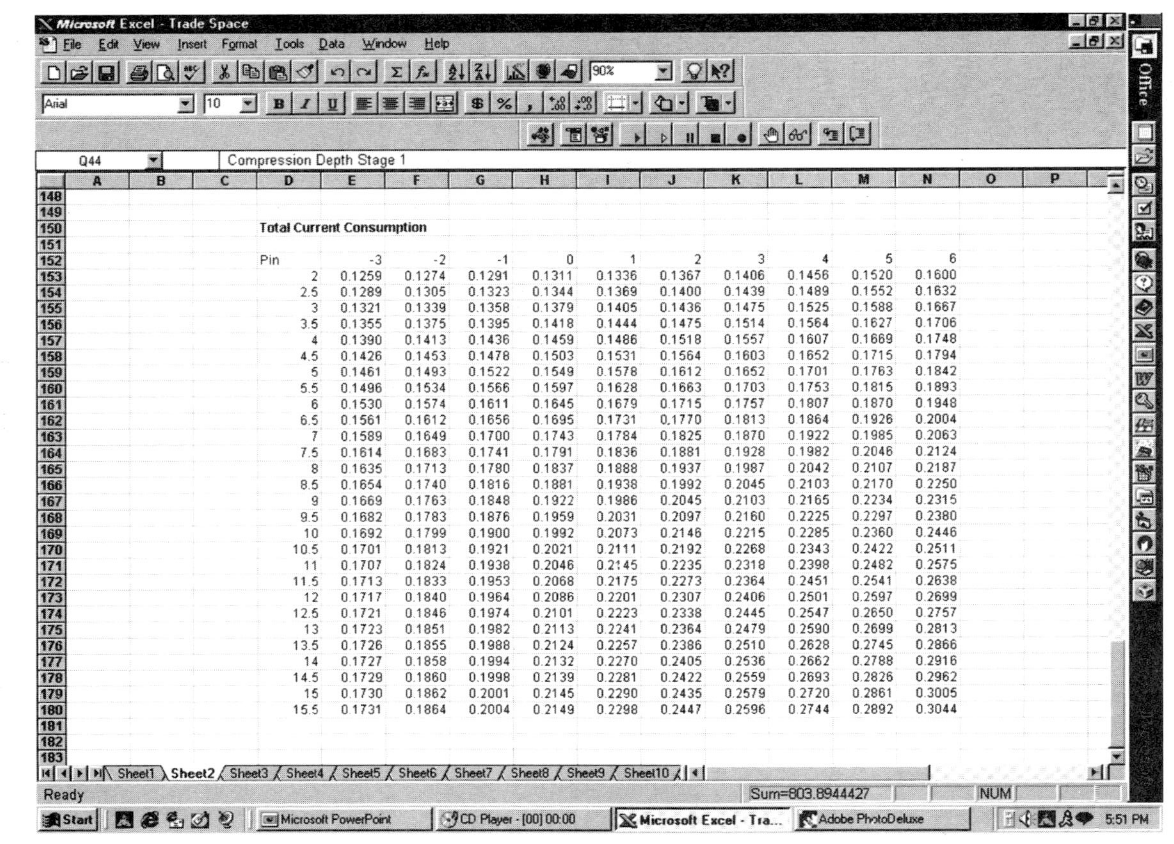

Total Current Consumption

Pin	-3	-2	-1	0	1	2	3	4	5	6
2	0.1259	0.1274	0.1291	0.1311	0.1336	0.1367	0.1406	0.1456	0.1520	0.1600
2.5	0.1289	0.1305	0.1323	0.1344	0.1369	0.1400	0.1439	0.1489	0.1552	0.1632
3	0.1321	0.1339	0.1358	0.1379	0.1405	0.1436	0.1475	0.1525	0.1588	0.1667
3.5	0.1355	0.1375	0.1395	0.1418	0.1444	0.1475	0.1514	0.1564	0.1627	0.1706
4	0.1390	0.1413	0.1436	0.1459	0.1486	0.1518	0.1557	0.1607	0.1669	0.1748
4.5	0.1426	0.1453	0.1478	0.1503	0.1531	0.1564	0.1603	0.1652	0.1715	0.1794
5	0.1461	0.1493	0.1522	0.1549	0.1578	0.1612	0.1652	0.1701	0.1763	0.1842
5.5	0.1496	0.1534	0.1566	0.1597	0.1628	0.1663	0.1703	0.1753	0.1815	0.1893
6	0.1530	0.1574	0.1611	0.1645	0.1679	0.1715	0.1757	0.1807	0.1870	0.1948
6.5	0.1561	0.1612	0.1656	0.1695	0.1731	0.1770	0.1813	0.1864	0.1926	0.2004
7	0.1589	0.1649	0.1700	0.1743	0.1784	0.1825	0.1870	0.1922	0.1985	0.2063
7.5	0.1614	0.1683	0.1741	0.1791	0.1836	0.1881	0.1928	0.1982	0.2046	0.2124
8	0.1635	0.1713	0.1780	0.1837	0.1888	0.1937	0.1987	0.2042	0.2107	0.2187
8.5	0.1654	0.1740	0.1816	0.1881	0.1938	0.1992	0.2045	0.2103	0.2170	0.2250
9	0.1669	0.1763	0.1848	0.1922	0.1986	0.2045	0.2103	0.2165	0.2234	0.2315
9.5	0.1682	0.1783	0.1876	0.1959	0.2031	0.2097	0.2160	0.2225	0.2297	0.2380
10	0.1692	0.1799	0.1900	0.1992	0.2073	0.2146	0.2215	0.2285	0.2360	0.2446
10.5	0.1701	0.1813	0.1921	0.2021	0.2111	0.2192	0.2268	0.2343	0.2422	0.2511
11	0.1707	0.1824	0.1938	0.2046	0.2145	0.2235	0.2318	0.2398	0.2482	0.2575
11.5	0.1713	0.1833	0.1953	0.2068	0.2175	0.2273	0.2364	0.2451	0.2541	0.2638
12	0.1717	0.1840	0.1964	0.2086	0.2201	0.2307	0.2406	0.2501	0.2597	0.2699
12.5	0.1721	0.1846	0.1974	0.2101	0.2223	0.2338	0.2445	0.2547	0.2650	0.2757
13	0.1723	0.1851	0.1982	0.2113	0.2241	0.2364	0.2479	0.2590	0.2699	0.2813
13.5	0.1726	0.1855	0.1988	0.2124	0.2257	0.2386	0.2510	0.2628	0.2745	0.2866
14	0.1727	0.1858	0.1994	0.2132	0.2270	0.2405	0.2536	0.2662	0.2788	0.2916
14.5	0.1729	0.1860	0.1998	0.2139	0.2281	0.2422	0.2559	0.2693	0.2826	0.2962
15	0.1730	0.1862	0.2001	0.2145	0.2290	0.2435	0.2579	0.2720	0.2861	0.3005
15.5	0.1731	0.1864	0.2004	0.2149	0.2298	0.2447	0.2596	0.2744	0.2892	0.3044

Figure 9.23 Total amplifier average current as a function of power input is calculated.

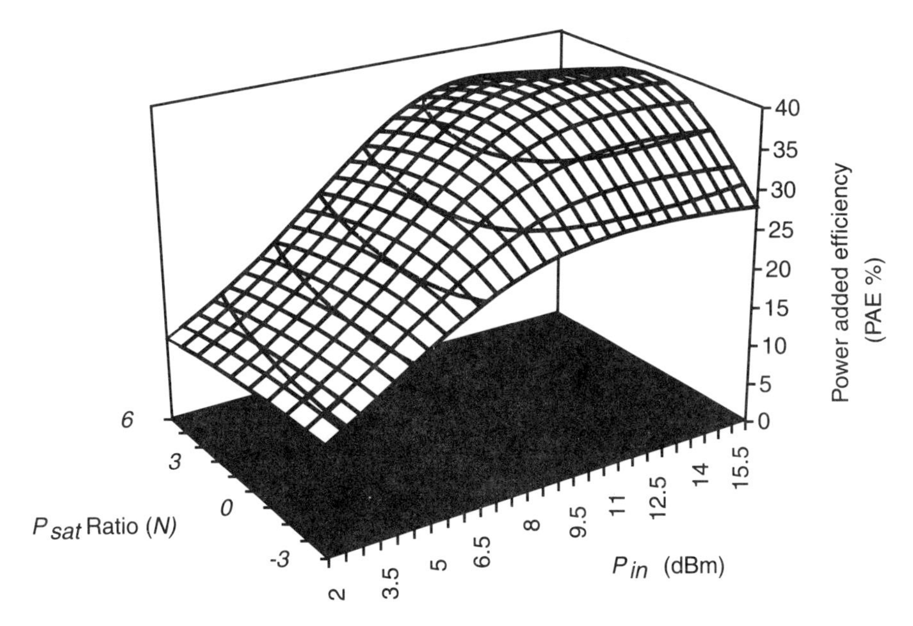

Figure 9.24 Power-added efficiency values that correspond to power output values shown in Figure 9.22.

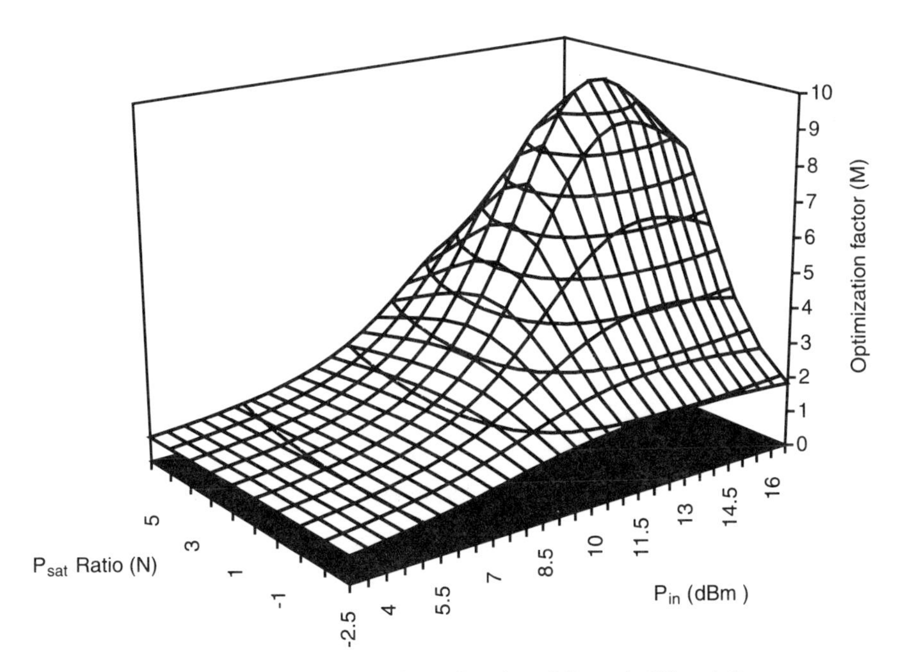

Figure 9.25 Optimization factor *M* plotted as a function of P_{sat} ratio (N) and P_{in}.

9.4.2 Compression Depth in Multistage Power Amplifiers

Amplifier stage compression depth is defined by (4.5), which can be simplified to

$$Compression = P_{in} + G - P_{out}$$

where P_{out} is nonlinear, compressed power output. Substitute (9.14) into this expression for compression depth and a new definition for cascaded amplifier stage compression depth is obtained.

$$Compression = P_{in} + G_1 + G_2 + K\log_{10}\left[10^{\frac{-P_{sat2}}{K}} + 10^{\frac{-(P_{sat1}+G_2)}{K}} + 10^{\frac{-(P_{in}+G_1+G_2)}{K}}\right]$$

which reduces to

$$Compression = +K\log_{10}\left[1 + 10^{\frac{(P_{in}+G_1+G_2-P_{sat2})}{K}} + 10^{\frac{(P_{in}+G_1-P_{sat1})}{K}}\right] \tag{9.18}$$

The center term in the logarithm argument is the compression depth of the final amplifier stage, and the right most term in the argument is compression depth of the first amplifier stage. Extension of (9.16) to compression depth of amplifiers having more than two stages is easily realized. Compression depth of a three-stage amplifier is expressed by

$$Compression = +K * \log_{10}\left[1 + 10^{\frac{(P_{in}+G_1+G_2+G_3-P_{sat3})}{K}} + 10^{\frac{(P_{in}+G_1+G_2-P_{sat2})}{K}} + 10^{\frac{(P_{in}+G_1-P_{sat1})}{K}}\right] \tag{9.19}$$

Proof of (9.19) is left to the reader.

9.4.3 Compression Phase Shift in Multistage Power Amplifiers

Amplifier stage compression phase shift is defined by (4.14) to be the product of compression depth and AM to PM conversion sensitivity θ,

$$\varphi = \theta K\log_{10}\left[1 + 10^{\frac{P_{in}+G_{ss}-P_{sat}}{K}}\right]$$

This definition can now be expanded to multiple stages. Multiply the multistage compression depth definition (9.16), or an expanded version of (9.16) covering a greater number of amplifier stages, by the AM to PM conversion ratio θ to obtain

$$\varphi = \theta K \log_{10}\left[1 + 10^{\frac{(P_{in}+G_1+G_2-P_{sat2})}{K}} + 10^{\frac{(P_{in}+G_1-P_{sat1})}{K}}\right] \tag{9.20}$$

9.5 Summary

Trade spaces have been developed for multiple stage low-noise amplifiers, and power amplifiers by applying the behavioral models for class A and class AB amplifiers developed in Chapters 4 and 5. All of these amplifier trade spaces give system and subsystem engineers the ability to quickly perform a sanity check for parameters their hardware designs require in order to meet overall system performance. The trade spaces provide a means of developing realistic flow-down specifications for design and development of circuits and transistors needed to meet subsystem requirements. The low-noise amplifier trade space gives the designer the ability to intelligently select the right combinations of amplifier stage parameters that simultaneously satisfy current consumption, noise figure, and input third-order intercept requirements. It shows the practical limits on noise figure imposed by requiring a specific input third-order intercept value.

A power amplifier trade space is developed which enables the designer to determine saturated power output values for each amplifier stage that result in optimum *PAE* and power output. Equations developed for the power amplifier give invaluable insight into the factors that limit cascaded amplifier compressed power output. The equations give intuitive understanding that saturated power output of each amplifier stage has an impact on final stage power output as a function of the gain following in later amplifier stages. Cascaded amplifier compression depth is simply the sum of compression depths of all of the amplifier stages. Equations (9.18) and (9.19) define total amplifier compression depth.

Cascaded amplifier stage compression phase shift is also affected by compression in each of the amplifier stages. Equation (9.20) is proposed as an extension into multiple stages of the model for phase shift of a single amplifier stage.

Behavioral characteristics of single amplifier stages have been extended into cascaded stages, methods have been developed for the modeling of population statistics, the models have been extended to cover the frequency domain, and amplifier parameter sensitivity to temperature has been modeled. All of these modeling techniques are pulled together in Chapter 10 to reveal the impact that all of these factors in combination can have on performance of a multiple-staged amplifier with respect to specifications that flow down from subsystem and system requirements.

9.6 Problems

Problem 9.1

Re-work Example 9.2 but partition gain as 8 dB for stage 1, 7 dB for stage 2, and 6 dB for stage 3.

Problem 9.2

Re-work Example 2 using 7 dB gain for each stage, but partition the problem such that stages 1 and 2 are combined in circuit element A and stage 3 is circuit element B. An infinite combination of solutions can be obtained for values of circuit element A and circuit element B. Use the solution that gives circuit element B the same current consumption that was derived for amplifier stage 3 in Example 9.2. The solution obtained when decomposing circuit element A into amplifier stages 1 and 2 should be identical to that obtained in Example 9.2.

Problem 9.3

Given: A three-stage amplifier having small signal gain per stage of 8 dB, (G_c = 24 dB), operates from a 9-volt power source, the first-stage transistor has i_{dss} = 0.015A, second-stage transistor has i_{dss} = 0.08A, and third-stage transistor has i_{dss} = 0.5A. All amplifier stages are biased at 20 percent i_{dss}. Assume transistor v_k = 1V.

Determine: Amplifier power output, compression depth, and phase shift with +10 dBm power input.

Problem 9.4

Prove that equation (9.15) can be written as

$$P_{out} = P_{in} + G_1 + G_2 + G_3$$

$$- K\log_{10}\left[1 + 10^{\frac{P_{in}+G_1+G_2+G_3-P_{sat3}}{K}} + 10^{\frac{P_{in}+G_1+G_2-P_{sat2}}{K}} + 10^{\frac{P_{in}+G_1-P_{sat1}}{K}}\right]$$

10

Models Upon Models

10.1 The Sum of All Models

Transistor output current versus applied signal and bias voltage was modeled using new curve-fit techniques involving right-hand functions (RHF) and left-hand functions (LHF) in Chapter 3. The new curve-fit equations model transistor saturation characteristics with accuracy that is lacking in models based on polynomial expansions. Combined bias voltage and signal voltage as a function of normalized time were applied as inputs to the new nonlinear models using spreadsheet analysis. Transistor output current waveforms for different combinations of bias voltage and signal voltage level were generated. These waveforms were then analyzed using the spreadsheet's Fourier analysis capability. The Fourier transform gives spectral content, providing average DC current, and fundamental and harmonic amplitude data as a function of input signal level. Unusual behavior of second, third, and higher harmonics as a function of input signal voltage and DC bias voltage is predicted by this nonlinear analysis. Variation of transistor gain as a function of DC bias and signal voltage is also revealed.

The curve-fit techniques developed in Chapter 2 are applied in Chapter 4 to develop a nonlinear class A amplifier behavioral model. Definitions and equations for typical amplifier parameters of bias coefficient b_q, saturated power output P_{sat}, output third-order intercept point P_{OIP}, compression coefficient K, and one dB compression point P_{1dB} are established for class A amplifiers in Chapter 4. Methods of computing third-order intermodulation IM components are also developed. A rule of thumb for estimating amplifier stage noise figure is offered. These definitions and models are expanded to include class AB amplifiers in Chapter 5. New models that compute average bias current and small signal gain as a function of input signal level in class AB amplifiers are developed in Chapter 5. These models give dynamic function to power-added efficiency *PAE* as a function of input signal power level. Harmonic output as a function of input signal level is based on data generated by discrete Fourier transform (DFT) analysis performed in Chapter 3.

Frequency as a variable was added to the modeling technique in Chapter 6. Amplifier small signal gain G_{ss}, signal input power P_{in}, compression coefficient K, and saturated power output P_{sat} are all functions of frequency, and as they vary compression depth changes, and so does compression phase shift. Methods of modeling these parameters over frequency by curve-fitting measured data or by simulating Butterworth or Chebishev bandpass responses were explained. All of the parameters associated with class A and class AB amplifiers can now be expressed as a function of frequency adding a new dimension to the dynamic models.

Temperature sensitivity of amplifier parameters was added to the inventory of modeling tools in Chapter 7. Methods of modeling nonlinear temperature sensitivity as a function of frequency are described and applied to the amplifier behavioral model. Temperature sensitivity of three parameters, small signal gain G_{ss}, saturated power output P_{sat}, and noise figure *NF* is discussed, and typical temperature sensitivities are offered from experience.

Statistical variability of transistor parameters and amplifier parameters within a population is discussed in Chapter 8. Gaussian and Weibull probability density functions are proposed as appropriate models for the estimation of parameter variability range, standard deviation of population values about the mean, and yield analysis. Typical values for standard deviation of amplifier stage small signal gain, current consumption, and saturated power output are offered from experience. Amplifier assembly small signal gain variability range as a function of the number of cascaded stages is determined in Chapter 8. Amplifier architectures that use independent and uncorrelated amplifier stages are shown to have less variation of cascaded small signal gain compared to monolithic amplifiers having the same number of amplifier stages, which have highly correlated parameters. An example applying population statistics to the amplifier frequency response behavioral model shows how simulation of frequency response variation within a population can be modeled. The use of commercial risk analysis spreadsheet add-in software to estimate amplifier population yields is illustrated.

The amplifier behavioral model is applied to circuit architectures of multiple amplifier stages of low-noise and power amplifier designs in Chapter 9. A process for configuring amplifier parameter trade spaces is presented, and trade spaces for low-noise amplifiers and power amplifiers are developed. Low-noise amplifier trade space results clearly show there is a minimum achievable noise figure associated with a combination of total amplifier gain and required input third-order intercept point. Furthermore, there is an optimum circuit parameter value allocation associated with a desired input third-order intercept point which results in lowest power consumption. Trade space parameters that show advantages of using fewer high gain amplifier stages in place of a greater number of low gain stages are developed. There is also an optimum selection of gain per stage and saturated power output per stage that maximizes power amplifier compressed power output and power-added efficiency. Techniques for configuring spreadsheets to perform multidimensional trades are described and examples are given.

Concepts from all of these tools are pulled together in this Chapter 10 to illustrate the total impact on amplifier subsystem performance. The object here is to show that subsystem specifications need to be based on the total range of component parameter variability. Variability of subsystem performance parameters as affected by population statistics, temperature, frequency, and circuit architecture can be unbelievably large. Quite often the only way to manage these large variances is to use gain control devices such as attenuators or voltage controlled variable gain amplifiers. Spreadsheet techniques are developed here to help determine the range of gain control needed to cover the full range of component parameter variability.

10.2 Suballocating System Parameters to Subsystem Requirements

The best way to teach the technique of suballocating system parameters to subsystem element requirements is by example. A receiver system is used here as an example.

Example 10.1 Suballocating Receive System Parameters

System requirements are flowed down to subsystem components by suballocating acceptable system level parameters and tolerances through a root-sum-squares process. For instance, consider a double conversion heterodyne receive system that outputs a digital signal having 12 bits plus sign. The analog to digital A/D converter output saturates (all bits become ones) when power into the A/D converter is +10 dBm. Dynamic range of the 12 bit A/D converter is

$$20 \log_{10}[2^{12}] = 72.2 \text{ dB} \tag{10.1}$$

With 72.2 dB dynamic range and +10 dBm saturation level, sinusoidal signal input to the A/D converter smaller than −62.2 dBm will not toggle the least significant bit (LSB). Signal level −62.2 dBm can be either sinusoidal or average noise power integrated over a defined bandwidth. The receive channel noise bandwidth is 1 MHz. Total integrated noise power at the A/D converter input has to be −62.2 dBm in order to cause the LSB to flicker. No signal can pass through the A/D converter unless the LSB is toggling. Signal levels much lower than −62.2 dBm can pass through the A/D converter if the LSB is toggling on total integrated noise. Most designers set average integrated noise level 3 dB higher (in this example −59.2 dBm) than the minimum needed to cause the LSB to toggle. Average integrated noise at the A/D converter input, −59.2 dBm, results when noise power spectral density of −119.2 dBm per Hz is integrated over a 1 MHz receive channel noise bandwidth (+60 dB). This noise power spectral density at the A/D converter is referenced to the receiver front end by reducing its value by receive channel total gain. This noise power spectral density is directly related to excess noise generated in the receiver front end, which is related to receiver noise figure. If the system designer determines that system noise figure has to be 4.0 dB (−170 dBm per Hz noise power spectral density) in order to obtain maximum receive range, then the receive channel total gain has to be the difference between noise power spectral density at the A/D converter input and noise power spectral density at the receiver front end; in this example that difference is 50.8 dB. Receive channel gain needs to be 50.8 dB. Figure 10.1 illustrates the receiver just described.

Statistical variation of three receive channel parameter values, total receive channel gain, receive channel noise figure, and receive channel bandwidth has

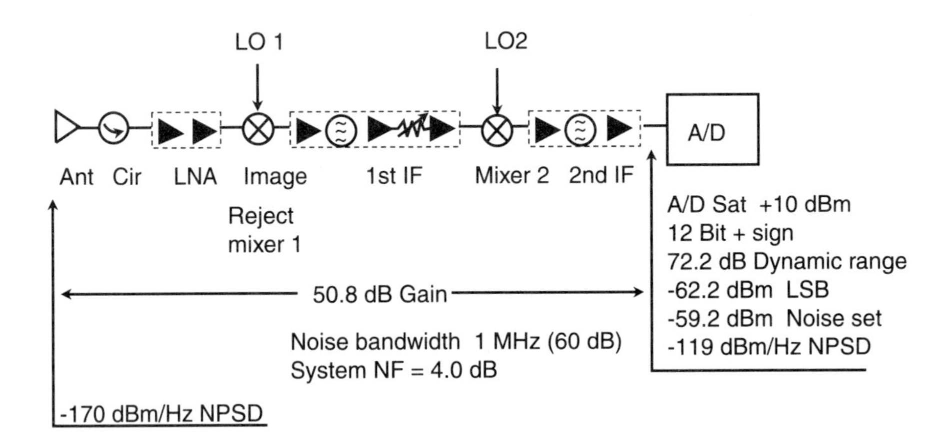

Figure 10.1 A hypothetical receive channel's total gain is gain set by system parameters.

an impact on the noise power density that appears at the A/D converter. Receive channel element temperature also impacts noise power spectral density at the A/D converter input. The system will work well as long as the LSB continues to flicker on average noise power appearing at the A/D converter input. The system designer might require average noise power at the A/D converter input to fall within ±2 dB of the desired 3 dB over that needed to toggle the LSB under all conditions of temperature ranging from 0 degrees C to 50 degrees C. Every system assembled needs to meet these standards for maximum yield.

Suballocate Total System Gain to Understand the Extent of the Problem

The duplex receive channel uses an antenna with 0.2 dB insertion loss and a circulator with insertion loss of 0.5 dB. The receiver is a double conversion system which uses an image reject mixer for Mixer 1 to reject image noise and signals that pass through the wide band LNA. Mixer 2 is doubly balanced. The image reject mixer has 10 dB conversion loss and +30.5 dBm input third-order intercept. Mixer 2 introduces 8 dB conversion loss and has an input third-order intercept point of +27 dBm. There are two local oscillator (LO) frequencies and two intermediate frequency (IF) sections. Bandpass filtering is used in each IF amplifier section, and noise gain control is performed in the first IF section. Bandpass filters in the IF amplifiers are passive, and have insertion loss of 2 dB each. Gain control is achieved with a binary attenuator having 0.5 dB per step, 0 to 15.5 dB total range, input third-order intercept +40 dBm, and 2.5 dB insertion loss at 0 dB setting. The gain trim attenuator can be set to within ±0.25 dB of any desired attenuation value. The second IF output is applied directly to the A/D converter, which undersamples the second IF frequency. Passive components are assumed to have input third-order intercept value at least +100 dBm. That is not to say that they can handle power levels that high, but that they are extremely linear at the power levels actually running through the receiver, and that they have no contribution to total receive channel input third-order intercept. Table 10.1 lists receive channel elements, their gains, losses and input third-order intercept points known at this point in the architecture development. The gain trim attenuator is commanded to the 0 dB setting. Only its insertion loss value appears in Table 10.1. Power supply voltage available to operate the receive channel is +5 volts. Available transistors develop 12 dB small signal gain in LNA and IF amplifier stages.

A receive channel architecture is configured. Receive channel noise figure of 4.0 dB over temperature and for all assembled receive channels is required. High receive channel input third-order intercept and low-power consumption is desired.

Table 10.1
Receiver Channel Parameters Known Before Design Iteration

Component	Gain/Loss	Gain/Loss 1σ	Current	Noise Figure	Input 3rd OIP
Antenna	−0.2 dB	0.01 dB	—	0.2 dB	+100 dBm
Circulator	−0.5 dB	0.01 dB	—	0.5 dB	+100 dBm
LNA	24.0 dB	1.0 dB	TBD	TBD	TBD
Mixer 1	−10.0 dB	0.04 dB	—	10.0 dB	+30.5 dBm
IF Amp 1	12.0 dB	0.5 dB*	TBD	TBD	TBD
IF Filter 1	−2.0 dB	0.04 dB	—	2.5 dB	+100 dBm
IF Amp 2	12.0 dB	0.5 dB*	TBD	TBD	TBD
Gain Cont	−2.5 dB	0.03 dB	—	2.5 dB	+40 dBm
IF Amp 3	12.0 dB	0.5 dB*	TBD	TBD	TBD
Mixer 2	−8.0 dB	0.04 dB	—	8.0 dB	+27.0 dBm
IF Amp 4	12.0 dB	0.5 dB**	TBD	TBD	TBD
IF Filter 2	−2.0 dB	0.04 dB	—	2.5 dB	+100 dBm
IF Amp 5	12.0 dB	0.5 dB**	TBD	TBD	TBD
Rx Channel Totals	58.8 dB	TBD	TBD	TBD	TBD

*IF Amp stages 1, 2, and 3 are on one monolithic chip and together exhibit $1\sigma = 1.5$ dB.

**IF Amp stages 4, and 5 are on one monolithic chip and together exhibit $1\sigma = 1.0$ dB.

Transistor gate periphery or class A saturated power output for each of the receive channel amplifier stages needs to be determined. Total receive channel input third-order intercept and power consumption needs to be determined. The composition and values of Table 10.1 will change as statistical models and thermal sensitivity models are applied to the understanding of the receive channel performance.

Average receive channel gain at +25 degrees C in a population of assemblies is 58.8 dB with the gain trim attenuator set at 0.0 dB. The desired receive channel gain of 50.8 dB, which establishes noise power spectral density of −119 dBm/Hz at the A/D converter input, is realized by setting the gain control attenuator for 8 dB. Setting the attenuator at 8 dB to compensate for excess gain for the average assembly at room temperature uses approximately half of the adjustment range in the attenuator.

Is There Sufficient Attenuator Control Range?

The extent of variability of receive channel gain needs to be determined to assure there is sufficient gain control range. If total receive channel gain in any assembly of randomly selected circuit components exceeds 66.55 dB (50.8 dB, +15.5 dB, +0.25 dB maximum gain trim), there is not sufficient gain trim range. Standard deviation of total receive channel gain for a population of receivers is determined by computing the square root of the sum of the squares (root-sum-square, RSS) of standard deviations of individual contributors. The low-noise amplifier (LNA) is monolithic and exhibits standard deviation of small signal gain $1\sigma = 1.0$ dB. Monolithic IF amplifier stages 1, 2, and 3 are lumped into a single standard deviation $1\sigma = 1.5$ dB, and monolithic IF amp stages 4 and 5 are lumped into a single $1\sigma = 1.0$ dB. Estimates are made for passive component standard deviation of loss in dB. Receive channel standard deviation

$$\sigma_{RX} = \sqrt{\underset{\text{Ant}}{(0.01)^2} + \underset{\text{Circ}}{(0.01)^2} + \underset{\text{LNA}}{(1)^2} + \underset{\text{Mxr}}{(0.04)^2} + \underset{1^{st}\text{ IF}}{(1.5)^2} + \underset{\text{Filter}}{(0.04)^2} + \underset{\text{Atten}}{(0.03)^2} + \underset{\text{Mxr}}{(0.04)^2} + \underset{2^{nd}\text{ IF}}{(1)^2} + \underset{\text{Filter}}{(0.04)^2}}$$

$$\sigma_{RX} = 2.06 \text{ dB}$$

Receive channel $3\sigma = 6.18$ dB, indicating that total small signal gain can be as low as 52.62 dB in some assemblies and as high as 64.98 dB in others at ambient room temperature of +25 degrees C. Small signal gain temperature sensitivity of each amplifier stage is assumed to be −0.015 dB/degrees C. (see Chapter 7). Seven cascaded amplifier stages will exhibit a combined gain sensitivity of −0.105 dB/degrees C. Ambient temperature of +50 degrees C will cause population mean value of receiver gain to drop 2.63 dB. Minimum (-3σ) small signal gain in a population will be 49.99 dB at ambient temperature of +50 degrees C. Maximum ($+3\sigma$) small signal gain in a population of receive channels at +50 degrees C will be 62.35 dB. Gaussian probability density predicts that with a gain mean value of 56.17 dB at +50 degrees C and a $3\sigma = 6.18$ dB, the gain trim attenuator setting will be at 0 dB in 0.4 percent of the assemblies and no more than 11.5 dB in any of the assemblies. Receive channel gain will be low by −0.8 dB in the worst case. Mean attenuator setting for the population at +50 degrees C will be 6.0 dB. Attenuator setting error for 99.6 percent of the population at +50 degrees C will be ±0.25 dB.

At ambient temperature 0 degrees C, receiver gain will rise 2.63 dB and maximum gain ($+3\sigma$) in some receiver assemblies will be as high as 67.61 dB, 1.31 dB greater than can be controlled by the gain trim attenuator range. Gaussian probability density predicts that with a gain mean value of 61.43 dB at 0 degrees C and a $3\sigma = 6.18$ dB, the gain trim attenuator setting will be at maximum value of 15.5 dB in 0.9 percent of the assemblies. Worst case receive channel gain will be 67.6 dB. Maximum attenuation of 15.5 dB will reduce the gain to 52.1 dB, which is 1.31 dB greater than the desired 50.8 dB. Only 0.5

percent of the receive channels in a population will have attenuator setting error 1.06 dB greater than the ideal ±0.25 dB at 0 degrees C. Minimum gain in any one assembly will be 55.3 dB at 0 degrees C ambient temperature. Minimum attenuator setting for any unit in the population will be at least 4.5 dB. Mean value of attenuator setting for the population will be 10.5 dB. The net range of noise power spectral density that the gain set attenuator needs to compensate over temperature is 17.6 dB. The total gain trim attenuator control range is 15.5 dB. There is a 2.1 dB total mismatch which is balanced between hot and cold ambient temperature and should not impact product yield significantly. Figure 10.2 illustrates the receive channel gain range that will be experienced in randomly selected assemblies caused by random selection of components for assembly, and temperature extremes.

The Gain Trim Range Mismatch Will Be Partially Compensated by Noise Figure Sensitivity to Temperature

Amplifier noise figure sensitivity to temperature is described in Chapter 7 to be +0.015 dB/degrees C. At ambient temperature of +50 degrees C, noise power spectral density rises 0.38 dB/Hz. At the same temperature small signal gain drops by 2.63 dB. Net change in noise power spectral density at the A/D converter input is only −2.25 dB. A similar situation occurs at 0 degrees C. Receive channel gain increases by 2.63 dB, while noise power spectral density drops by 0.38 dB causing a net increase in noise power spectral density of only

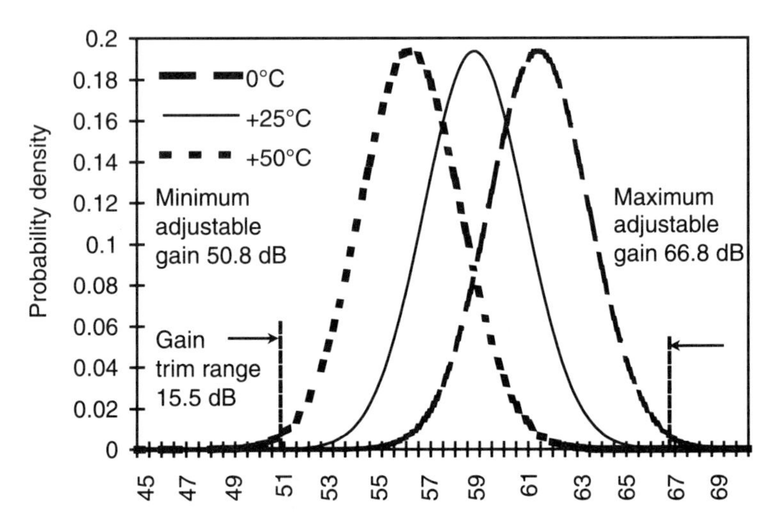

Figure 10.2 Receiver assembly small signal gain extremes due to temperature and statistical variances.

+2.25 dB. Net total change in noise power spectral density at the A/D converter input as a function of temperature is 16.8 dB. The gain trim attenuator is required to compensate this gain change. The shortfall in gain trim control range of noise power spectral density is 1.3 dB, which is split equally (0.65 dB) at each end of the control range at temperature extremes. There is also an uncertainty of ±0.25 dB in attenuator setting due to the binary step gain trim attenuator resolution of 0.5 dB per step. Total error in noise power spectral density setting at the A/D converter allowed by the system engineer is 2 dB. The maximum error in gain trim setting will be 0.9 dB and that will occur only at temperature extremes in about 0.9 percent of the units.

Is Receive Channel 4.0 dB Noise Figure Obtainable?

An architecture has been selected as shown in Figure 10.1 above. Gains and losses have been distributed throughout the receive channel based on realizable transistor gain at the applied frequency and passive component parameters gleaned from data sheets. The gain control attenuator will have a nominal setting of 8.0 dB for most of the assemblies at room temperature, 6.0 dB at +50 degrees C, and 10.5 dB at 0 degrees C.

Is a noise figure of 4.0 dB obtainable from the proposed architecture at any temperature between 0 degrees C and +50 degrees C, for any randomly populated receive channel assembly? What range of noise figure values are found from one receiver to the next as randomly selected components are assembled, and the gain trim attenuator is adjusted to compensate for gain variances and for temperature variation?

Noise figure values are easily allocated to subsystem elements by understanding and using the familiar cascaded element noise figure (9.11). Realize that the noise figure value in decibels of passive elements, such as the antenna and the circulator, is simply the magnitude of the insertion loss of that element. The antenna and circulator are the first elements in the receive channel, have total insertion loss of −0.7 dB, and develop the first 0.7 dB of receive channel total noise figure. The noise figure looking into the low-noise amplifier cannot be greater than 3.3 dB at any temperature if 4.0 dB noise figure is to be achieved over temperature for all randomly assembled receive channel assemblies. Several factors affect the noise figure value looking into the LNA. The noise figure rule of thumb (4.17) gives an estimate of noise figure mean value for each amplifier stage in the LNA. Bias current through each stage will vary with standard deviation of 15 percent from monolithic chip to monolithic chip (see Chapter 8) causing noise figure value to vary. Noise figure of the first LNA amplifier stage will be affected by current variation and that has direct impact on receive channel noise figure. Noise figure of the second stage will add to the value of receive channel noise figure looking into the LNA by an amount determined by gain of the first LNA amplifier stage. Gain of the first stage will vary with standard deviation of gain $1\sigma = 0.5$ dB. Likewise noise figure value looking into the first mixer will add to the receive channel noise figure looking into the LNA by an amount determined by total gain of the LNA, which has a standard deviation of gain $1\sigma = 1$ dB.

The LNA monolithic circuit will need to have a noise figure value significantly less than 3.3 dB at +50 degrees C in order to combine with the loss of the antenna and circulator and with the noise figure of the first IF amplifiers, to achieve less than a 4.0 dB total receive channel noise figure, a worst case in any receive channel assembly. The chart shown in Figure 10.3 is a handy tool for estimating the degradation of noise figure ΔNF that occurs looking into circuit element A when cascaded with the noise figure value looking into circuit element B. Derivation of the Figure 10.3 chart is found in Appendix C. Circuit elements A and B can be sequential cascaded amplifier stages or can be sequential cascaded circuit elements consisting of combinations of mixers, amplifier stages, filters, and attenuators. Applying what is known about the LNA at this point to Figure 10.3, gain 24 dB, noise figure approximately 2.5 dB, and first IF amplifier noise figure in the range of 14 dB, realize that noise figure looking into the LNA will be degraded by approximately 0.3 dB by the IF amplifiers.

Develop a table of values as illustrated in Table 10.2 of noise figure degradation expected at 50 degrees C as a result of cascading the circuit elements (use

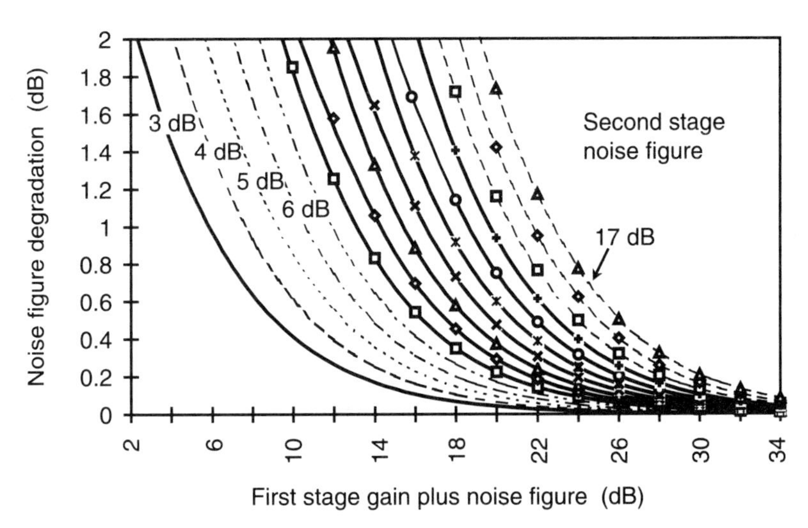

Figure 10.3 The effect of latter stage noise figure on cascaded stage noise figure is easily determined.

Table 10.2
Noise Figure Degradation With Cascaded Element Mean Values at +50 Degrees C

Circuit Element	Expected Gain	Expected *NF*	Expected ΔNF
LNA Stage 1	11.62 dB	1.9 dB	—
LNA Stage 2	11.62 dB	4.0 dB	0.35 dB
1st IF Channel	13.5 dB	14 dB*	0.3 dB
2nd IF Channel	13.5 dB	12 dB**	0.2 dB
Expected *NF* Degradation Due to Cascading Stages			0.85 dB

*Noise figure of 10 dB is attributed to the image reject mixer and 4 dB to the first IF amplifier.

**Noise figure of 8 dB is attributed to the doubly balanced mixer and 4 dB to the fourth IF amplifier.

the curves of Figure 10.3 to determine each element's contribution to reduction of receive channel noise figure). Use mean values of gain at +50 degrees C for estimates in the table. Estimate circuit element noise figures based on data sheet values and experience. The probability that all circuit elements will simultaneously have worst case values is remote.

Table 10.2 leads to the conclusion that degradation in LNA first-stage noise figure will be on the order of 0.85 dB due to circuit elements after the first-stage. This means first stage noise figure has to be no worse than 2.45 dB at +50 degrees C. Since noise figure temperature sensitivity is 0.015 dB/degrees C, worst case value of first-stage noise figure at 25 degrees C will be 2.45 − 0.015 * 25 = 2.07 dB.

First-stage noise figure value will vary within a population as a function of transistor bias current variation. Bias current variation can exhibit standard deviation of 15 percent, $3\sigma = 45$ percent. Noise figure 3σ variation due to this current variability is a function of the bias current level as estimated by rule of thumb (4.17). Figure 10.4 shows estimated noise figure 3σ variation as a function of bias current mean value. From Figure 10.4 conclude that if first-stage worst case noise figure at +25 degrees C can be no worse than 2.07 dB (mean value $+3\sigma$), first-stage noise figure mean value has to be 1.72 dB and that worst case noise figure ($+3\sigma$) is 0.35 dB greater than the mean.

First-stage noise figure of 1.72 dB can be achieved with average bias current of $i_q = 0.015$ A, determined by solving (4.17) for bias current. Table 10.2 estimates degradation of LNA first-stage noise figure of 0.35 dB caused by noise figure of amplifier stage 2. The low-noise amplifier monolithic circuit

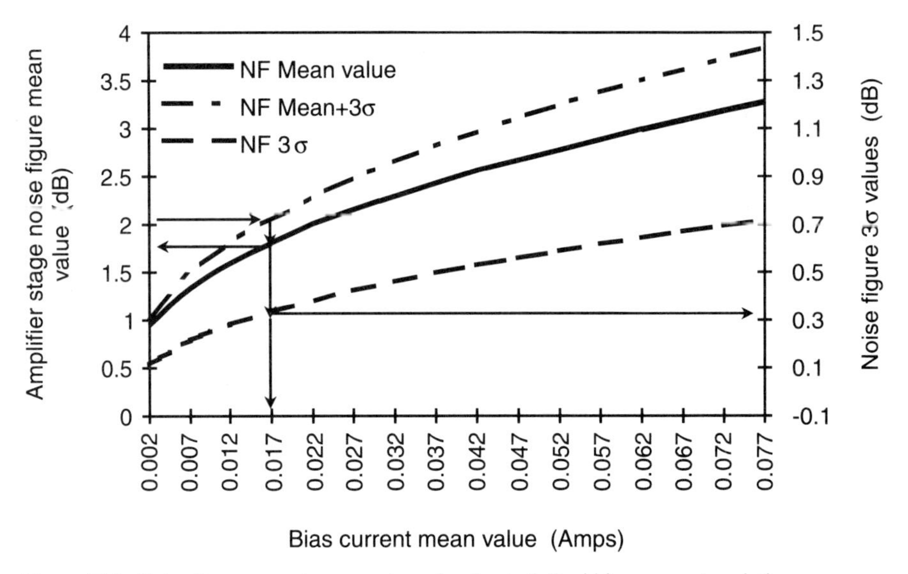

Figure 10.4 Noise figure worst case values due to statistical bias current variation.

element of two amplifier stages will need to have population mean value noise figure 2.07 dB at +25 degrees C, that's 0.35 dB greater than 1.72 dB first-stage noise figure.

What LNA Circuit Element Input Third-Order Intercept Point Can Be Supported With 2.07 dB LNA Noise Figure?

The fastest way to answer this question is to open the "Amplifier Analysis" Excel spreadsheet workbook included in this book's software package and select the "Two Stage" worksheet. This workbook uses trade space techniques developed in Chapter 9. Enter the proposed gain per stage at +25 degrees C (12 dB each stage), enter the operating voltage (5 volts), and transistor knee voltage (assume 1 volt), then iterate amplifier input third-order intercept value until noise figure value of 2.07 dB is obtained. Figure 10.5 shows the results from the workbook page.

Low-noise amplifier circuit element input third-order intercept point $P_{IIP} = +6.55$ dBm with noise figure $NF = 2.07$ dB. Design parameters for the two class A amplifier stages in the LNA are listed showing current consumption, noise figure, saturated power output, output third-order intercept point, and 1 dB compression point. Total current expected to be drawn by the LNA is 0.178 amperes. The LNAs output third-order intercept point is +27.5 dBm. Note that the LNA second stage amplifier's noise figure is higher than the guess

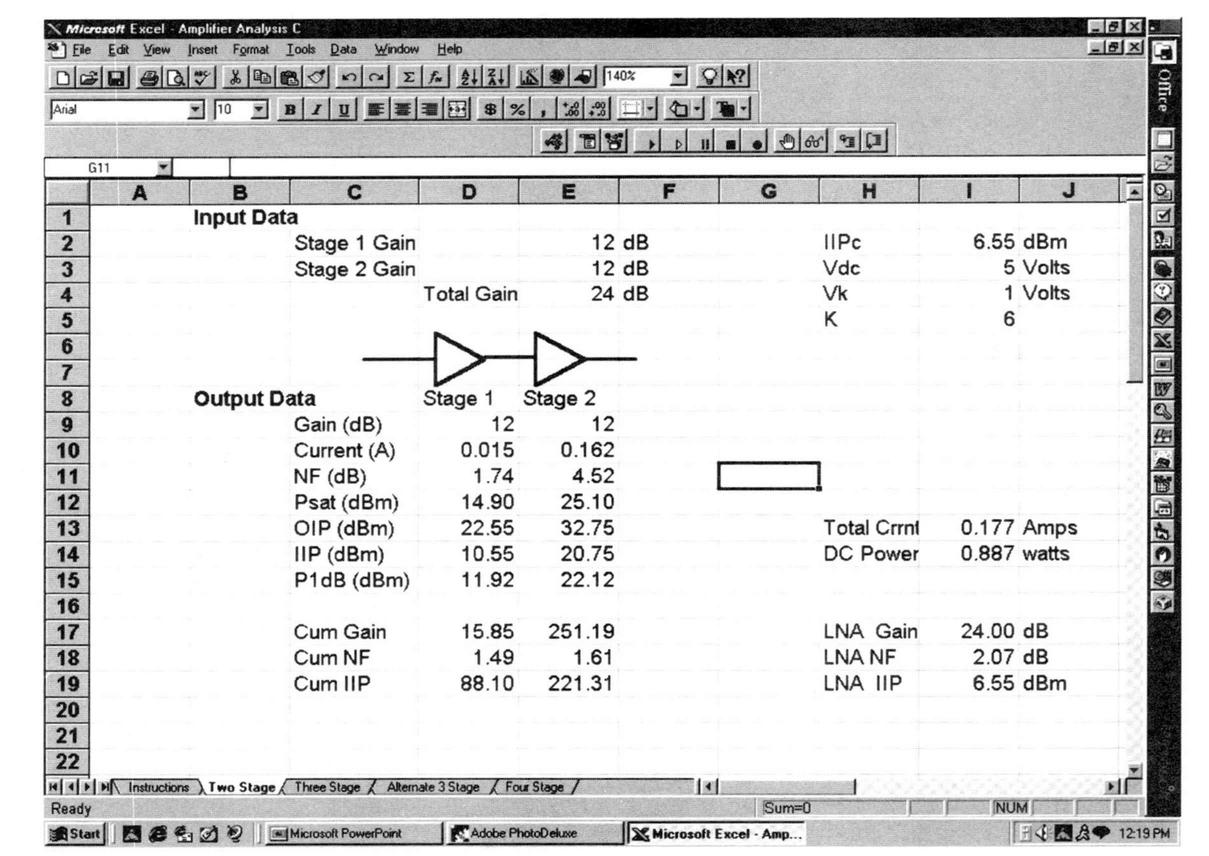

Figure 10.5 Page from "Amplifier Analysis" workbook utility showing two-stage amplifier solution.

entered into Table 10.2. The first IF amplifier's noise figure will also turn out to be higher than is estimated in Table 10.2.

The most desirable approach for designing the IF portion of the receive channel would be to support the input third-order intercept established by this LNA design through the entire receive channel. That approach will prove to be too costly in terms of noise figure degradation and current consumption. The LNA provides +6.55 dBm input third-order intercept, which is increased by the loss in the circulator and antenna to a level of +7.25 dBm third-order intercept at the antenna input. The receive channel has this level of signal linearity for signal frequencies which fall in the bandpass of the LNA, but do not down convert into the first IF. A practical IF channel design will suffer degradation of in-band input third-order intercept by 20 dB less than the LNA *IIP*.

Design the IF Amplifiers to Preserve Receive Channel Noise Figure

The receive channel input third-order intercept looking into the LNA can never be any greater than the input intercept point of the LNA, +6.55 dBm. The passive insertion loss ahead of the LNA increases receive channel input intercept value by the value of insertion loss to +7.25 dBm. The LNA output third-order intercept point is +30.55 dBm. Input third-order intercept point of the image reject mixer is +30.5 dBm (see Table 10.1). Joint intercept point at the LNA-image reject mixer interface is +27.5 dBm. This reflects back through the LNA to the antenna input and limits receive channel input third-order intercept to +4.2 dBm. Third-order intercept of IF amplifier stages will degrade this +4.2 dBm at the antenna even more. Receiver bandwidth up to the image reject mixer is wide, and signals enter the LNA that do not down convert into the IF amplifier. These unwanted signals, if large, will cause compression in the LNA and result in cross modulation. Cross modulation appears as amplitude and phase distortion in the desired signal and passes through both IF amplifiers on the received signal. Intermodulation results when two large signals that are separated by a frequency difference equal to the first IF enter the LNA, mix, and are down converted. Intermodulation signals usually cause interference and confusion, not distortion. The desired received signal frequency is either higher than or less than the first local oscillator frequency by a difference equal to the first IF. The image reject mixer is designed to reject the frequency band that is either lower than or higher than the first local oscillator frequency, whichever frequency band (the image band) is opposite that of the desired signal. The image reject mixer also cancels the noise in the image frequency band, keeping it out of the IF amplifier where it would add to the in-band noise power spectral density reaching the A/D converter input. Receiver noise figure is automatically 3 dB higher if image frequency noise is allowed to pass into the IF amplifiers.

The impact that IF amplifier stage parameters has on final receiver noise figure, and on final receiver input third-order intercept is best understood if a spreadsheet receiver model is developed.

Develop a Spreadsheet Receive Channel Model to Facilitate Iteration of IF Amplifier Parameters

The architecture selected for the IF amplifier stages in Figure 10.1 distributes gain and loss in the receive channel to maximize receive channel input third-order intercept. Loss in the form of mixer conversion loss, filter loss, and gain trim attenuator loss is inserted between IF amplifier stages. The high input third-order intercept values of the lossy components does not significantly degrade the lower input third-order intercept value of each of the IF amplifier stages. Develop a spreadsheet model of the receive channel that includes gain and loss, saturated power output, current, noise figure and input third-order intercept values of each element. Use the approach illustrated in Figure 10.6.

The first three rows contain reference data of temperature, voltage, and compression coefficient used in the data field of the sheet. Row 6 contains population mean insertion loss or amplifier gain values in dB for each element in the receive channel. Row 7 contains mean gain values entered in Row 6 that are modified by temperature sensitivity (7.2) using the ambient temperature value entered in cell H1. The assumption is made that temperature sensitivity of passive circuit elements is insignificant. Row 9 contains values of room temperature noise figure for each of the circuit elements. Passive element noise figure value is simply the negative of the element's insertion loss value in decibels. LNA amplifier stage noise figure values are taken from the LNA trade study performed by the Amplifier Analysis tool illustrated in Figure 10.5. Room temperature noise figure estimates are entered into cells for each of the IF amplifier stages across Row 9. These noise figure values will be iterated to arrive at an acceptable overall receive channel noise figure, total current consumption, and input third-order intercept. Row 10 contains values of noise figure at the ambient temperature entered into cell H1, calculated by using (7.2) and noise figure temperature sensitivity of +0.015 dB/degrees C. Row 11 contains room temperature mean value of current drawn by each amplifier stage, which is calculated by solving the noise figure rule of thumb (4.17) for current, knowing the amplifier stage's room temperature noise figure value in Row 9. Values for amplifier stage saturated power output at room temperature in Row 14 are calculated using (4.3), which draws values of current consumption from Row 11 and voltage values from cells C1 and C2. The assumption is made that all amplifier stages are class A, and i_{ds} values in Row 11 are equal to 50 percent of i_{dss}. Saturated power output at ambient temperature, Row 15, is calculated using (7.2) with P_{sat} temperature sensitivity of -0.015 dB/degrees C and ambient

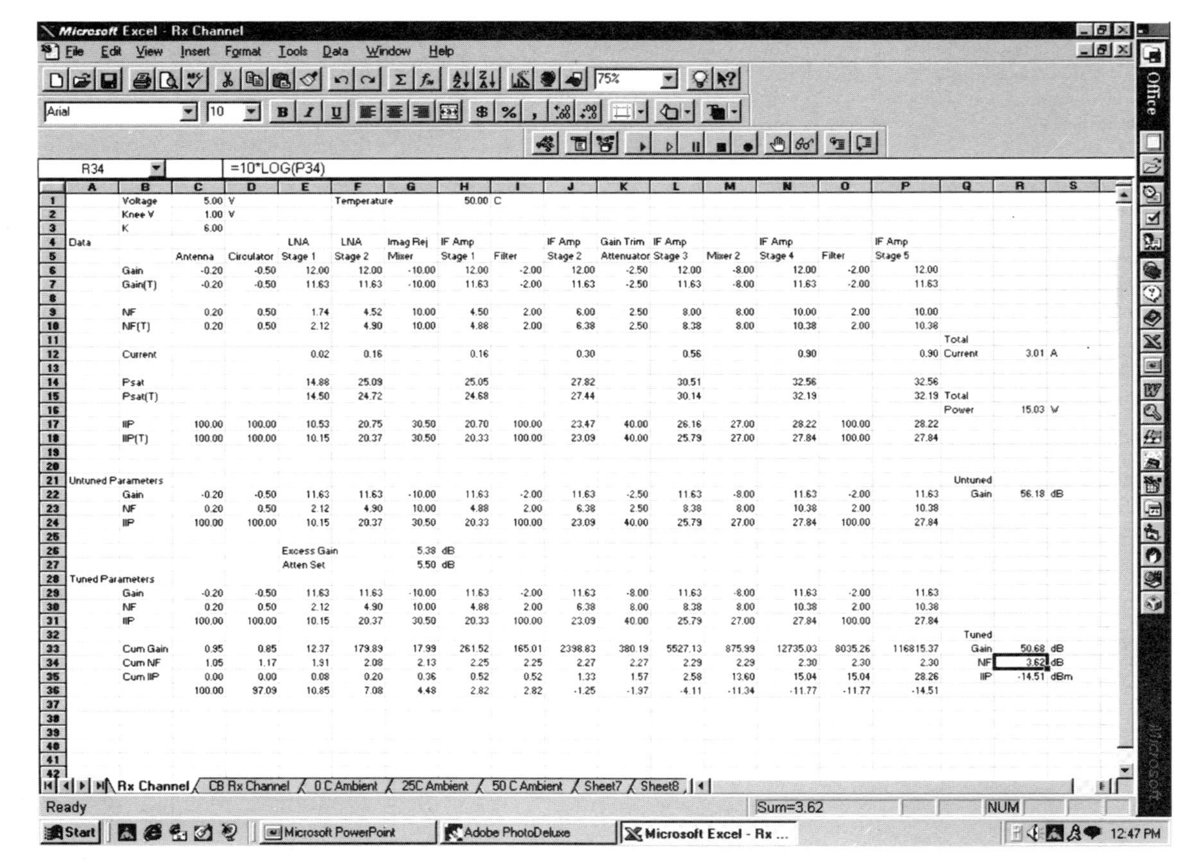

	A	B	C	D	E	F	G	H	I	J	K	L	M	N	O	P	Q	R	S
1		Voltage	5.00	V		Temperature		50.00	C										
2		Knee V	1.00	V															
3		K	6.00																
4	Data				LNA	LNA	Imag Rej	IF Amp		IF Amp	Gain Trim	IF Amp		IF Amp		IF Amp			
5			Antenna	Circulator	Stage 1	Stage 2	Mixer	Stage 1	Filter	Stage 2	Attenuator	Stage 3	Mixer 2	Stage 4	Filter	Stage 5			
6		Gain	-0.20	-0.50	12.00	12.00	-10.00	12.00	-2.00	12.00	-2.50	12.00	-8.00	12.00	-2.00	12.00			
7		Gain(T)	-0.20	-0.50	11.63	11.63	-10.00	11.63	-2.00	11.63	-2.50	11.63	-8.00	11.63	-2.00	11.63			
9		NF	0.20	0.50	1.74	4.52	10.00	4.50	2.00	6.00	2.50	8.00	8.00	10.00	2.00	10.00			
10		NF(T)	0.20	0.50	2.12	4.90	10.00	4.88	2.00	6.38	2.50	8.38	8.00	10.38	2.00	10.38			
11																	Total		
12		Current			0.02	0.16		0.16		0.30		0.56		0.90		0.90	Current	3.01	A
14		Psat			14.88	25.09		25.05		27.82		30.51		32.56		32.56			
15		Psat(T)			14.50	24.72		24.68		27.44		30.14		32.19		32.19	Total		
16																	Power	15.03	W
17		IIP	100.00	100.00	10.53	20.75	30.50	20.70	100.00	23.47	40.00	26.16	27.00	28.22	100.00	28.22			
18		IIP(T)	100.00	100.00	10.15	20.37	30.50	20.33	100.00	23.09	40.00	25.79	27.00	27.84	100.00	27.84			
21	Untuned Parameters																Untuned		
22		Gain	-0.20	-0.50	11.63	11.63	-10.00	11.63	-2.00	11.63	-2.50	11.63	-8.00	11.63	-2.00	11.63	Gain	56.18	dB
23		NF	0.20	0.50	2.12	4.90	10.00	4.88	2.00	6.38	2.50	8.38	8.00	10.38	2.00	10.38			
24		IIP	100.00	100.00	10.15	20.37	30.50	20.33	100.00	23.09	40.00	25.79	27.00	27.84	100.00	27.84			
26					Excess Gain		5.38	dB											
27					Atten Set		5.50	dB											
28	Tuned Parameters																		
29		Gain	-0.20	-0.50	11.63	11.63	-10.00	11.63	-2.00	11.63	-8.00	11.63	-8.00	11.63	-2.00	11.63			
30		NF	0.20	0.50	2.12	4.90	10.00	4.88	2.00	6.38	8.00	8.38	8.00	10.38	2.00	10.38			
31		IIP	100.00	100.00	10.15	20.37	30.50	20.33	100.00	23.09	40.00	25.79	27.00	27.84	100.00	27.84			
32																	Tuned		
33		Cum Gain	0.95	0.85	12.37	179.89	17.99	261.52	165.01	2398.83	380.19	5527.13	875.99	12735.03	8035.26	116815.37	Gain	50.68	dB
34		Cum NF	1.05	1.17	1.91	2.08	2.13	2.25	2.25	2.27	2.27	2.29	2.29	2.30	2.30	2.30	NF	3.62	dB
35		Cum IIP	0.00	0.00	0.08	0.20	0.36	0.52	0.52	1.33	1.57	2.58	13.60	15.04	15.04	28.26	IIP	-14.51	dBm
36			100.00	97.09	10.85	7.08	4.48	2.82	2.82	-1.25	-1.97	-4.11	-11.34	-11.77	-11.77	-14.51			

Figure 10.6 Design a receive channel spreadsheet to study IF amplifier parameters.

temperature entered into cell H1. Amplifier stage input third-order intercept value in Row 17 is calculated using (4.13), which calls values of P_{sat} from Row 14, gain from Row 6, and compression coefficient from cell C3. Input third-order intercept values at ambient temperature, Row 18, also use (4.13), but call P_{sat} values from Row 15, gain values from Row 7, and compression coefficient from cell C3.

Untuned receive channel parameters of gain, noise figure, and input third-order intercept appearing in Rows 22, 23, and 24 are copied from ambient temperature values appearing in Rows 7, 10, and 18 respectively. Gain values across Row 22 are summed to obtain the total untuned gain value appearing in cell R22. Excess gain, the difference between total untuned gain at ambient temperature, cell R22, and 50.8 dB small signal gain needed to set noise power spectral density at the A/D converter input, is calculated in cell G26.

Gain trim is a 5 bit binary attenuator, with step size of 0.5 dB, and a total range of 0 to 15.5 dB. The excess gain value calculated in cell G26 is used to determine which of the 32 states the gain trim attenuator needs to be commanded. Cell G27 contains an EXCEL decision statement

=IF(G26<=0,0,IF(G26>15.5,15.5,IF((G26-INT(G26))>0.75,
INT(G26)+1,IF((G26-INT(G26))>0.25,INT(G26)+0.5,INT(G26))))).

(10.2)

This decision statement equation tests the value in cell G26 and determines the closest binary attenuator setting value that tunes the receive channel for 50.8 dB. If untuned gain is less than 50.8 dB, Equation (10.1) inserts 0 dB into cell C27. If untuned gain is greater than 66.3 dB, Equation (10.1) inserts 15.5 dB into cell G27. Excess gain at ambient temperature 25 degrees C is 8.0 dB requiring a gain trim attenuator setting of 8.0 dB. The error in gain setting will be less than ± 0.25 dB for all cases where untuned receive channel gain is in the range of 50.55 dB to 66.55 dB. Receive channel gain lower than 50.55 dB could occur in 0.4 percent of the assemblies at ambient temperature of +50 degrees C, causing the gain trim attenuator to be set at 0 dB. Receive channel gain higher than 66.55 dB could occur in 0.9 percent of the assemblies at ambient temperature of 0 degrees C, causing the gain trim attenuator to be set at 15.5 dB.

Tuned receive channel parameters appear in Rows 29, 30, and 31. The only difference between tuned and untuned parameters is the gain trim attenuator set value insertion loss and noise figure. Rows 33, 34, and 35 contain cumulative values of gain, noise figure, and input third-order intercept from which final receive channel values in cells R33, R34, and R35 are determined.

Cell R12 contains the sum of all currents used in the LNA and IF amplifier stages.

Set ambient temperature at +50 degrees C, cell H1, and insert values for IF amplifier stage noise figure. The goal is to obtain a high input third-order intercept value when tuned, while holding tuned noise figure value at a value of 3.62 dB. The LNA first stage is biased at 0.016 amperes. Figure 10.4 indicates the LNA first stage noise figure $3\sigma = 0.35$ dB. Worst case ($+3\sigma$) receive channel noise figure at +50 degrees C would then be 3.97 dB. Figure 10.6 shows the receive channel spreadsheet model at +50 degrees C ambient temperature with IF amplifier noise figure values that support receive channel noise figure of 3.62 dB. The receiver input third-order intercept value is −14.51 dBm for signals that pass through the IF amplifiers. Input third-order intercept point is +7.08 dBm for signals that enter the LNA but do not down convert into the IF amplifiers. Current consumption for the entire receive channel is 3.01 amperes. Power consumption at 5 volts is 15.05 watts. This spreadsheet is a useful tool for developing the amplifier parameter values that satisfy the subsystem requirements. Values developed here can be passed down confidently as specifications for circuit assemblies to be developed in greater detail with circuit design CAD software. Additional confidence is obtained by using risk analysis spreadsheet add-in software such as CRYSTAL BALL to test the amplifier stage parameter mean value selection. Risk analysis add-in software performs the building of thousands of subassemblies, tunes them at ambient temperature extremes, and displays desired parameters as statistics.

Modify the Spreadsheet Receive Channel Model to Use CRYSTAL BALL

The spreadsheet receive channel model shown in Figure 10.6 is modified by the addition of cells used to generate random values of gain, saturated power output, and current at 25 degrees C ambient. Values of gain, saturated power output, noise figure, and input third-order intercept at ambient temperature are then calculated from the statistical values generated. The modified spreadsheet calculates receive channel mean values of current, saturated power output, and input third-order intercept as is done in the spreadsheet shown in Figure 10.6. Figure 10.7 shows organization of the new spreadsheet.

Rows 9 through 18 from Figure 10.6 are moved up, filling in empty rows, to occupy Rows 8 through 14. Standard deviation of gain is added as an input in Row 16. CBGain in Row 15 is random value of gain generated by CRYSTAL BALL using mean value of temperature compensated gain from Row 7, and standard deviation of gain from Row 16. LNA stage 1 gain value is randomly generated. LNA stage 2 gain is set equal to the value generated for stage 1. This simulates the unity covariance in gain found in multiple amplifier stages within

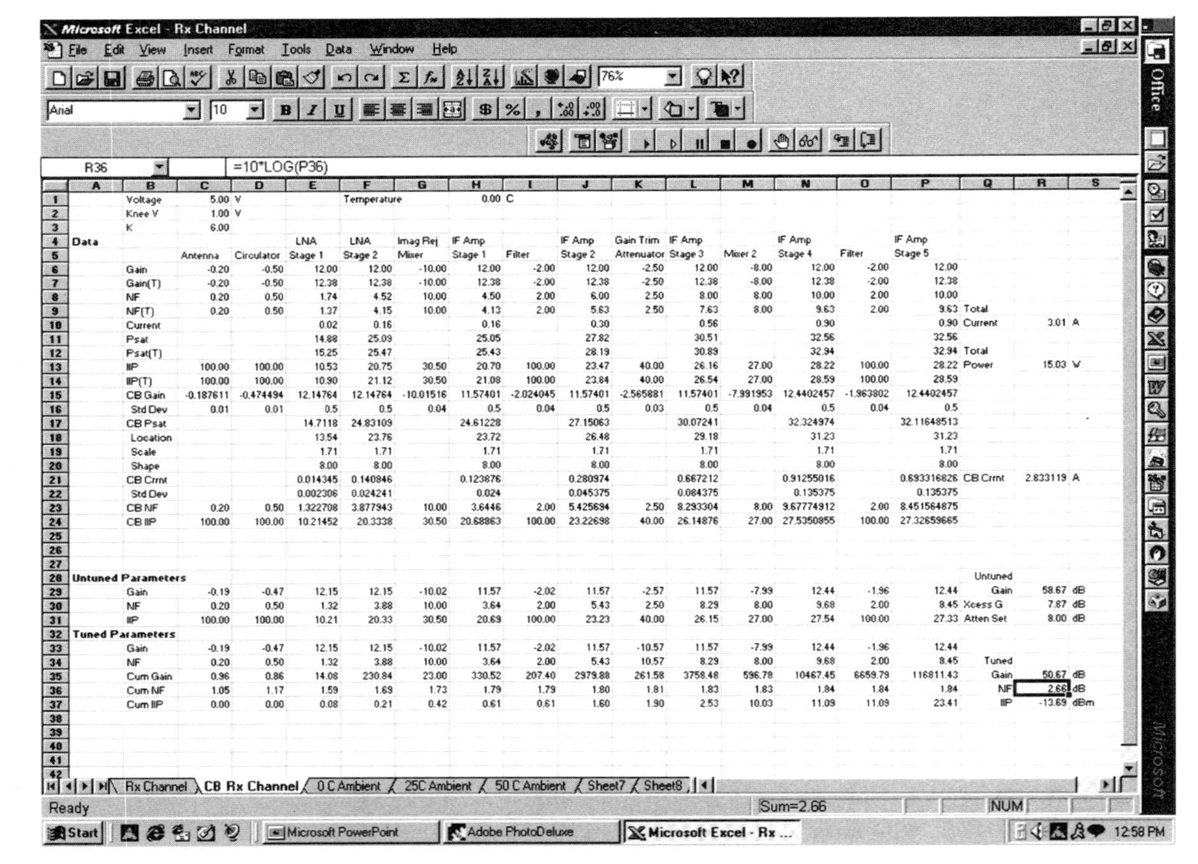

	A	B	C	D	E	F	G	H	I	J	K	L	M	N	O	P	Q	R	S
1		Voltage	5.00	V		Temperature		0.00	C										
2		Knee V	1.00	V															
3		K	6.00																
4	Data				LNA	LNA	Imag Rej	IF Amp		IF Amp	Gain Trim	IF Amp		IF Amp		IF Amp			
5			Antenna	Circulator	Stage 1	Stage 2	Mixer	Stage 1	Filter	Stage 2	Attenuator	Stage 3	Mixer 2	Stage 4	Filter	Stage 5			
6		Gain	-0.20	-0.50	12.00	12.00	-10.00	12.00	-2.00	12.00	-2.50	12.00	-8.00	12.00	-2.00	12.00			
7		Gain(T)	-0.20	-0.50	12.38	12.38	-10.00	12.38	-2.00	12.38	-2.50	12.38	-8.00	12.38	-2.00	12.38			
8		NF	0.20	0.50	1.74	4.52	10.00	4.50	2.00	6.00	2.50	8.00	8.00	10.00	2.00	10.00			
9		NF(T)	0.20	0.50	1.37	4.15	10.00	4.13	2.00	5.63	2.50	7.63	8.00	9.63	2.00	9.63	Total		
10		Current			0.02	0.16		0.16		0.30		0.56		0.90		0.90	Current	3.01	A
11		Psat			14.88	25.09		25.05		27.82		30.51		32.56		32.56			
12		Psat(T)			15.25	25.47		25.43		28.19		30.89		32.94		32.94	Total		
13		IIP	100.00	100.00	10.53	20.75	30.50	20.70	100.00	23.47	40.00	26.16	27.00	28.22	100.00	28.22	Power	15.03	V
14		IIP(T)	100.00	100.00	10.90	21.12	30.50	21.08	100.00	23.84	40.00	26.54	27.00	28.59	100.00	28.59			
15		CB Gain	-0.187611	-0.474494	12.14764	12.14764	-10.01516	11.57401	-2.024045	11.57401	-2.565881	11.57401	-7.991953	12.4402457	-1.963802	12.4402457			
16		Std Dev	0.01	0.01	0.5	0.5	0.04	0.5	0.04	0.5	0.03	0.5	0.04	0.5	0.04	0.5			
17		CB Psat			14.7118	24.83109		24.61228		27.15063		30.07241		32.324974		32.11648513			
18		Location			13.54	23.76		23.72		26.48		29.18		31.23		31.23			
19		Scale			1.71	1.71		1.71		1.71		1.71		1.71		1.71			
20		Shape			8.00	8.00		8.00		8.00		8.00		8.00		8.00			
21		CB Crrnt			0.014345	0.140846		0.123876		0.280974		0.667212		0.91255016		0.693316826	CB Crrnt	2.833119	A
22		Std Dev			0.002306	0.024241		0.024		0.045375		0.084375		0.135375		0.135375			
23		CB NF	0.20	0.50	1.322708	3.877943	10.00	3.6446	2.00	5.425694	2.50	8.293304	8.00	9.67774912	2.00	8.451564875			
24		CB IIP	100.00	100.00	10.21452	20.3338	30.50	20.68863	100.00	23.22698	40.00	26.14876	27.00	27.5350855	100.00	27.32659665			
25																			
26																			
27																			
28	Untuned Parameters																Untuned		
29		Gain	-0.19	-0.47	12.15	12.15	-10.02	11.57	-2.02	11.57	-2.57	11.57	-7.99	12.44	-1.96	12.44	Gain	58.67	dB
30		NF	0.20	0.50	1.32	3.88	10.00	3.64	2.00	5.43	2.50	8.29	8.00	9.68	2.00	8.45	Xcess G	7.87	dB
31		IIP	100.00	100.00	10.21	20.33	30.50	20.69	100.00	23.23	40.00	26.15	27.00	27.54	100.00	27.33	Atten Set	8.00	dB
32	Tuned Parameters																		
33		Gain	-0.19	-0.47	12.15	12.15	-10.02	11.57	-2.02	11.57	-10.57	11.57	-7.99	12.44	-1.96	12.44			
34		NF	0.20	0.50	1.32	3.88	10.00	3.64	2.00	5.43	10.57	8.29	8.00	9.68	2.00	8.45	Tuned		
35		Cum Gain	0.96	0.86	14.08	230.84	23.00	330.52	207.40	2979.88	261.58	3758.48	596.78	10467.45	6659.79	116811.43	Gain	50.67	dB
36		Cum NF	1.05	1.17	1.59	1.69	1.73	1.79	1.79	1.80	1.81	1.83	1.83	1.84	1.84	1.84	NF	2.66	dB
37		Cum IIP	0.00	0.00	0.08	0.21	0.42	0.61	0.61	1.60	1.90	2.53	10.03	11.09	11.09	23.41	IIP	-13.69	dBm

Figure 10.7 The spreadsheet of Figure 10.6 is modified to include random variables.

a single monolithic circuit chip. The same procedure is used for gain in the two IF amplifier monolithic circuits.

Row 17 contains CRYSTAL BALL generated saturated power output values CBP_{sat} using location, scale, and shape values from Rows 18, 19, and 20. Location values in Row 18 are determined by subtracting 1.71 dB from temperature compensated P_{sat} values in Row 12. Scale values are 1.71, and shape values are 8. Row 21 contains CRYSTAL BALL generated values of current, CBcrrnt, that use Row 10 current values as mean value and standard deviation values from Row 22. Row 23 calculates temperature compensated noise figure values using noise figure rule of thumb (4.17) and CBcrrnt values, and adding temperature sensitivity of noise figure (7.11). Row 24 calculates amplifier stage input third-order intercept point using temperature compensated, CB-generated values of saturated power output, gain, and compression coefficient.

CRYSTAL BALL generated parameters are then transferred to the untuned parameter portion of the spreadsheet model, Rows 28 through 31, where excess gain is calculated and gain trim attenuator set value is determined. Gain trim attenuator value is then transferred to the tuned parameter section, Rows 32 through 37, where tuned gain, noise figure, and input third-order intercept values are calculated. Cells identified as CRYSTAL BALL forecast values are CB Total Current (cell R22), Attenuator Set (cell R31), Tuned Gain (cell R35), Tuned *NF* (cell R36), and Tuned IIP (cell R37).

Build 2000 Receive Channels Using Mean Value Parameters Shown in Figure 10.6 and Test Statistics at Temperature Extremes

Using the IF amplifier stage parameter values entered into the spreadsheet shown in Figure 10.6 as mean values, build 2000 receive channel assemblies, and study the performance statistics. Trials are run at ambient temperatures of 0 degrees C, +25 degrees C, and +50 degrees C. Forecast values, CRYSTAL BALL output data, are Current, Attenuator Set, Tuned Gain, Tuned *NF*, and Tuned IIP.

Tuned gain results are shown in Figure 10.8 for all three temperatures. Notice that tuned gain at 0 degrees C shows twelve assemblies that had too much gain for the gain set attenuator to trim. Data for these twelve assemblies fell outside of the 51.1 dB range displayed. Predictions made earlier from Gaussian probability distribution functions were for 0.9 percent with too high gain. This CRYSTAL BALL data shows 1.2 percent with too high gain. At ambient temperature of +25 degrees C tuned gain always falls within the range of 50.55 dB and 51.05 dB. Probability density is reasonably uniform for 2000 trials, probability is nearly uniform that receive channels will tune to any gain within the range. At +50 degrees C five assemblies lacked sufficient gain to tune within the range, even with zero gain trim attenuator setting. Those five

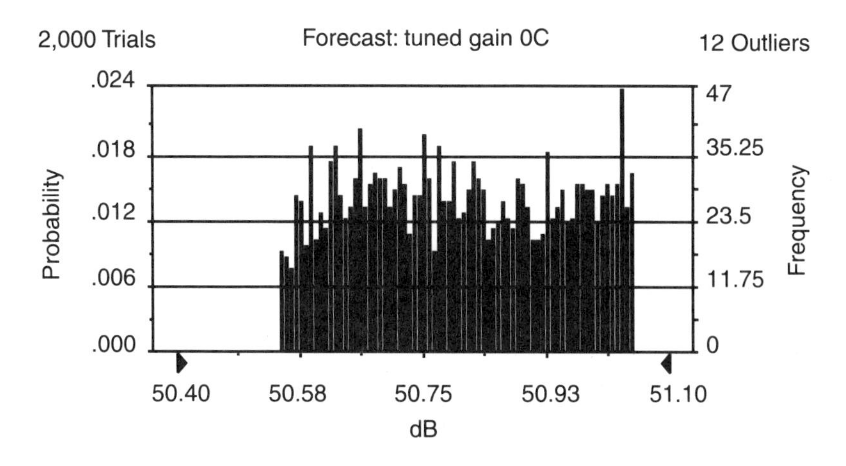

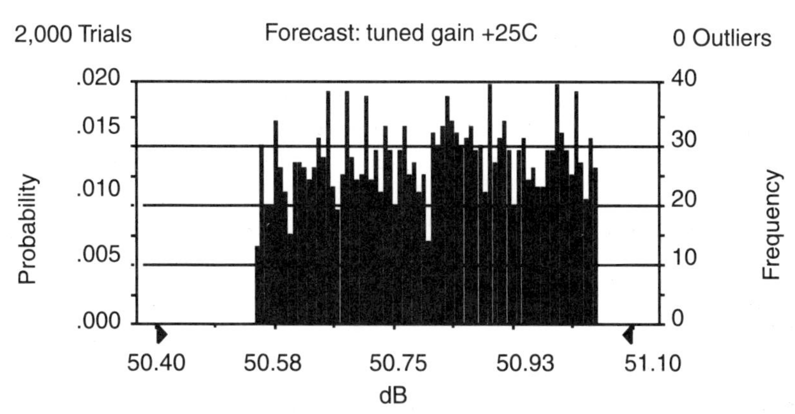

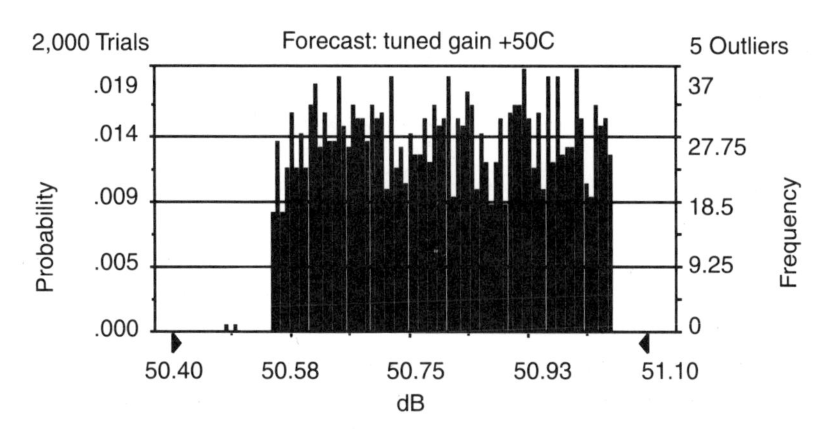

Figure 10.8 Receive channel tuned gain for 2000 trials at three ambient temperatures.

receive channels will still provide sufficient gain to cause the A/D converter LSB to toggle on noise.

Forecast values for gain trim attenuator settings at ambient temperatures of 0 degrees C, +25 degrees C, and +50 degrees C for the 2000 trials are shown in Figure 10.9. Receive channel gain is highest at 0 degrees C, and tuned gain trim attenuator settings move to the upper end of the attenuator range. When untuned gain is greater than 66.3 dB, the gain set attenuator is set at maximum of 15.5 dB. The twelve receive channels that had too much gain in Figure 10.8 at 0 degrees C cause the gain trim attenuator to go to 15.5 dB. It appears that more than twelve receive channels have high enough gain to cause the gain trim attenuator to go to 15.5 dB (see Figure 10.9, 0 degrees C). The forecast for 0 degrees C indicates that 0.4 percent of the assemblies have untuned gain so high as to cause more noise power spectral density than desired at the A/D converter input. This is more desirable than less noise power spectral density, which might cause the A/D converter LSB to stop toggling.

At +25 degrees C ambient temperature the gain trim attenuator setting range moves down and becomes centered. All of the receive channels tune properly. At +50 degrees C untuned gain in five of the receive channels drops below 50.55 dB (see Figure 10.8 +50 degrees C). A small number of receive channels have gain values less than 51.05 dB. Figure 10.9 for +50 degrees C shows approximately 12 units having 0 dB gain trim setting.

Tuned noise figure varies over a range of 2.3 dB minimum at 0 degrees C to 4.0 dB maximum at +50 degrees C. The design goal of providing 50.8 dB tuned receive channel gain with 4.0 dB maximum noise figure at any temperature between 0 degrees C and +50 degrees C for any assembly of randomly selected components has been satisfied. The maximum noise figure seen in 2000 assemblies at +50 degrees C is just 4.0 dB as shown in Figure 10.10 for +50 degrees C.

Input third-order intercept point mean value of −14.0 dBm for signals that pass into the IF amplifiers is realized for the receive channel population. Temperature sensitivity is 0.02 dB per degree centigrade. Input third-order intercept value for signals that do not get into the IF amplifier is +4.48 dB. Figure 10.11 shows forecast values of input third-order intercept for signals that do pass into the IF amplifier at three ambient temperatures of 0 degrees C, +25 degrees C, and +50 degrees C.

Mean value of current consumption is 3.0 amperes. Figure 10.12 shows the forecast for receive channel current ranging from 2.3 amperes to 3.7 amperes. Standard deviation of current consumption for the population is 0.23 amperes. Mean value of power consumption is 15.05 watts.

This example has shown an approach to suballocating receive channel parameters to achieve desired noise figure performance over a specified

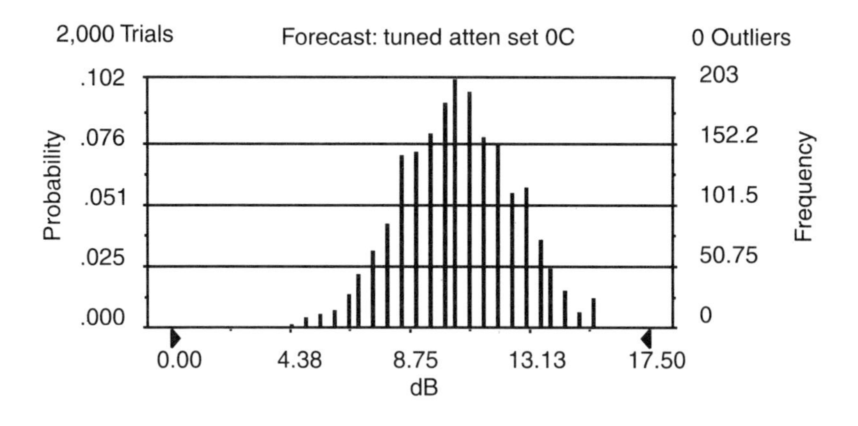

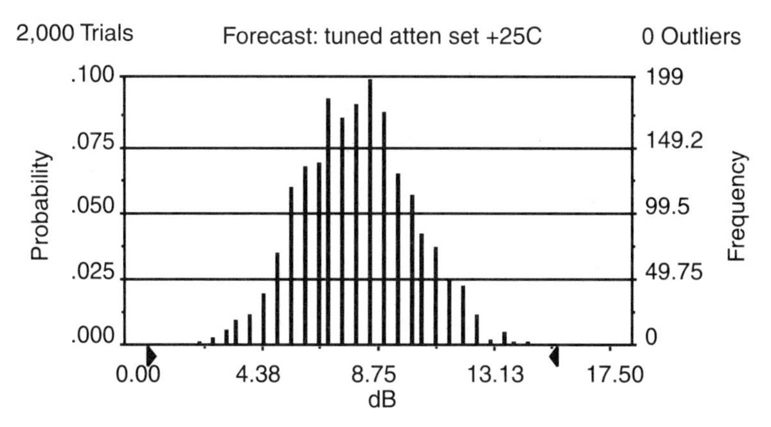

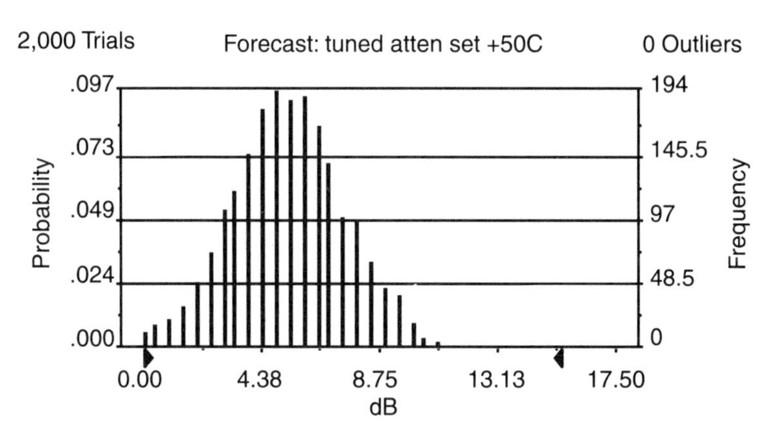

Figure 10.9 Receive channel tuned gain trim attenuator setting for 2000 trials at three ambient temperatures.

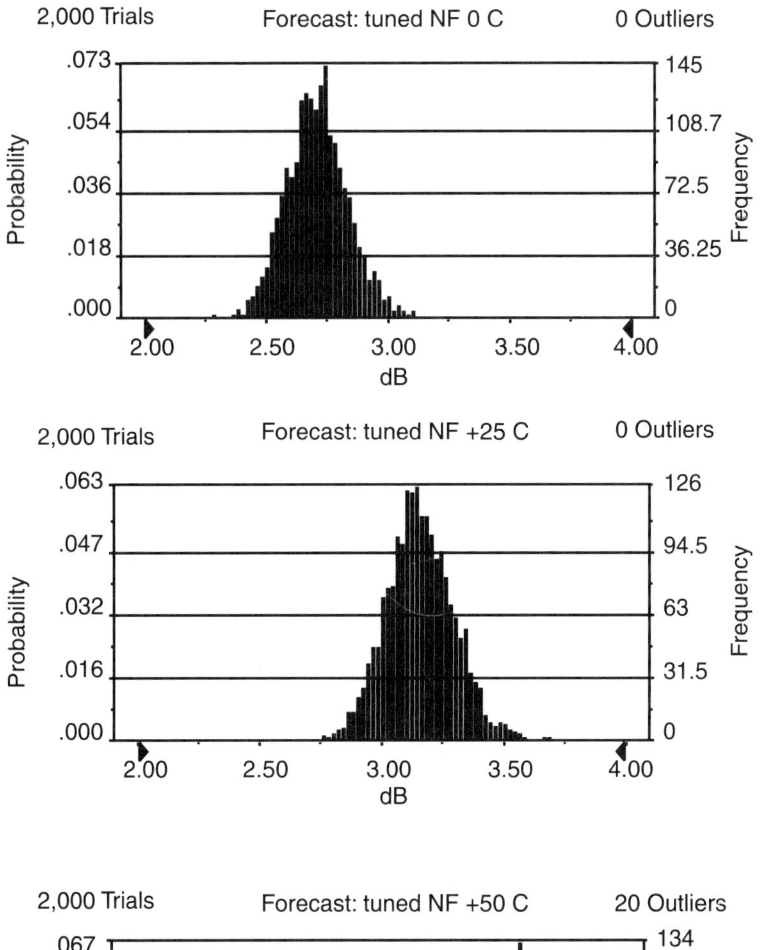

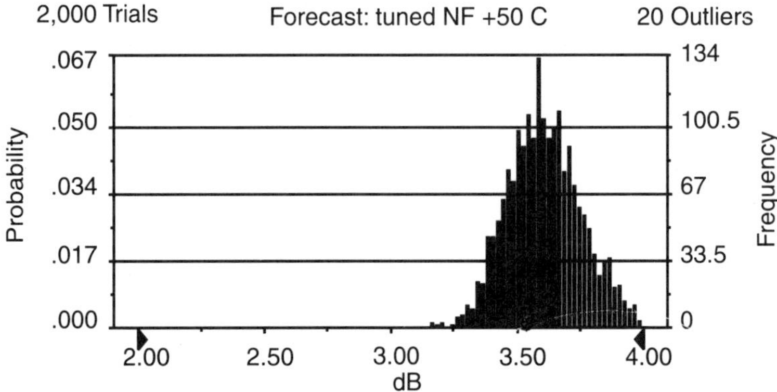

Figure 10.10 Receive channel tuned noise figure for 2000 trials at three ambient temperatures.

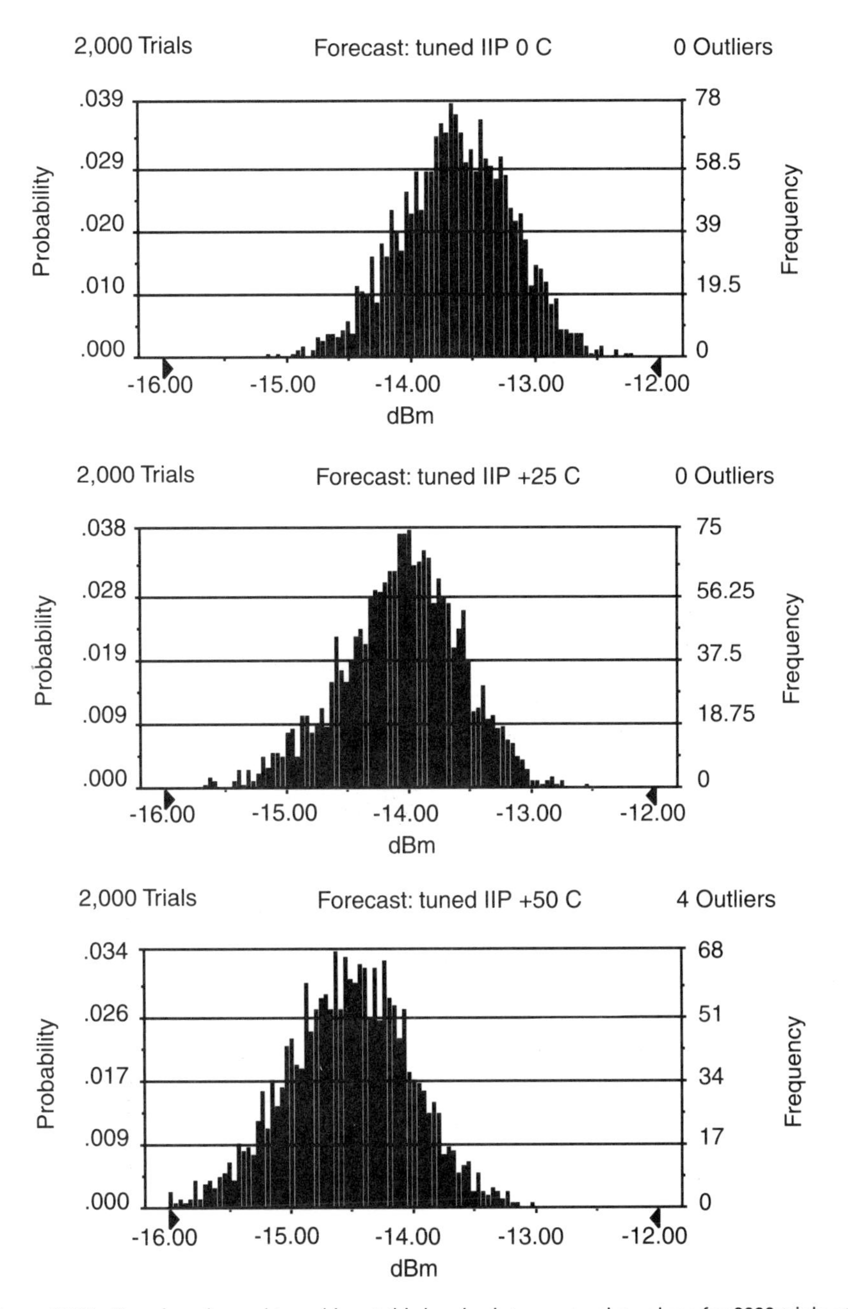

Figure 10.11 Receive channel tuned input third-order intercept point values for 2000 trials at three ambient temperatures.

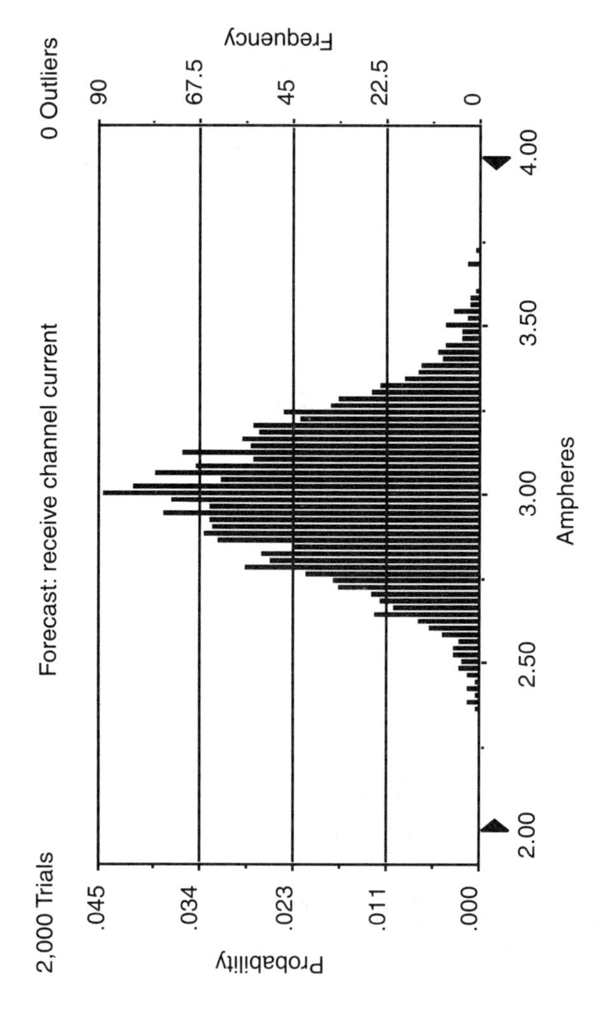

Figure 10.12 Receive channel population current consumption forecast.

temperature range for a population of 2000 receiver assemblies. The parameters flowed down to subassembly levels of LNA, and IF amplifiers can be confidently included in a specification for those subassemblies. Furthermore, data has been generated that flows down to the amplifier stage level to help define transistor parameters needed for the individual amplifier circuit designs.

An approach for suballocating transmit channel parameters to obtain desired power output and power-added efficiency over an operating temperature range for a population of transmitter assemblies is described in the next example. The approach is different because efficient transmitter amplifier stages have to be driven into compression.

Example 10.2 Suballocating Transmit Channel Parameters

The 0 dBm, X band frequency, signal out of a frequency up-converter mixer is to be amplified and transmitted. Transmitter amplifier power output needs to be +33 dBm minimum at any temperature 0 degrees C to +50 degrees C over a population of transmitter amplifier assemblies. Transistors available for the power amplifier can develop 10 dB small signal gain per amplifier stage. Voltage available to power the transmitter power amplifier is +9.0 V. Maximum current available is 1.2 amperes. Architect a transmitter power amplifier to satisfy these requirements.

Begin With the Final Amplifier Stage and Work Backwards Through the Transmitter

Assume for the moment that five amplifier stages will be needed to develop sufficient tuned amplifier compressed gain to amplify the 0 dBm to +33 dBm at +50 degrees C. The actual number of stages needed will be determined logically as the model is developed. First, look at parameters for the final power amplifier stage 5 and the stage 4 that drives it. Temperature sensitivity of saturated power output is −0.015 dB/degrees C. Saturated power output of each stage will change ±0.375 dB at temperature extremes with respect to room temperature. Worst-case decrease in saturated power output as a function of temperature will be at +50 degrees C. Final amplifier stage 5 saturated power output is expected to exhibit standard deviation of 0.25 dB over a population of transistors. Refer to Figure 8.10, which shows worst case saturated power output is 1.05 dB lower than the population mean in a population with $1\sigma = 0.25$ dB. Population mean value of transmitter power output at +25 degrees C needs to be at least 1.425 dB (0.375 dB + 0.79 dB) greater than the required +33 dBm just to allow for temperature and population statistics. The mean will need to be even higher to allow for loss of power output due to amplifier stage compression.

Develop a trade space for the two final amplifier stages using the method described in Section 9.4.1 to determine the loss of power output due to amplifier stage compression. The trade space will also reveal the ratio of amplifier stage saturated power outputs that give optimum power-added efficiency for amplifiers with 10 dB small signal gain and mean compressed power output in the range of +34.4 dBm. Use (9.14) to develop an array of power output versus power input for values of P_{sat} ratio N as defined by (9.16). Figure 10.13 shows the ranges of power input to the two final stages P_{in} and N that are useful for the trade space.

Row 1 in the spreadsheet contains compression coefficient values calculated using (5.11), which depends on bias coefficient b_q that is entered in row 5. Row 2 in the spreadsheet contains the above estimated value P_{sat2} = +34.4 dBm. This value may have to be adjusted later as the trade space reveals optimum P_{sat} ratios, and power loss due to compression at the optimum operating point. Row 3 contains a value for gain of the final amplifier stage G_2 = 10 dB. Row 4 contains values for P_{sat1} corresponding to values of the variable $N = (P_{sat1} + G_2) - P_{sat2}$ where $-3 \leq N \leq +6$. Rows 6 and 7 contain values for the parameter $P_2(b)$ and $P_1(b)$ (see (5.18) and (5.19)) needed to determine average current consumption of a class AB amplifier as a function of power input using (5.15). Rows 12 through 33 contain power output data as a function of power input and variable N.

Plot the data generated in the spreadsheet shown in Figure 10.13 as a surface as shown in Figure 10.14. This surface shows the reduction in power output with respect to saturated power output of the final amplifier stage P_{sat2} = +34.4 dBm, due to P_{sat} ratio, and compression depth of the two last stages in the transmitter amplifier. The object now is to determine where the optimum operating point is located for best efficiency and power output.

Using (5.15) and referring to rows 5, 6, and 7 for values of b_q, P_2, and P_1, calculate average current as a function of power input for amplifier stage 4. Assemble the calculated data into a spreadsheet array as illustrated in Figure 10.15.

The power output of amplifier stage 4 is the power input of amplifier stage 5 and will determine the average current drawn by stage 5. Create an array that calculates power output of amplifier stage 4 using (5.1). Power output of stage 4 is a function of P_{in} and variable N as illustrated in Figure 10.16. Saturated power output values for amplifier stage 4 are calculated in row 4 of the spreadsheet (see Figure 10.13). Amplifier stage 4 gain is 10 dB and compression coefficient K is listed in row 1 of the spreadsheet.

Using amplifier stage 4 power output data shown in Figure 10.16 as power input data for amplifier stage 5, calculate average current consumed by amplifier stage 5 as a function of power input to that stage and variable N by creating another spreadsheet array as illustrated in Figure 10.17.

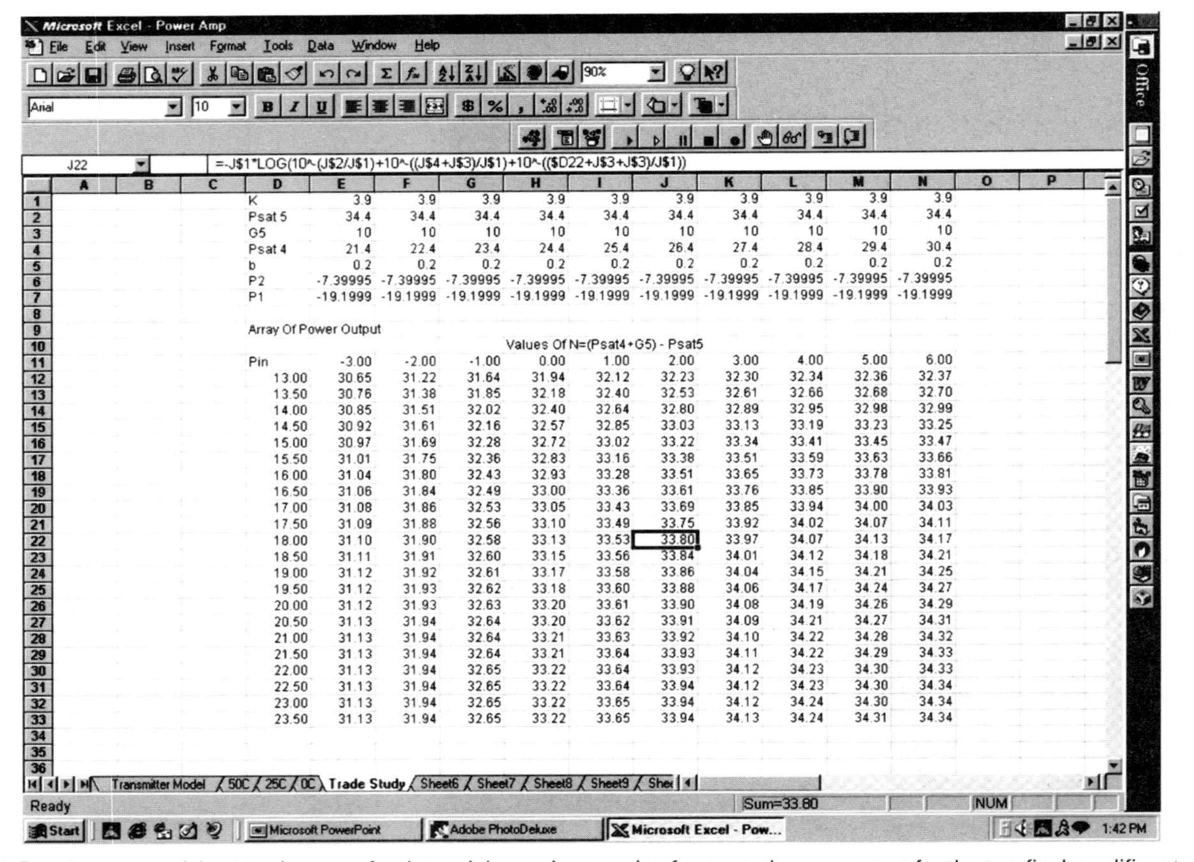

	E	F	G	H	I	J	K	L	M	N
K	3.9	3.9	3.9	3.9	3.9	3.9	3.9	3.9	3.9	3.9
Psat 5	34.4	34.4	34.4	34.4	34.4	34.4	34.4	34.4	34.4	34.4
G5	10	10	10	10	10	10	10	10	10	10
Psat 4	21.4	22.4	23.4	24.4	25.4	26.4	27.4	28.4	29.4	30.4
b	0.2	0.2	0.2	0.2	0.2	0.2	0.2	0.2	0.2	0.2
P2	-7.39995	-7.39995	-7.39995	-7.39995	-7.39995	-7.39995	-7.39995	-7.39995	-7.39995	-7.39995
P1	-19.1999	-19.1999	-19.1999	-19.1999	-19.1999	-19.1999	-19.1999	-19.1999	-19.1999	-19.1999

Array Of Power Output

Values Of N=(Psat4+G5) - Psat5

Pin	-3.00	-2.00	-1.00	0.00	1.00	2.00	3.00	4.00	5.00	6.00
13.00	30.65	31.22	31.64	31.94	32.12	32.23	32.30	32.34	32.36	32.37
13.50	30.76	31.38	31.85	32.18	32.40	32.53	32.61	32.66	32.68	32.70
14.00	30.85	31.51	32.02	32.40	32.64	32.80	32.89	32.95	32.98	32.99
14.50	30.92	31.61	32.16	32.57	32.85	33.03	33.13	33.19	33.23	33.25
15.00	30.97	31.69	32.28	32.72	33.02	33.22	33.34	33.41	33.45	33.47
15.50	31.01	31.75	32.36	32.83	33.16	33.38	33.51	33.59	33.63	33.66
16.00	31.04	31.80	32.43	32.93	33.28	33.51	33.65	33.73	33.78	33.81
16.50	31.06	31.84	32.49	33.00	33.36	33.61	33.76	33.85	33.90	33.93
17.00	31.08	31.86	32.53	33.05	33.43	33.69	33.85	33.94	34.00	34.03
17.50	31.09	31.88	32.56	33.10	33.49	33.75	33.92	34.02	34.07	34.11
18.00	31.10	31.90	32.58	33.13	33.53	33.80	33.97	34.07	34.13	34.17
18.50	31.11	31.91	32.60	33.15	33.56	33.84	34.01	34.12	34.18	34.21
19.00	31.12	31.92	32.61	33.17	33.58	33.86	34.04	34.15	34.21	34.25
19.50	31.12	31.93	32.62	33.18	33.60	33.88	34.06	34.17	34.24	34.27
20.00	31.12	31.93	32.63	33.20	33.61	33.90	34.08	34.19	34.26	34.29
20.50	31.13	31.94	32.64	33.20	33.62	33.91	34.09	34.21	34.27	34.31
21.00	31.13	31.94	32.64	33.21	33.63	33.92	34.10	34.22	34.28	34.32
21.50	31.13	31.94	32.64	33.21	33.64	33.93	34.11	34.22	34.29	34.33
22.00	31.13	31.94	32.65	33.22	33.64	33.93	34.12	34.23	34.30	34.33
22.50	31.13	31.94	32.65	33.22	33.64	33.94	34.12	34.23	34.30	34.34
23.00	31.13	31.94	32.65	33.22	33.65	33.94	34.12	34.24	34.30	34.34
23.50	31.13	31.94	32.65	33.22	33.65	33.94	34.13	34.24	34.31	34.34

Figure 10.13 Develop a spreadsheet trade space for determining optimum ratio of saturated power output for the two final amplifier stages.

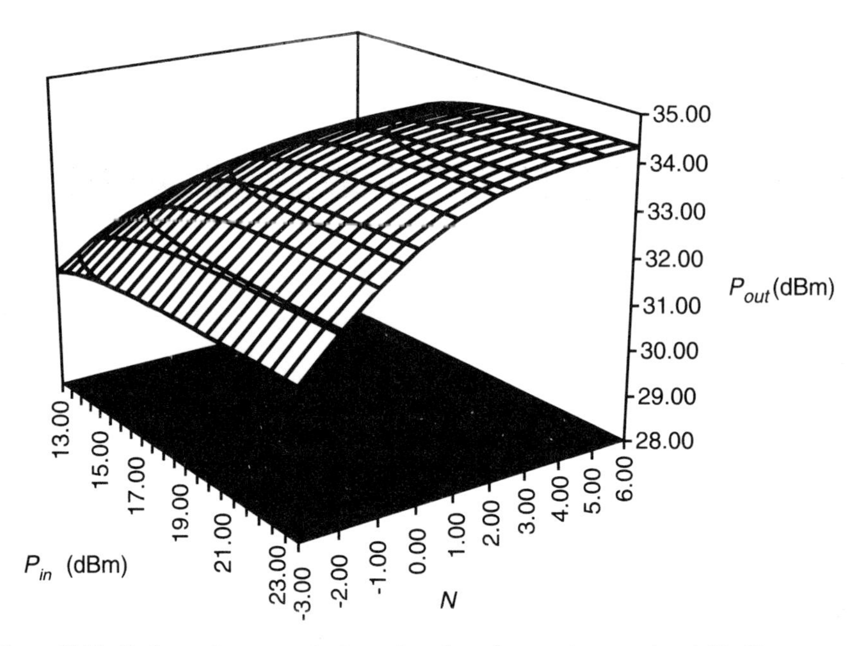

Figure 10.14 Surface of power output as a function of power input and variable *N*.

Using power output data calculated in the array shown in Figure 10.13, power input data in the first column of the array, and total current consumption of the two amplifier stages (the sum of the two current consumption arrays Figures 10.15 and 10.17), calculate power-added efficiency of the two amplifier stages as a function of power input and variable N as illustrated in Figure 10.18. Equation (4.16) is used to calculate *PAE*.

Plot data created in the array of Figure 10.18 as a surface shown in Figure 10.19. This surface shows there is a soft peak in power-added efficiency for $2 \leq N \leq 3$, and for power input in the range $+17.5$ dBm $\leq P_{in} \leq +18.0$ dBm. Study the amplifier power output surface Figure 10.14 and notice there is a soft roll-off of power output in the same region of P_{in} and N. A definitive technique is needed to focus on the optimum combination of power output and power-added efficiency.

Determine the maximum *PAE* value in the array of Figure 10.18. Add 3 to that value. Create another array where values in the array are power output from Figure 10.13 in units of dBm divided by $(PAE_{max} + 3 - PAE)$ as illustrated in Figure 10.20. This calculation of figure of merit M (see Section 9.4.1) combines power output and *PAE* into a single quantity and magnifies the optimum combination of power input and variable N, giving the result that optimum power input to the two final amplifier stages is $P_{in} = +18.0$ dBm, and the best ratio of saturated power output values is obtained when $N = 2$.

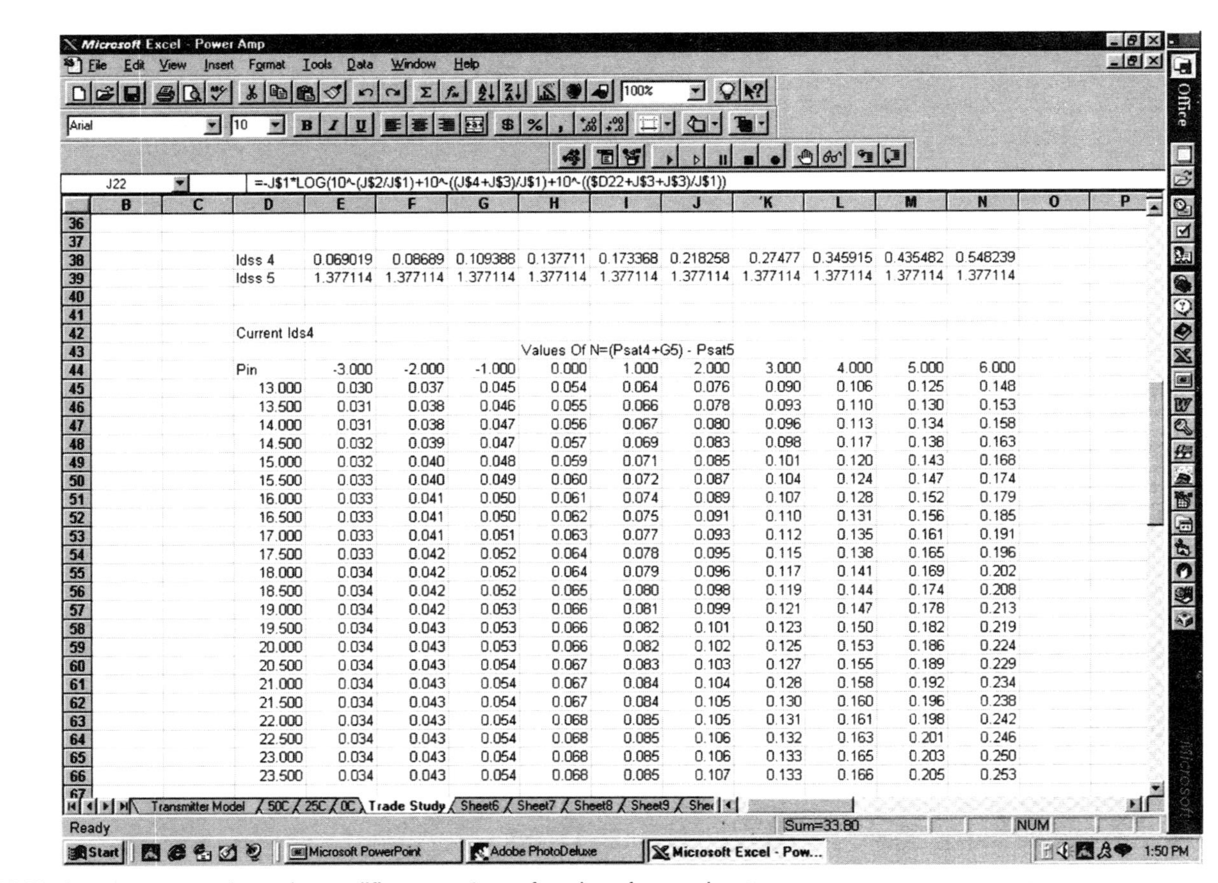

	D	E	F	G	H	I	J	K	L	M	N
38	Idss 4	0.069019	0.08689	0.109388	0.137711	0.173368	0.218258	0.27477	0.345915	0.435482	0.548239
39	Idss 5	1.377114	1.377114	1.377114	1.377114	1.377114	1.377114	1.377114	1.377114	1.377114	1.377114
42	Current Ids4										
43					Values Of N=(Psat4+G5) - Psat5						
44	Pin	-3.000	-2.000	-1.000	0.000	1.000	2.000	3.000	4.000	5.000	6.000
45	13.000	0.030	0.037	0.045	0.054	0.064	0.076	0.090	0.106	0.125	0.148
46	13.500	0.031	0.038	0.046	0.055	0.066	0.078	0.093	0.110	0.130	0.153
47	14.000	0.031	0.038	0.047	0.056	0.067	0.080	0.096	0.113	0.134	0.158
48	14.500	0.032	0.039	0.047	0.057	0.069	0.083	0.098	0.117	0.138	0.163
49	15.000	0.032	0.040	0.048	0.059	0.071	0.085	0.101	0.120	0.143	0.168
50	15.500	0.033	0.040	0.049	0.060	0.072	0.087	0.104	0.124	0.147	0.174
51	16.000	0.033	0.041	0.050	0.061	0.074	0.089	0.107	0.128	0.152	0.179
52	16.500	0.033	0.041	0.050	0.062	0.075	0.091	0.110	0.131	0.156	0.185
53	17.000	0.033	0.041	0.051	0.063	0.077	0.093	0.112	0.135	0.161	0.191
54	17.500	0.033	0.042	0.052	0.064	0.078	0.095	0.115	0.138	0.165	0.196
55	18.000	0.034	0.042	0.052	0.064	0.079	0.096	0.117	0.141	0.169	0.202
56	18.500	0.034	0.042	0.052	0.065	0.080	0.098	0.119	0.144	0.174	0.208
57	19.000	0.034	0.042	0.053	0.066	0.081	0.099	0.121	0.147	0.178	0.213
58	19.500	0.034	0.043	0.053	0.066	0.082	0.101	0.123	0.150	0.182	0.219
59	20.000	0.034	0.043	0.053	0.066	0.082	0.102	0.125	0.153	0.186	0.224
60	20.500	0.034	0.043	0.054	0.067	0.083	0.103	0.127	0.155	0.189	0.229
61	21.000	0.034	0.043	0.054	0.067	0.084	0.104	0.128	0.158	0.192	0.234
62	21.500	0.034	0.043	0.054	0.067	0.084	0.105	0.130	0.160	0.196	0.238
63	22.000	0.034	0.043	0.054	0.068	0.085	0.105	0.131	0.161	0.198	0.242
64	22.500	0.034	0.043	0.054	0.068	0.085	0.106	0.132	0.163	0.201	0.246
65	23.000	0.034	0.043	0.054	0.068	0.085	0.106	0.133	0.165	0.203	0.250
66	23.500	0.034	0.043	0.054	0.068	0.085	0.107	0.133	0.166	0.205	0.253

Figure 10.15 Average current drawn by amplifier stage 4 as a function of power input.

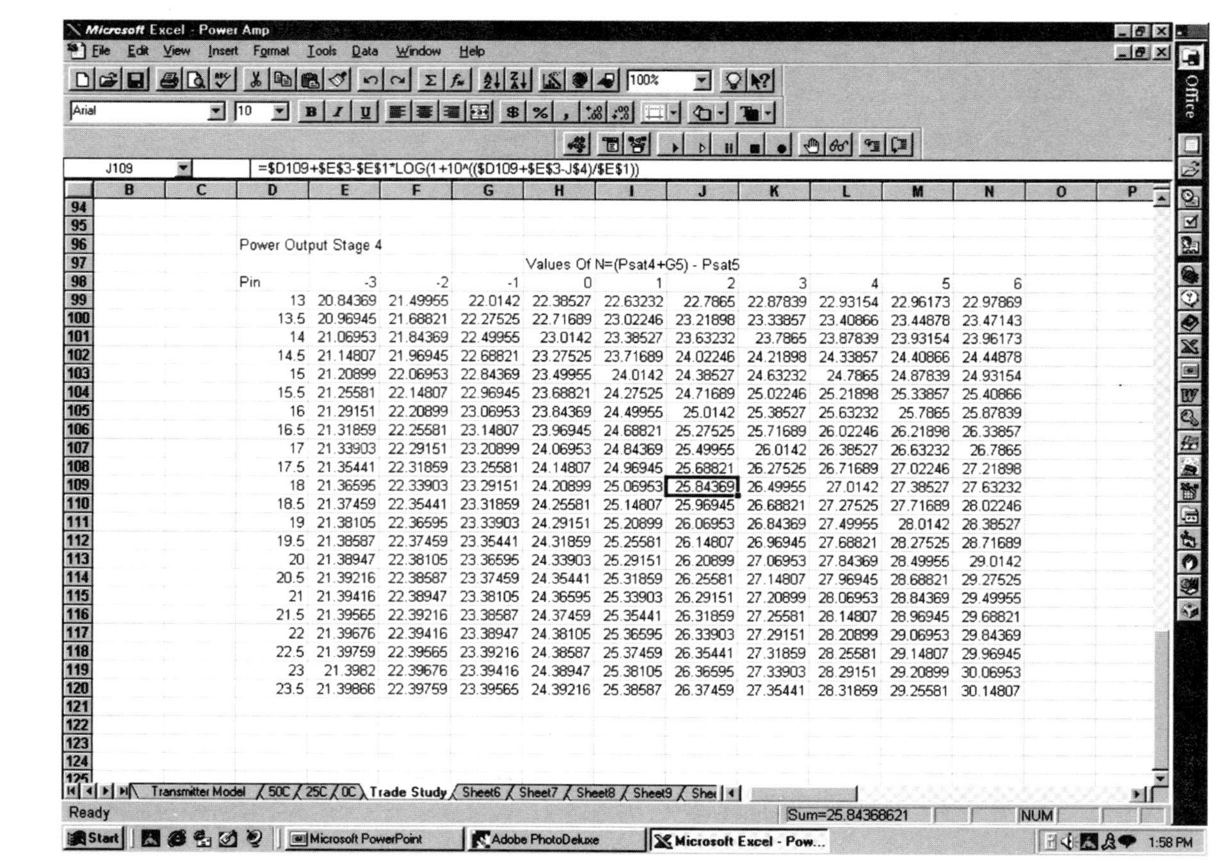

Microsoft Excel - Power Amp

J109 =$D109+$E$3-$E$1*LOG(1+10^(($D109+$E$3-J$4)/$E$1))

Power Output Stage 4

Values Of N=(Psat4+G5) - Psat5

Pin	-3	-2	-1	0	1	2	3	4	5	6
13	20.84369	21.49955	22.0142	22.38527	22.63232	22.7865	22.87839	22.93154	22.96173	22.97869
13.5	20.96945	21.68821	22.27525	22.71689	23.02246	23.21898	23.33857	23.40866	23.44878	23.47143
14	21.06953	21.84369	22.49955	23.0142	23.38527	23.63232	23.7865	23.87839	23.93154	23.96173
14.5	21.14807	21.96945	22.68821	23.27525	23.71689	24.02246	24.21898	24.33857	24.40866	24.44878
15	21.20899	22.06953	22.84369	23.49955	24.0142	24.38527	24.63232	24.7865	24.87839	24.93154
15.5	21.25581	22.14807	22.96945	23.68821	24.27525	24.71689	25.02246	25.21898	25.33857	25.40866
16	21.29151	22.20899	23.06953	23.84369	24.49955	25.0142	25.38527	25.63232	25.7865	25.87839
16.5	21.31859	22.25581	23.14807	23.96945	24.68821	25.27525	25.71689	26.02246	26.21898	26.33857
17	21.33903	22.29151	23.20899	24.06953	24.84369	25.49955	26.0142	26.38527	26.63232	26.7865
17.5	21.35441	22.31859	23.25581	24.14807	24.96945	25.68821	26.27525	26.71689	27.02246	27.21898
18	21.36595	22.33903	23.29151	24.20899	25.06953	25.84369	26.49955	27.0142	27.38527	27.63232
18.5	21.37459	22.35441	23.31859	24.25581	25.14807	25.96945	26.68821	27.27525	27.71689	28.02246
19	21.38105	22.36595	23.33903	24.29151	25.20899	26.06953	26.84369	27.49955	28.0142	28.38527
19.5	21.38587	22.37459	23.35441	24.31859	25.25581	26.14807	26.96945	27.68821	28.27525	28.71689
20	21.38947	22.38105	23.36595	24.33903	25.29151	26.20899	27.06953	27.84369	28.49955	29.0142
20.5	21.39216	22.38587	23.37459	24.35441	25.31859	26.25581	27.14807	27.96945	28.68821	29.27525
21	21.39416	22.38947	23.38105	24.36595	25.33903	26.29151	27.20899	28.06953	28.84369	29.49955
21.5	21.39565	22.39216	23.38587	24.37459	25.35441	26.31859	27.25581	28.14807	28.96945	29.68821
22	21.39676	22.39416	23.38947	24.38105	25.36595	26.33903	27.29151	28.20899	29.06953	29.84369
22.5	21.39759	22.39565	23.39216	24.38587	25.37459	26.35441	27.31859	28.25581	29.14807	29.96945
23	21.3982	22.39676	23.39416	24.38947	25.38105	26.36595	27.33903	28.29151	29.20899	30.06953
23.5	21.39866	22.39759	23.39565	24.39216	25.38587	26.37459	27.35441	28.31859	29.25581	30.14807

Transmitter Model / 50C / 25C / 0C / Trade Study / Sheet6 / Sheet7 / Sheet8 / Sheet9

Ready Sum=25.84368621 NUM

Figure 10.16 Amplifier stage 4 power output as a function of P_{in} and N is calculated.

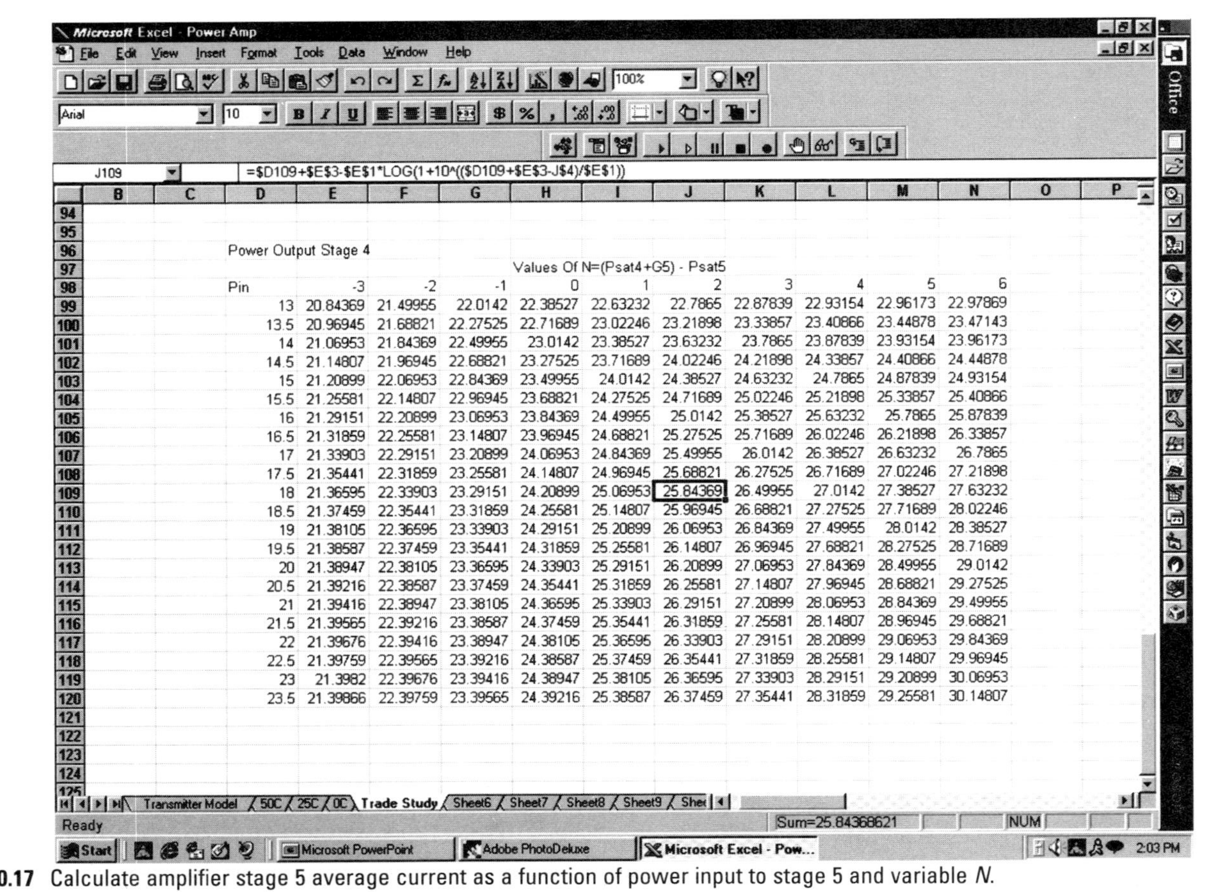

Microsoft Excel - Power Amp

J109 =$D109+$E$3-$E$1*LOG(1+10^(($D109+$E$3-J$4)/$E$1))

Power Output Stage 4

Values Of N=(Psat4+G5) - Psat5

Pin	-3	-2	-1	0	1	2	3	4	5	6
13	20.84369	21.49955	22.0142	22.38527	22.63232	22.7865	22.87839	22.93154	22.96173	22.97869
13.5	20.96945	21.68821	22.27525	22.71689	23.02246	23.21898	23.33857	23.40866	23.44878	23.47143
14	21.06953	21.84369	22.49955	23.0142	23.38527	23.63232	23.7865	23.87839	23.93154	23.96173
14.5	21.14807	21.96945	22.68821	23.27525	23.71689	24.02246	24.21898	24.33857	24.40866	24.44878
15	21.20899	22.06953	22.84369	23.49955	24.0142	24.38527	24.63232	24.7865	24.87839	24.93154
15.5	21.25581	22.14807	22.96945	23.68821	24.27525	24.71689	25.02246	25.21898	25.33857	25.40866
16	21.29151	22.20899	23.06953	23.84369	24.49955	25.0142	25.38527	25.63232	25.7865	25.87839
16.5	21.31859	22.25581	23.14807	23.96945	24.68821	25.27525	25.71689	26.02246	26.21898	26.33857
17	21.33903	22.29151	23.20899	24.06953	24.84369	25.49955	26.0142	26.38527	26.63232	26.7865
17.5	21.35441	22.31859	23.25581	24.14807	24.96945	25.68821	26.27525	26.71689	27.02246	27.21898
18	21.36595	22.33903	23.29151	24.20899	25.06953	25.84369	26.49955	27.0142	27.38527	27.63232
18.5	21.37459	22.35441	23.31859	24.25581	25.14807	25.96945	26.68821	27.27525	27.71689	28.02246
19	21.38105	22.36595	23.33903	24.29151	25.20899	26.06953	26.84369	27.49955	28.0142	28.38527
19.5	21.38587	22.37459	23.35441	24.31859	25.25581	26.14807	26.96945	27.68821	28.27525	28.71689
20	21.38947	22.38105	23.36595	24.33903	25.29151	26.20899	27.06953	27.84369	28.49955	29.0142
20.5	21.39216	22.38587	23.37459	24.35441	25.31859	26.25581	27.14807	27.96945	28.68821	29.27525
21	21.39416	22.38947	23.38105	24.36595	25.33903	26.29151	27.20899	28.06953	28.84369	29.49955
21.5	21.39565	22.39216	23.38587	24.37459	25.35441	26.31859	27.25581	28.14807	28.96945	29.68821
22	21.39676	22.39416	23.38947	24.38105	25.36595	26.33903	27.29151	28.20899	29.06953	29.84369
22.5	21.39759	22.39565	23.39216	24.38587	25.37459	26.35441	27.31859	28.25581	29.14807	29.96945
23	21.3982	22.39676	23.39416	24.38947	25.38105	26.36595	27.33903	28.29151	29.20899	30.06953
23.5	21.39866	22.39759	23.39565	24.39216	25.38587	26.37459	27.35441	28.31859	29.25581	30.14807

Transmitter Model / 50C / 25C / 0C / Trade Study / Sheet6 / Sheet7 / Sheet8 / Sheet9

Ready Sum=25.84368621 NUM

Figure 10.17 Calculate amplifier stage 5 average current as a function of power input to stage 5 and variable *N*.

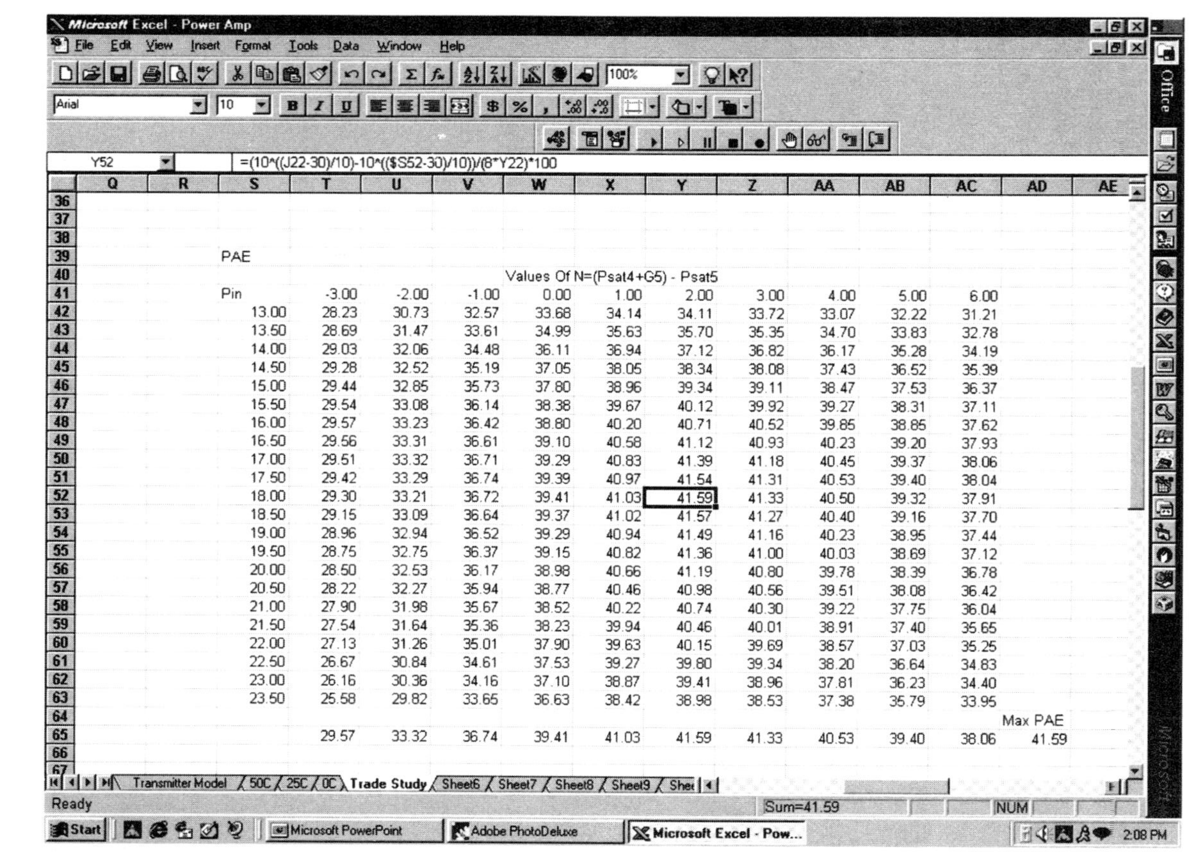

Pin	-3.00	-2.00	-1.00	0.00	1.00	2.00	3.00	4.00	5.00	6.00
13.00	28.23	30.73	32.57	33.68	34.14	34.11	33.72	33.07	32.22	31.21
13.50	28.69	31.47	33.61	34.99	35.63	35.70	35.35	34.70	33.83	32.78
14.00	29.03	32.06	34.48	36.11	36.94	37.12	36.82	36.17	35.28	34.19
14.50	29.28	32.52	35.19	37.05	38.05	38.34	38.08	37.43	36.52	35.39
15.00	29.44	32.85	35.73	37.80	38.96	39.34	39.11	38.47	37.53	36.37
15.50	29.54	33.08	36.14	38.38	39.67	40.12	39.92	39.27	38.31	37.11
16.00	29.57	33.23	36.42	38.80	40.20	40.71	40.52	39.85	38.85	37.62
16.50	29.56	33.31	36.61	39.10	40.58	41.12	40.93	40.23	39.20	37.93
17.00	29.51	33.32	36.71	39.29	40.83	41.39	41.18	40.45	39.37	38.06
17.50	29.42	33.29	36.74	39.39	40.97	41.54	41.31	40.53	39.40	38.04
18.00	29.30	33.21	36.72	39.41	41.03	41.59	41.33	40.50	39.32	37.91
18.50	29.15	33.09	36.64	39.37	41.02	41.57	41.27	40.40	39.16	37.70
19.00	28.96	32.94	36.52	39.29	40.94	41.49	41.16	40.23	38.95	37.44
19.50	28.75	32.75	36.37	39.15	40.82	41.36	41.00	40.03	38.69	37.12
20.00	28.50	32.53	36.17	38.98	40.66	41.19	40.80	39.78	38.39	36.78
20.50	28.22	32.27	35.94	38.77	40.46	40.98	40.56	39.51	38.08	36.42
21.00	27.90	31.98	35.67	38.52	40.22	40.74	40.30	39.22	37.75	36.04
21.50	27.54	31.64	35.36	38.23	39.94	40.46	40.01	38.91	37.40	35.65
22.00	27.13	31.26	35.01	37.90	39.63	40.15	39.69	38.57	37.03	35.25
22.50	26.67	30.84	34.61	37.53	39.27	39.80	39.34	38.20	36.64	34.83
23.00	26.16	30.36	34.16	37.10	38.87	39.41	38.96	37.81	36.23	34.40
23.50	25.58	29.82	33.65	36.63	38.42	38.98	38.53	37.38	35.79	33.95
	29.57	33.32	36.74	39.41	41.03	41.59	41.33	40.53	39.40	38.06

Figure 10.18 Power-added efficiency is calculated for the two amplifier stages as a function of P_{in} and variable N.

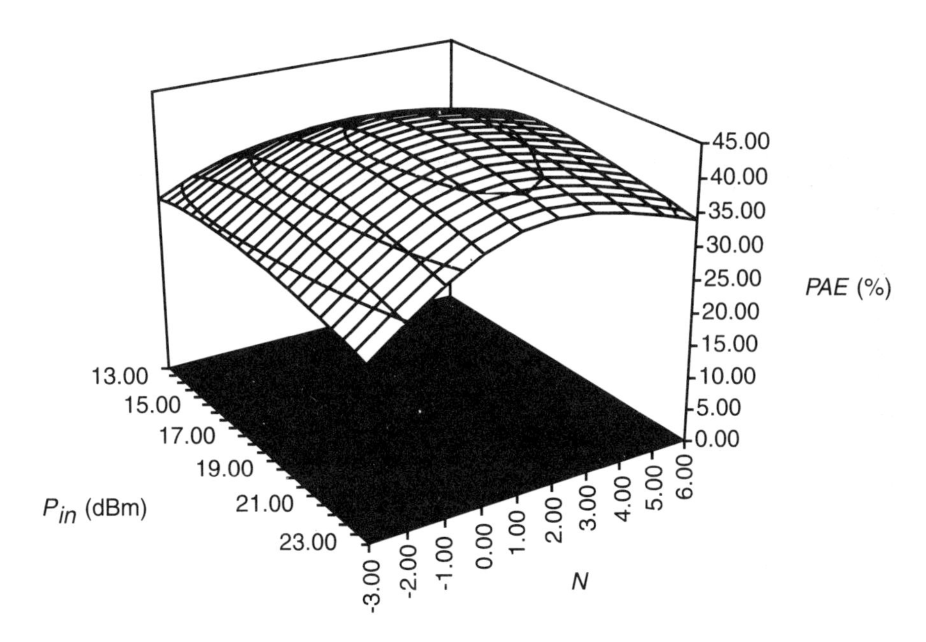

Figure 10.19 Two-stage amplifier *PAE* surface as a function of power input and variable *N*.

The data generated in Figure 10.20 is plotted as a surface in Figure 10.21 showing the magnification realized by using the figure of merit M to determine the optimum operating point. With $N = 2$, $G_2 = 10$, and $P_{sat5} = +34.4$ dBm, the optimum value for $P_{sat4} = +26.4$ dBm is determined. More important than the actual value of P_{sat4} is the ratio $(P_{sat5} - P_{sat4}) = 8.0$ dB for this combination of amplifier stage small signal gain G_{ss}, compression coefficient K, and bias coefficient b_q.

Another parameter that can be determined by extending the trade study is compression depth of the last two stages at the optimum operating point. Create another spreadsheet array that uses (9.18) to calculate compression depth as a function of power input P_{in} and variable value N. Figure 10.22 illustrates the data calculated. Locate the cell which corresponds to $P_{in} = +18.0$ dBm and $N = 2$, and determine that total compression depth is 4.2 dB for the two stages. That is roughly 2.1 dB per stage. The addition of another class AB stage 3 adds another 2.1 dB for a total of 6.3 dB compression in the last three stages at peak *PAE*.

The values for P_{sat4} and P_{sat5} that will satisfy the requirements of $P_{out} = +33.0$ dBm minimum at +50 degrees C have to be determined by studying temperature, population statistics, compression characteristics, and degradation of P_{sat} in cascaded stages.

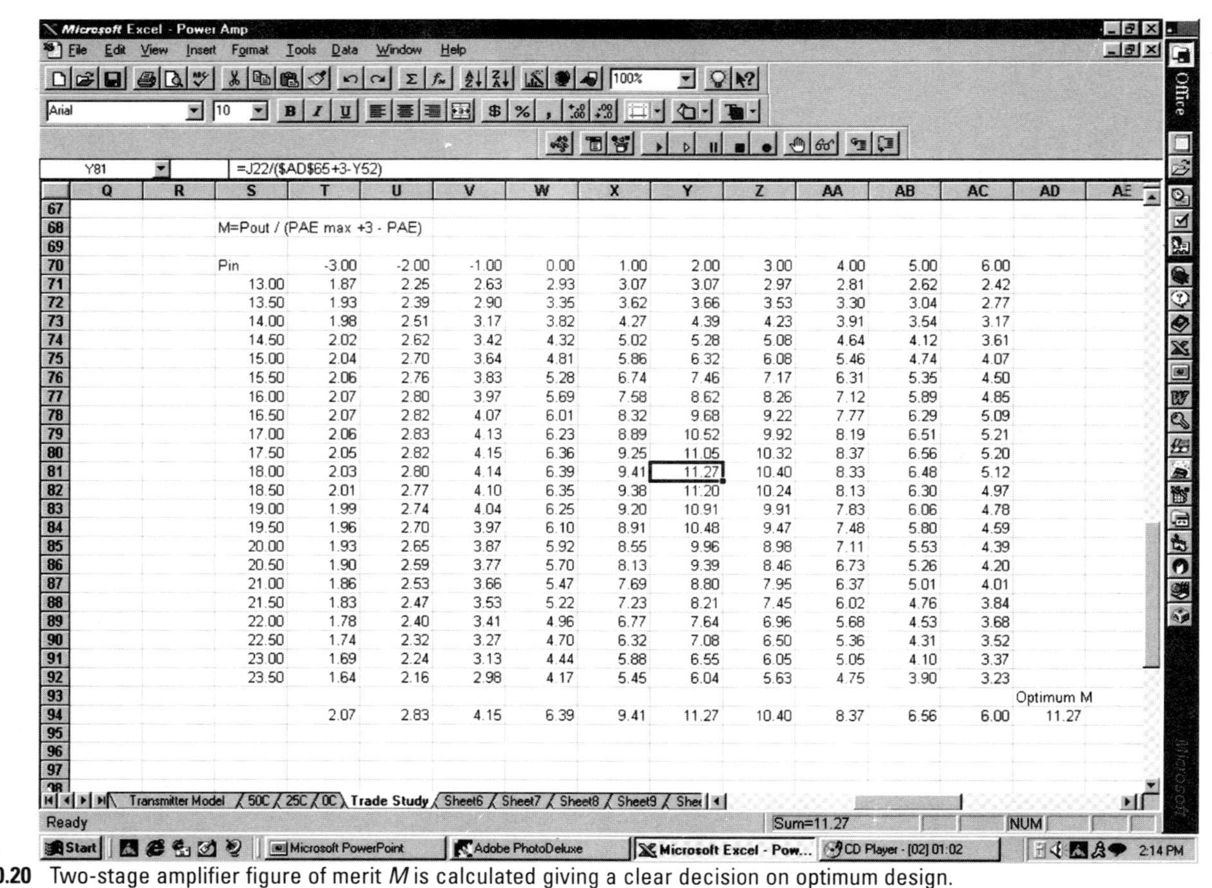

=J22/($AD$65+3-Y52)

M=Pout / (PAE max +3 - PAE)

S	T	U	V	W	X	Y	Z	AA	AB	AC	AD
Pin	-3.00	-2.00	-1.00	0.00	1.00	2.00	3.00	4.00	5.00	6.00	
13.00	1.87	2.25	2.63	2.93	3.07	3.07	2.97	2.81	2.62	2.42	
13.50	1.93	2.39	2.90	3.35	3.62	3.66	3.53	3.30	3.04	2.77	
14.00	1.98	2.51	3.17	3.82	4.27	4.39	4.23	3.91	3.54	3.17	
14.50	2.02	2.62	3.42	4.32	5.02	5.28	5.08	4.64	4.12	3.61	
15.00	2.04	2.70	3.64	4.81	5.86	6.32	6.08	5.46	4.74	4.07	
15.50	2.06	2.76	3.83	5.28	6.74	7.46	7.17	6.31	5.35	4.50	
16.00	2.07	2.80	3.97	5.69	7.58	8.62	8.26	7.12	5.89	4.85	
16.50	2.07	2.82	4.07	6.01	8.32	9.68	9.22	7.77	6.29	5.09	
17.00	2.06	2.83	4.13	6.23	8.89	10.52	9.92	8.19	6.51	5.21	
17.50	2.05	2.82	4.15	6.36	9.25	11.05	10.32	8.37	6.56	5.20	
18.00	2.03	2.80	4.14	6.39	9.41	11.27	10.40	8.33	6.48	5.12	
18.50	2.01	2.77	4.10	6.35	9.38	11.20	10.24	8.13	6.30	4.97	
19.00	1.99	2.74	4.04	6.25	9.20	10.91	9.91	7.83	6.06	4.78	
19.50	1.96	2.70	3.97	6.10	8.91	10.48	9.47	7.48	5.80	4.59	
20.00	1.93	2.65	3.87	5.92	8.55	9.96	8.98	7.11	5.53	4.39	
20.50	1.90	2.59	3.77	5.70	8.13	9.39	8.46	6.73	5.26	4.20	
21.00	1.86	2.53	3.66	5.47	7.69	8.80	7.95	6.37	5.01	4.01	
21.50	1.83	2.47	3.53	5.22	7.23	8.21	7.45	6.02	4.76	3.84	
22.00	1.78	2.40	3.41	4.96	6.77	7.64	6.96	5.68	4.53	3.68	
22.50	1.74	2.32	3.27	4.70	6.32	7.08	6.50	5.36	4.31	3.52	
23.00	1.69	2.24	3.13	4.44	5.88	6.55	6.05	5.05	4.10	3.37	
23.50	1.64	2.16	2.98	4.17	5.45	6.04	5.63	4.75	3.90	3.23	
											Optimum M
	2.07	2.83	4.15	6.39	9.41	11.27	10.40	8.37	6.56	6.00	11.27

Figure 10.20 Two-stage amplifier figure of merit M is calculated giving a clear decision on optimum design.

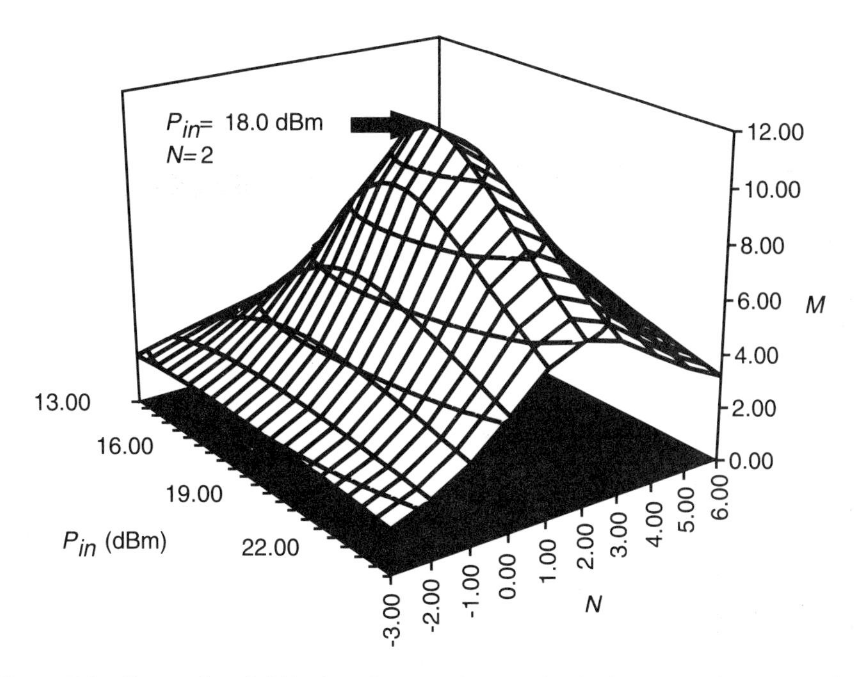

Figure 10.21 Figure of merit M is plotted as a surface to clearly show an optimum operating point exists at P_{in} = +18.0 dBm and $N = 2$.

Determine P_{sat4} and P_{sat5} Values

The trade study indicates saturated power output difference between class AB amplifier stages with 10 dB small signal gain, $b_q = 2$ and $K = 3.6$, needs to be 8.0 dB for optimum efficiency. The final parameter that needs to be confirmed from this trade study is the population mean value of saturated power output for P_{sat4} and P_{sat5} that will give +33.0 dBm minimum power output at +50 degrees C for any transmitter assembled by random selection of components. Refer to Figure 10.13 and read the spreadsheet cell P_{out} value where P_{in} = +18.0 dBm and $N = 2$. The value is P_{out} = +33.8 dBm at ambient temperature +25 degrees C. Note that the value of the final amplifier stage saturated power output was arbitrarily set at P_{sat5} = +34.4 dBm. The 0.6 dB difference between compressed power output and final stage saturated power output value is due to two factors. Maximum saturated power output of these two cascaded amplifier stages with 8 dB difference in P_{sat} values is not +34.4 dBm but is calculated by (9.15) to be

$$P_{max} = -3.9 \log_{10}\left[10^{\frac{-34.4}{3.9}} + 10^{\frac{-(26.4+10)}{3.9}}\right] = 33.94 \text{ dBm},$$

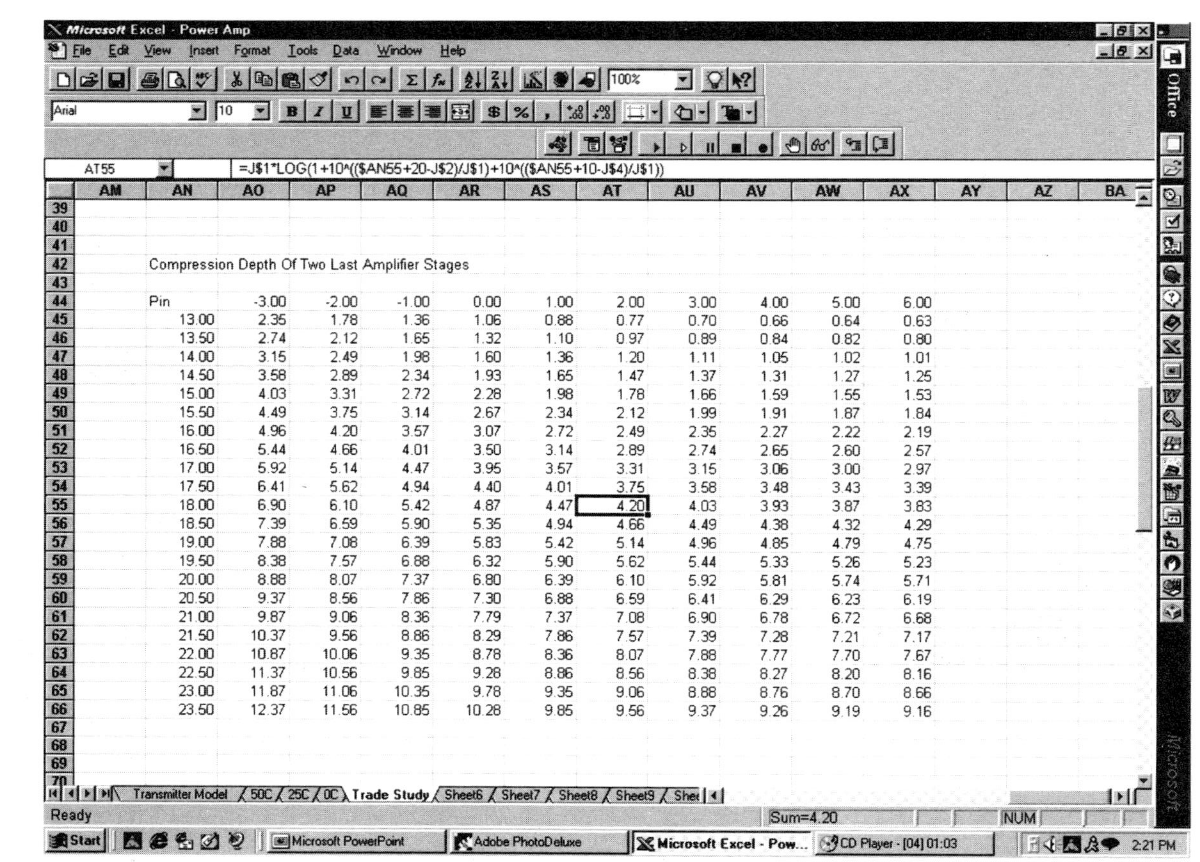

=J$1*LOG(1+10^(($AN55+20-J$2)/J$1)+10^(($AN55+10-J$4)/J$1))

Compression Depth Of Two Last Amplifier Stages

Pin	-3.00	-2.00	-1.00	0.00	1.00	2.00	3.00	4.00	5.00	6.00
13.00	2.35	1.78	1.36	1.06	0.88	0.77	0.70	0.66	0.64	0.63
13.50	2.74	2.12	1.65	1.32	1.10	0.97	0.89	0.84	0.82	0.80
14.00	3.15	2.49	1.98	1.60	1.36	1.20	1.11	1.05	1.02	1.01
14.50	3.58	2.89	2.34	1.93	1.65	1.47	1.37	1.31	1.27	1.25
15.00	4.03	3.31	2.72	2.28	1.98	1.78	1.66	1.59	1.55	1.53
15.50	4.49	3.75	3.14	2.67	2.34	2.12	1.99	1.91	1.87	1.84
16.00	4.96	4.20	3.57	3.07	2.72	2.49	2.35	2.27	2.22	2.19
16.50	5.44	4.66	4.01	3.50	3.14	2.89	2.74	2.65	2.60	2.57
17.00	5.92	5.14	4.47	3.95	3.57	3.31	3.15	3.06	3.00	2.97
17.50	6.41	5.62	4.94	4.40	4.01	3.75	3.58	3.48	3.43	3.39
18.00	6.90	6.10	5.42	4.87	4.47	4.20	4.03	3.93	3.87	3.83
18.50	7.39	6.59	5.90	5.35	4.94	4.66	4.49	4.38	4.32	4.29
19.00	7.88	7.08	6.39	5.83	5.42	5.14	4.96	4.85	4.79	4.75
19.50	8.38	7.57	6.88	6.32	5.90	5.62	5.44	5.33	5.26	5.23
20.00	8.88	8.07	7.37	6.80	6.39	6.10	5.92	5.81	5.74	5.71
20.50	9.37	8.56	7.86	7.30	6.88	6.59	6.41	6.29	6.23	6.19
21.00	9.87	9.06	8.36	7.79	7.37	7.08	6.90	6.78	6.72	6.68
21.50	10.37	9.56	8.86	8.29	7.86	7.57	7.39	7.28	7.21	7.17
22.00	10.87	10.06	9.35	8.78	8.36	8.07	7.88	7.77	7.70	7.67
22.50	11.37	10.56	9.85	9.28	8.86	8.56	8.38	8.27	8.20	8.16
23.00	11.87	11.06	10.35	9.78	9.35	9.06	8.88	8.76	8.70	8.66
23.50	12.37	11.56	10.85	10.28	9.85	9.56	9.37	9.26	9.19	9.16

Figure 10.22 Compression depth of two amplifier stages as a function of P_{in} and variable N.

a reduction in P_{sat} of 0.46 dB, and the final amplifier stage is not achieving full saturated power output at 2.1 dB compression causing P_{out} to be 0.14 dB less than the degraded P_{sat} value. Add a third class AB amplifier stage as a pre-driver having P_{sat3} 8.0 dB lower than P_{sat4}. Adding the third class AB amplifier stage to the trade study at this point is valid if consideration is given to the additional reduction in power output that it will cause. Calculate this additional reduction using (9.15)

$$P_{sat5} - P_{max} = 35.14 - 3.9 \log_{10}\left[10^{\frac{-35.14}{3.9}} + 10^{\frac{-(27.14+10)}{3.9}} + 10^{\frac{-(19.14+20)}{3.9}}\right]$$

$$= 0.57 \text{ dB}$$

This is an additional 0.11 dB degradation over the 0.46 dB calculated above for two cascaded stages. If the transmitter power output is to be $P_{out} =$ +33.0 dBm minimum at +50 degrees C for any assembly of randomly selected components, the final amplifier stage saturated power output at +25 degrees C needs to be no less than shown in Table 10.3.

Table 10.3 illustrates all of the factors that need to be considered to determine the value P_{sat5} at +25 degrees C ambient that will satisfy the +33.0 dBm minimum, any transmitter assembly at +50 degrees C.

From Table 10.3 P_{sat5} needs to be +35.13 dBm, P_{sat4} needs to be +27.13 dBm, and P_{sat3} needs to be +19.13 dBm.

How Many Amplifier Stages are Needed?

Now determine how many additional amplifier stages are needed. Use an architecture where the last three class AB amplifier stages are on a single monolithic chip. Insert a binary gain trim attenuator having 2.5 dB insertion loss at

Table 10.3
Factors That Affect Specification of P_{sat5} Value at +25 Degrees C

Parameter	Value
P_{out} Minimum at +50°C	+33.0 dBm
$P_{sat} - 4\sigma$ (Population Minimum)	1.05 dB
P_{sat} Degradation (2 Compressed Stages)	0.57 dB
P_{sat} Degradation Due to Temperature	0.375 dB
($P_{out} - P_{sat}$) ΔdB at 2.1 dB Compression	0.14 dB
P_{sat5} Specification at +25°C	+35.13 dBm

zero dB setting ahead of the three-stage power MMIC. The 32-bit binary attenuator MMIC has 0.5 dB step size and 0 dB to 15.5 dB control range. Absolute minimum amplifier stage small signal gain needed to amplify 0 dBm to +33 dBm with 6.3 dB compression in the last three transmitter power amplifier stages, and a gain trim insertion loss of −2.5 dB is 41.8 dB.

If five 10 dB gain amplifier stages with gain sensitivity to temperature of −0.015 dB/degrees C per stage are used, population mean small signal active gain will be 48.12 dB at +50 degrees C. Total transmitter compressed gain will be 39.3 dB accounting for compression depth and gain trim insertion loss. This is 6.3 dB more gain than needed to produce +33.0 dBm. The gain trim attenuator will be set at 6.5 dB in the average transmitter at +50 degrees C. The goal is to output +33.0 dBm minimum at +50 degrees C even for the worst case transmitter assembly.

What will transmitter yield be at +50 degrees C if this design is used? Determine amplifier assembly population statistics by root-sum-square calculation of standard deviations. Five-stage amplifier assembly standard deviation of gain is calculated as the root-sum-square of standard deviations of independent contributors in Table 10.4. The two-stage class A monolithic circuit has standard deviation of small signal gain of 1.0 dB. Change in gain of the two stages is correlated within each chip. The three-stage power amplifier chip has standard deviation of small signal gain of 1.5 dB. The gain trim attenuator has standard deviation of insertion loss value of 0.25 dB. See Section 8.3.2 and Figure 8.8 for a discussion of small signal gain variation in populations of multistage MMIC amplifiers.

Worst case amplifier small signal gain is mean value -3σ (−5.46 dB) at +50 degrees C. Mean value gain at +50 degrees C will be (50.0 − 0.015 * 5 * 25) = 48.12 dB. Reduce this value by -3σ of the five-stage transmitter amplifier (−5.46 dB), gain trim loss (−2.5 dB), and compression depth needed to obtain maximum *PAE* (6.3 dB) and worst case transmitter compressed gain at +50 degrees C will be +33.86 dB. This says that worst case compressed power

Table 10.4
Five-Stage Transmitter Assembly Standard Deviation of Small Signal Gain

Circuit Element	Standard Deviation of Gain
Class A MMIC	1.0 dB
Gain Trim	0.25 dB
Power MMIC	1.5 dB
Four Stage Transmitter Amplifier G_{ss} 1σ	1.82 dB

output with 0 dBm power input at +50 degrees C will still be 0.8 dB greater than minimum required. Place two 10 dB small signal gain class A amplifier stages ahead of the gain trim attenuator, and three 10 dB small signal gain class AB amplifier stages after the gain trim attenuator as illustrated in Figure 10.23.

The tuned transmitter gain trim setting at worst case +50 degrees C should be 0.5 dB to 1.0 dB. Mean value of gain trim setting at +50 degrees C should be 6.0 dB to 6.5 dB, 3σ above the worst case setting. When ambient temperature goes to 0 degrees C mean value of gain will increase by 3.75 dB relative to the +50 degrees C mean value. Mean gain trim setting should shift up to 10.0 to 10.5 dB. Worst case gain setting at 0 degrees C should be at 3σ above the mean, which is 15.5 to 16.0 dB.

Will Gain Trim Range of 0 dB to 15.5 dB be Sufficient?

Worst case high small signal gain in a population of transmitter assemblies is mean value $+3\sigma$ at 0 degrees C. Five-stage transmitter mean value gain, less gain trim insertion loss, less 6.3 dB compression, $+3\sigma$ at 0 degrees C is $50 - 2.5 - 6.3 + 1.875 + 5.46 = 48.54$ dB. Only 33.0 dB compressed gain is needed to output +33.0 dBm. The excess worst case gain will be controlled by setting the gain trim attenuator at 15.5 dB. It appears that gain trim attenuator range has no margin for error. It should have sufficient range on the lower end and may run up to the full 15.5 dB limit on a few transmitters when cold.

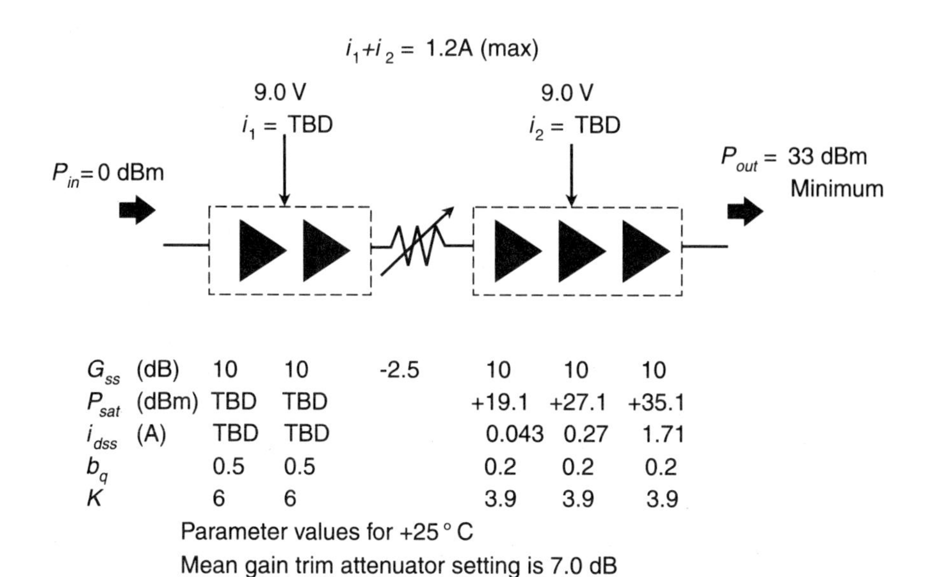

Figure 10.23 Five amplifier stages are used, two class A, three class AB.

The trade-off is the cost of adding greater range to all transmitters to save perhaps 1 percent of the transmitters at hot temperature.

What Saturated Power Output Value Is Needed for the First Two Class A Amplifier Stages?

At this point it is beneficial to develop a spreadsheet model of all five stages of the transmitter. The first two stages operate class A and should not be compressed more than a few tenths of a dB. Use the spreadsheet shown in Figure 10.24 as a model for the transmitter spreadsheet.

The top rows contain data inputs that are used by elements of the model. Cell C9 is the power input entry cell. Power input is copied into cell D9. Row 9 contains power input values in units of dBm for each stage of the transmitter. Power input of each stage is increased or decreased by gain or loss in that stage and stage power output is calculated and posted in row 23. Gain for each stage entered into row 10 is modified by temperature using sensitivity of −0.015 dB/degrees C. Saturated power output entered into row 13 is also modified by temperature sensitivity of −0.015 dB/degrees C. Power output for linear stages like the gain trim is simply the linear addition of power input plus gain or loss. Power output of nonlinear amplifier stages is calculated using (5.1). Compression coefficient K used in the nonlinear calculation is taken from row 22. The value of compression coefficient K is a function of quiescent bias coefficient b_q entered in row 18. Quiescent bias coefficient $b_q = 0.5$ for class A amplifiers where $b_q = 0.2$ for class AB amplifiers. Average bias current $i_{ds}/i_{dss} = 0.5$ for class A amplifiers, and is a variable quantity for class AB amplifiers. Class AB amplifier bias current is a function of quiescent bias coefficient b_q and power input P_{in} as described by (5.15). Values P_1 and P_2 in rows 19 and 20 are used in (5.15) and are a function of quiescent bias coefficient and i_{dss}. Values P_1 and P_2 are determined by (5.18) and (5.19). Transistor saturation current i_{dss} in row 17 is determined by solving (5.3) knowing values for saturated power output, applied DC voltage, and transistor knee voltage. Compression in nonlinear stages in row 24 is determined by subtracting power output from power input plus small signal gain. Compression depth of all stages is summed into cell K24. Gain trim value is manually iterated in cell F15 until total compression depth calculated in cell K24 is at or near 6.3 dB. Transmitter power output is then read in cell K23. Total current consumption is summed in cell K21, and transmitter power-added efficiency is calculated in cell K22.

P_{sat1} and P_{sat2} Values

Enter gain and saturated power output values into the spreadsheet Figure 10.24 for stages 3, 4, and 5 derived from the trade study performed above. Enter a mean value of −7 dB for the gain trim attenuator. Enter gain of 10 dB for

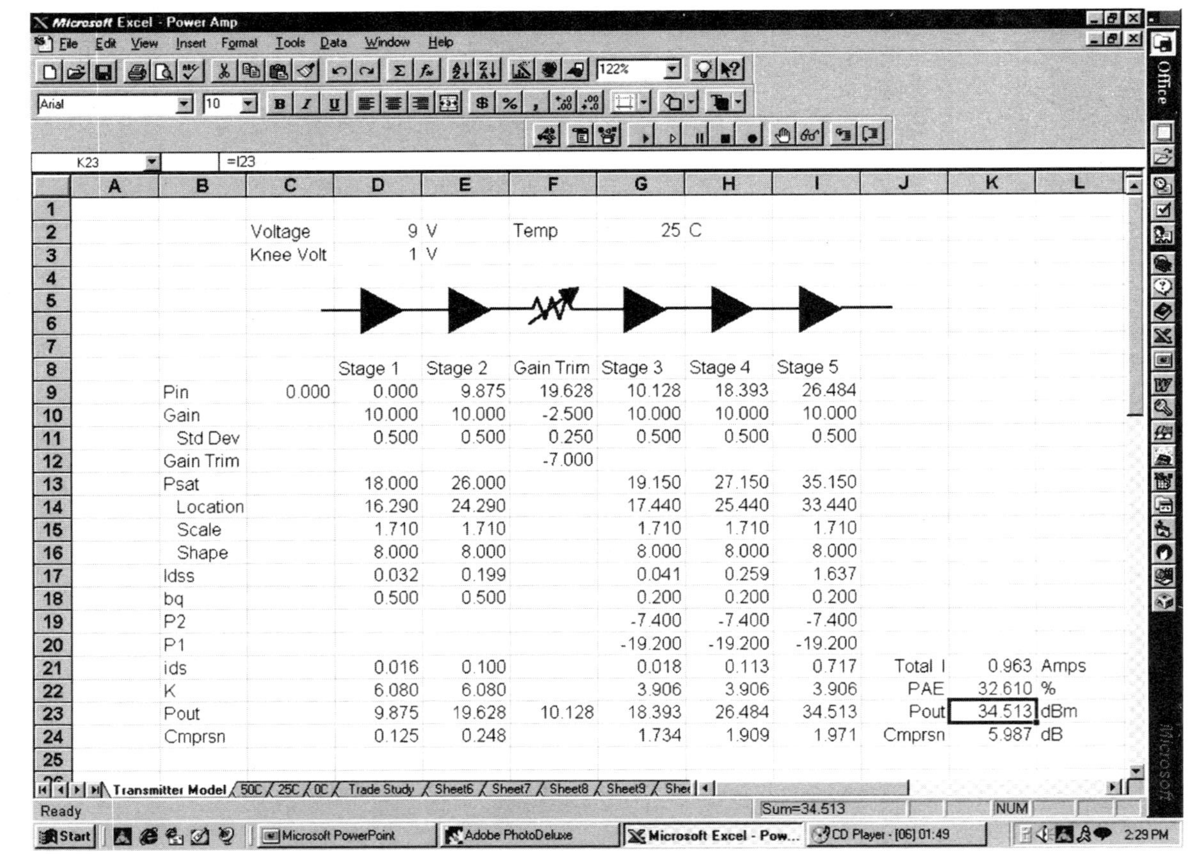

Voltage		9 V	Temp		25 C				
Knee Volt		1 V							
		Stage 1	Stage 2	Gain Trim	Stage 3	Stage 4	Stage 5		
Pin	0.000	0.000	9.875	19.628	10.128	18.393	26.484		
Gain		10.000	10.000	-2.500	10.000	10.000	10.000		
Std Dev		0.500	0.500	0.250	0.500	0.500	0.500		
Gain Trim				-7.000					
Psat		18.000	26.000		19.150	27.150	35.150		
Location		16.290	24.290		17.440	25.440	33.440		
Scale		1.710	1.710		1.710	1.710	1.710		
Shape		8.000	8.000		8.000	8.000	8.000		
Idss		0.032	0.199		0.041	0.259	1.637		
bq		0.500	0.500		0.200	0.200	0.200		
P2					-7.400	-7.400	-7.400		
P1					-19.200	-19.200	-19.200		
ids		0.016	0.100		0.018	0.113	0.717	Total I	0.963 Amps
K		6.080	6.080		3.906	3.906	3.906	PAE	32.610 %
Pout		9.875	19.628	10.128	18.393	26.484	34.513	Pout	34.513 dBm
Cmprsn		0.125	0.248		1.734	1.909	1.971	Cmprsn	5.987 dB

Figure 10.24 Develop a spreadsheet model of the transmitter amplifier assembly.

stages 1 and 2. Iterate values for saturated power output into P_{sat1} and P_{sat2}. A good ratio of P_{sat2} to P_{sat1} is the same 8 dB used in the class AB amplifiers. Arrive at acceptable values P_{sat2} = +26 dBm and P_{sat1} = 18 dBm. Amplifier stage 2 is compressed 0.25 dB, and stage 1 is compressed 0.12 dB. Total current is 0.96 amperes. Power-added efficiency is 32.6 percent. Total compression is 6.47 dB, and power output is +34.5 dBm at +25 degrees C. Parameter values for all amplifier stages are now known and are tabulated in Table 10.5.

Statistical behavior of the transmitter at temperature extremes is easily determined by applying the CRYSTAL BALL spreadsheet add-in application.

Modify the Spreadsheet for Statistical Analysis

Add standard deviation data to the spreadsheet as shown in Figure 10.24. Amplifier stages in the class A two-stage monolithic circuit experience the same gain variation so standard deviation for that MMIC is 1.0 dB. Likewise standard deviation of small signal gain for the three-stage class AB power amplifier MMIC is 1.5 dB. Small signal gain probability density function is Gaussian (see Section 8.3.1). Standard deviation of saturated power output for all stages is $1\sigma = 0.25$ dB. Standard deviation of saturated power output is independent from stage to stage. Saturated power output standard deviation of 0.25 dB relates to a Weibull function scale factor of 1.71 dB, a location 1.68 dB less than saturated power output population mean, and a shape factor of 8 (see Section 8.3.4).

Multiple transmitter models are set up in the spreadsheet. The first model accepts CRYSTAL BALL generated values of gain and saturated power output and calculates untuned transmitter parameters where the gain trim attenuator

Table 10.5
Transmitter Amplifier Stage Parameters at Ambient Temperature +25 Degrees C

Parameter	Stage 1	Stage 2	Gain Trim	Stage 3	Stage 4	Stage 5
G_{ss}	10 dB	10 dB	−2.5 dB	10 dB	10 dB	10 dB
Std Dev	0.5 dB	0.5 dB	0.25 dB	0.5 dB	0.5 dB	0.5 dB
Trim Range	—	—	15.5 dB	—	—	—
P_{sat}	18 dBm	26 dBm	—	19.15 dBm	27.15 dBm	35.15 dBm
Std Dev	0.25 dB	0.25 dB	—	0.25 dB	0.25 dB	0.25 dB
i_{dss}	0.031 A	0.2 A	—	0.041 A	0.27 A	1.71 A
b_q	0.5	0.5	—	0.2	0.2	0.2
K	6	6	—	3.9	3.9	3.9

value is set to 0 dB. Untuned compression depth is calculated and 6.3 dB subtracted from the value. Equation (10.1) is used to determine the binary attenuator setting value, and the gain trim setting is entered into a second transmitter model. The second transmitter model reduces drive power to the three final stages by the amount of gain trim applied. Compression depth then should be 6.3 dB mean value, and power output should be greater than +33 dBm. Results from building 2000 transmitters with CRYSTAL BALL are shown in the following figures.

Attenuator setting distributions used to tune the 2000 transmitters at temperatures of 0 degrees C, +25 degrees C, and +50 degrees C are shown in Figure 10.25. Mean attenuator setting ranges from 8.49 dB at 0 degrees C to 5.71 dB at +50 degrees C. None of the transmitters needed the full 15.5 dB attenuation; at least one transmitter used 15.0 dB. Three transmitters had insufficient gain at +50 degrees C to achieve the desired 6.3 dB compression depth and their attenuator settings were 0 dB as shown in Figure 10.25.

Mean value of power output ranges from +34.05 dBm at +50 degrees C to +35.06 dBm at 0 degrees C for 2000 transmitters. The worst case for power output is at +50 degrees C where 0.1 percent of the population (2 units) failed to output more than +33 dBm. Even so their power output is a respectable +32.7 dBm and +32.8 dBm. The low power output could have been due to insufficient compression or simply very low saturated power output value. The transmitter population standard deviation of power output is 0.25 dB, the same as standard deviation of saturated power output for the final amplifier stage. Figure 10.26 shows power output distribution of the 2000 transmitters at room temperature and at the temperature extremes.

Mean value of compression depth at all three temperatures is 6.35 dB. Compression depth range is ±0.25 dB as expected from the binary attenuator with 0.5 dB step size. Figure 10.27 shows compression depth distribution at room temperature and at temperature extremes. Distribution of compression depth is relatively uniform across the population of 2000 transmitters.

Total current consumption is less than the 1.2 amperes required at any temperature over the entire population of 2000 transmitters. Population mean value of current ranges from 0.9 amperes at +50 degrees C to 1.06 amperes at 0 degrees C. At least one transmitter required 1.17 amperes at 0 degrees C. Figure 10.28 shows distribution of current consumption in the population of 2000 transmitters at room temperature and at temperature extremes.

Population mean value of power-added efficiency ranged from 31.48 percent at +50 degrees C to 33.57 percent at 0 degrees C. One transmitter had *PAE* as low as 24 percent at +50 degrees C. Figure 10.29 shows distribution of *PAE* for the population of 2000 transmitters at room temperature and at temperature extremes.

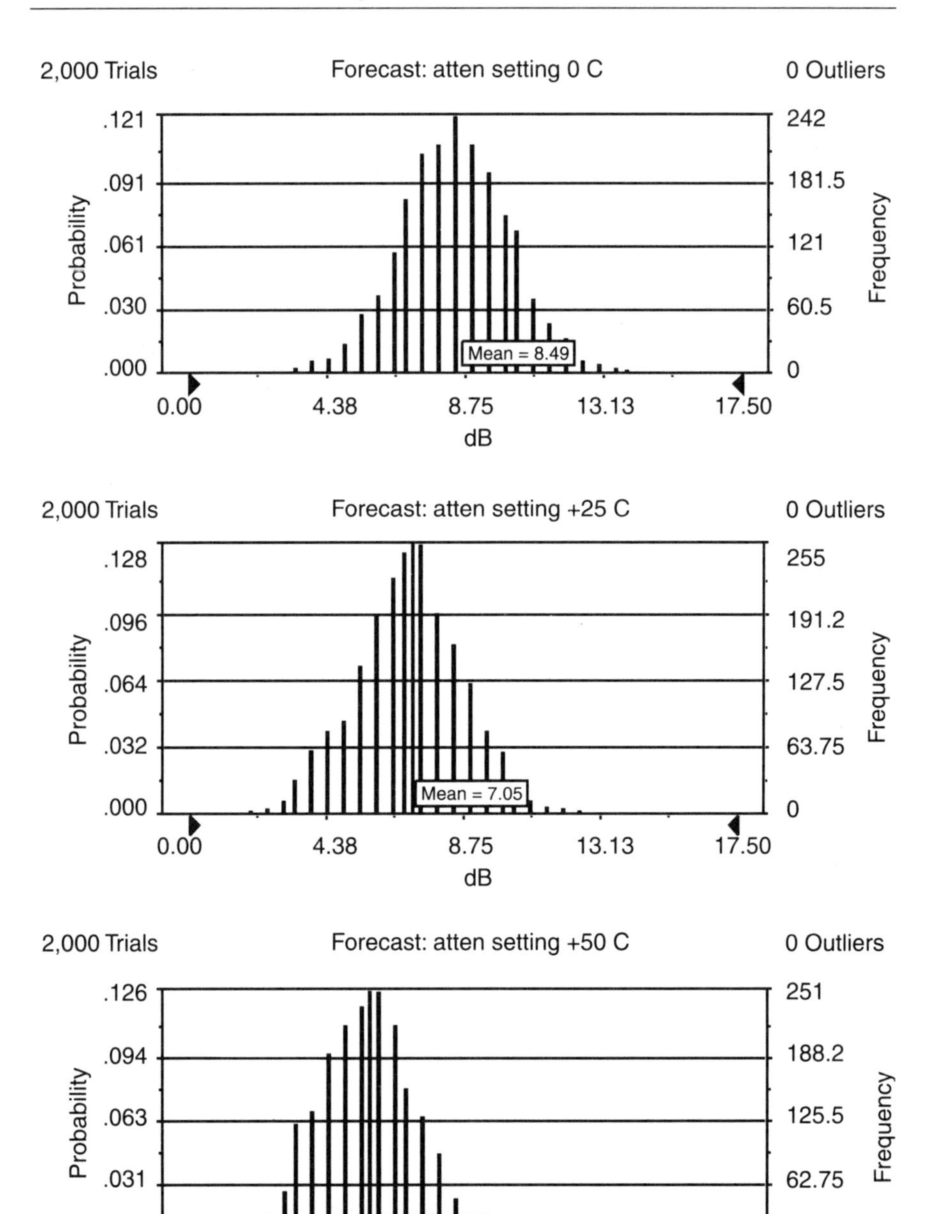

Figure 10.25 Gain trim attenuator setting distributions for 2000 transmitters at temperature extremes.

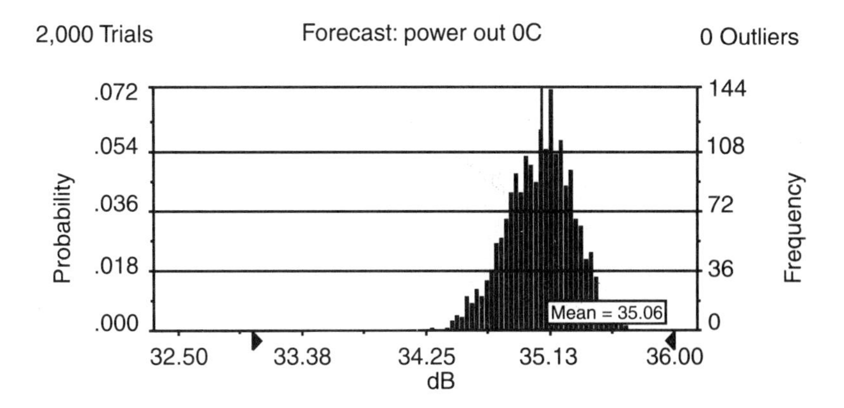

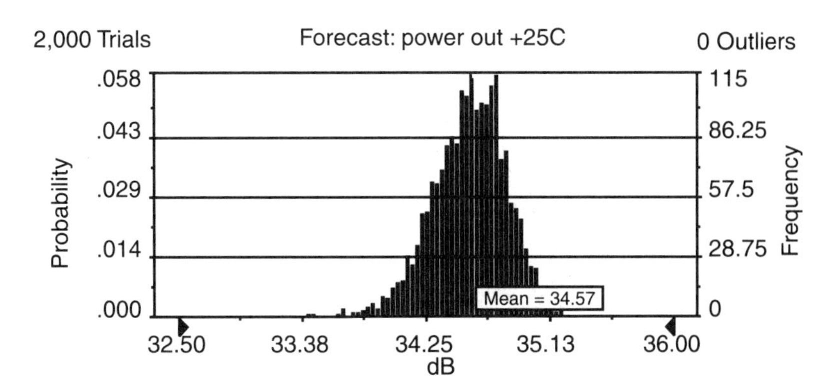

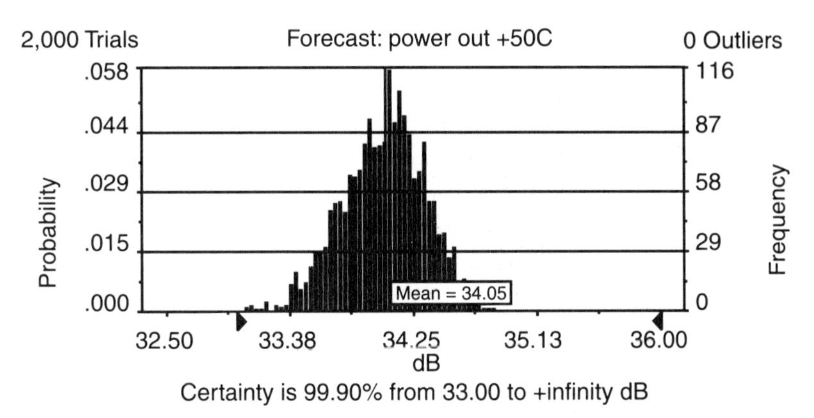

Figure 10.26 Power output distribution in 2000 transmitters at temperature extremes.

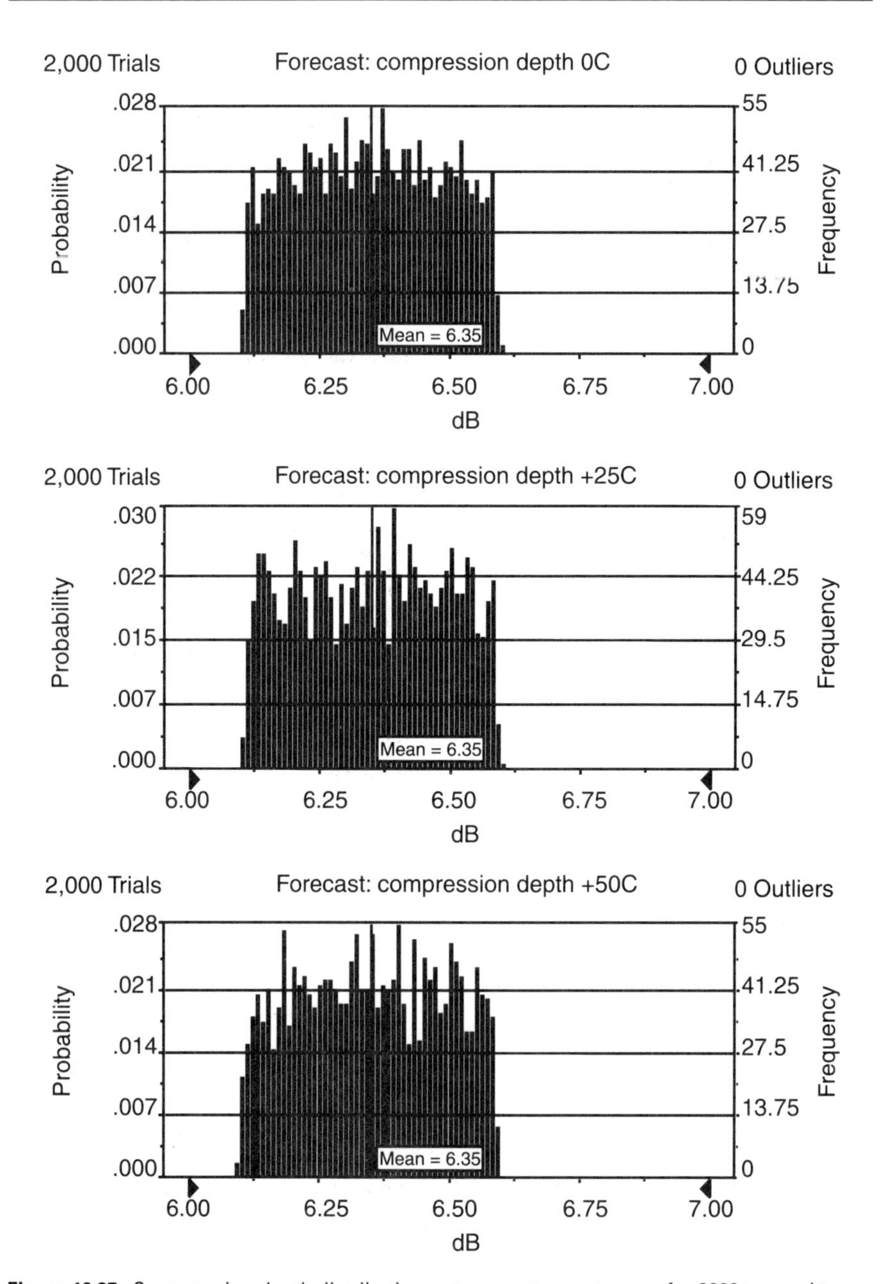

Figure 10.27 Compression depth distribution at temperature extremes for 2000 transmitters.

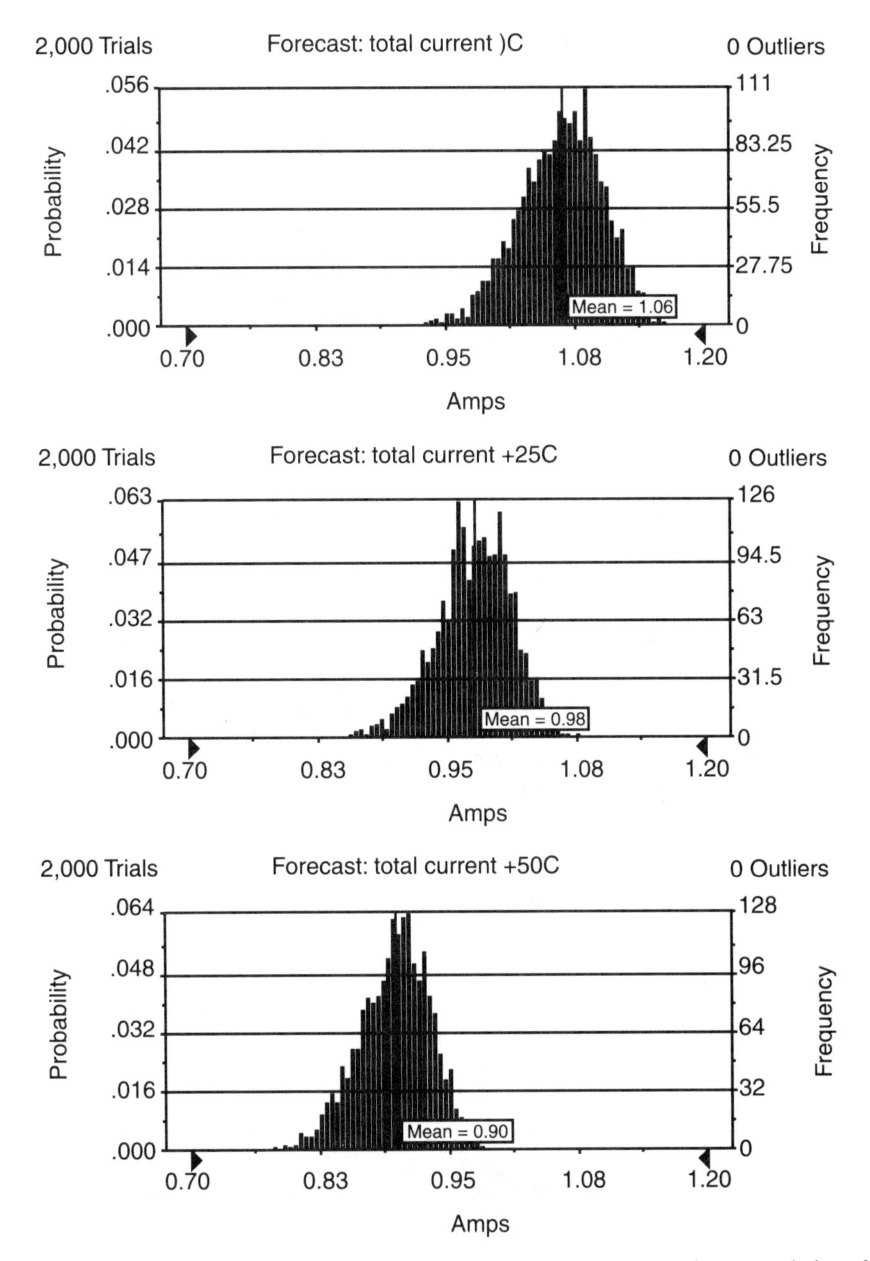

Figure 10.28 Current consumption distribution at temperature extremes for a population of 2000 transmitters.

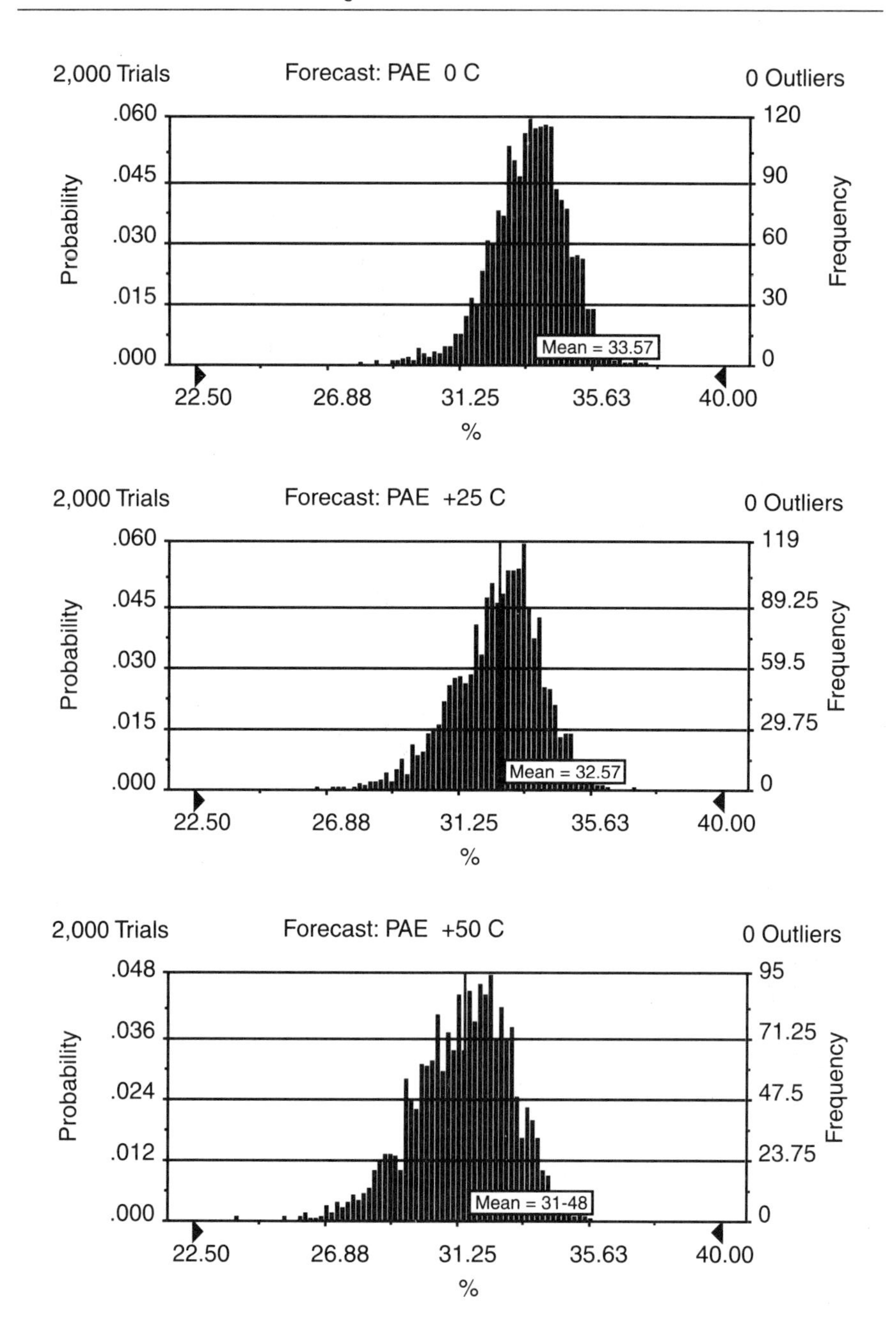

Figure 10.29 Power-added efficiency distribution at temperature extremes for a population of 2000 transmitters.

10.3 Summary

Models and methods developed throughout this book are applied in this chapter as a system design aid and a system analysis tool. The chapter consists of two examples. A receiver system is defined and methods to arrive at total small signal gain that satisfies system needs is developed. Trade space techniques for obtaining required receiver noise figure are described. Spreadsheet tools that lead to optimum input third-order intercept for a required receiver noise figure are explained and illustrated. Risk analysis software tools are applied to the receiver spreadsheet, and statistical results are obtained at room temperature and at temperature extremes.

The second example shows how system requirements for a transmitter power amplifier are flowed down to circuit level and to device level through the application of trade spaces that lead to optimum selection of saturated power output for each amplifier stage. A spreadsheet model of the transmitter is illustrated and risk analysis software is applied to generate statistical data for 2000 transmitters at room temperature and at temperature extremes.

10.4 Problem

Problem 10.1

Use data supplied in Example 10.2 and determine the performance obtained if the three class AB stages are biased with $b_q = 0.3$.

11

Odds and Ends

11.1 Odds and Ends

A new curve-fit technique has been described in this book that produces closed-form equations which do not generate excess harmonic content after Fourier transformation. A new method of modeling nonlinear circuit elements has been presented in this book based on the new curve-fit techniques. A new way of characterizing the behavior of devices, amplifiers, and circuit assemblies has been developed. Trade spaces have been developed for the purpose of guiding the designer and systems architect to optimum parameter relationships in circuits, circuit assemblies, and systems.

These new approaches to understanding circuit and subassembly behavior have been expanded into the frequency domain, with temperature variation, and with the uncertainty of stochastic variation of parameter values due to large volume production process variability. Examples of applications that did not seem to fit in the previous chapters are briefly discussed here primarily for the purpose to plant seeds of thought, to jog the imagination, and to lead to future applications that have yet to be considered. Consider what could be done with an S-parameter file that is a set of equations instead of a set of numbers. Complex variable S-parameter data files as a function of frequency typically represent amplifier or transistor behavior at a single power input, a single bias current, a single temperature, a specific variable gain control setting, and a single phase control setting. Great utility is added to the S-parameter file if it can be made to represent the object's behavior over frequency as a function of one or more of the above-mentioned variables. Imagine having a dynamic transistor S-parameter file that responds to change in bias current setting and could respond nonlinearly to change in input power. Circuit design software could be made to respond to dynamic change in variables that affect S-parameters.

Consider too the utility of having a single closed-form equation that represents a single impulse in time or a burst of several sine wave cycles that occur once in all time. Functions like these are easy to create with the techniques developed in this book. Also consider the advantage of being able to model junction capacitance as a function of junction voltage from the reverse bias region into the forward bias region, accurately reproducing measured capacitance peaking and roll-over with a single equation. These models are discussed in this chapter.

11.2 Modeling S-Parameters as a Function of Bias Current, RF Power, and Control Functions Over Frequency

An S-parameter dataset consists of four complex parameters, S11, S12, S21, and S22. Each parameter is a complex value consisting of a real and imaginary component or a magnitude and angle. An S-parameter file consists of hundreds, even thousands, of data sets. Each data set usually represents the S-parameters of a device at a specific frequency. An entire file might represent the behavior of a device at a specific bias current, a specific power level, a specific control voltage setting, or a specific temperature. Several files, perhaps tens of files, are required to represent the behavior of a device over a range of bias current. Another group of files are needed to represent the device behavior at different power levels, and yet another group of files to represent behavior over a range of temperatures. Huge lookup tables are needed to capture the device behavior as a function of all of the variables. Interpolation is required in order to arrive at values that fall between the recorded dataset values. Behavioral modeling techniques described thus far are very effective in converting device S-parameter files into sets of equations that account for multidimensioned, orthogonal variables, all included in one model. Furthermore, S-parameter values often trace out curves that have sharp discontinuities that polynomial expansions and other curve-fitting techniques fail to model correctly, or fail to model with a closed-form equation.

Any single-valued variable can be modeled with asymptotes, RHFs and LHFs and converted to a curve-fit equation having the form

$$y(x) = A + Bx + \sum_{1}^{i} C_{\mathrm{i}} * \log_{10}\left[\frac{10^{\frac{x-x_{\mathrm{i}}}{c_{\mathrm{i}}}}}{1 + 10^{\frac{x-x_{\mathrm{i}}}{c_{\mathrm{i}}}}}\right] + \sum_{1}^{k} D_{\mathrm{k}} * \log_{10}\left[1 + 10^{\frac{x-x_{\mathrm{k}}}{d_{\mathrm{k}}}}\right] \quad (11.1)$$

Equation (11.1) is used to model a specific parameter value ($|S11|$, $|S21|$, $|S12|$, $|S22|$, $/\underline{S11}$, $/\underline{S21}$, $/\underline{S12}$, $/\underline{S22}$) over frequency for a specific value of current, control voltage, power, or temperature. A second equation is generated for the same parameter at the second specific value of current, control voltage, power, or temperature, taking care to use the same number of asymptotes, RHFs and LHFs. A third equation is developed for the third specific value of current, control voltage, power, or temperature, again matching the number of asymptotes, RHFs and LHFs, and so on for as many equations as needed to characterize the device. This set of curve-fit equations is called the "primary set." Corresponding coefficients in the primary set of equations (A_i, B_i, C_i, D_i, x_i, c_i, and d_i) that represent behavior as a function of the primary variable (current, control voltage, power, or temperature) are then tabulated as a function of the primary variable values. Each coefficient (A_i, B_i, C_i, D_i, x_i, c_i, and d_i) is then modeled by an equation as a function of the primary variable. Each of these coefficient equations called the "secondary set" are then substituted back into the primary equation form (11.1) to create a single primary equation that represents behavior of the S-parameter variable over frequency and an additional variable (current, temperature, voltage, power input, and so forth). The secondary set of curve-fit equations needed to describe the variation of coefficient values can be either linear functions, polynomial expansions or another set of RHF-LHF curve-fit equations. Any common curve-fit technique described in Chapter 1 can be used to model a secondary equation. Procedures for developing polynomial equations have been outlined in examples in previous chapters of this book. Procedures for developing RHF-LHF functions are also described. The result of the process is a single, equation that represents an S-parameter value over frequency that includes the effect of one independent variable. This equation can then be used in CAD applications to provide dynamic control of circuits without having to rely on huge lookup tables and interpolation techniques. Additional variables can be added to the primary equation by expanding the process described above. Example 11.1 illustrates the modeling process described above and shows the differences that can result if bad choices are made in the curve-fit technique used to model the secondary set of coefficient variables.

Example 11.1 Modeling S-Parameters as a Function of Frequency and Bias Current

Consider a typical surface mount low-noise self-biased transistor amplifier such as the Hewlett Packard INA-12063. Published data for the device includes S-parameters as a function of frequency at bias current levels of 1.5mA, 2.5 mA,

5 mA, and 8 mA. The technique applied here can be used to develop equations for all of the S-parameters. For the sake of brevity, focus only on magnitude and angle of S21. Table 11.1 shows |S21| magnitude as a function of frequency and current, and Table 11.2 shows corresponding angle values.

The current values in these two tables are not equally spaced. Linear interpolation is used to develop a spreadsheet file that adds current values to cover the range 1.0 mA to 8.5 mA for frequencies 0.1 GHz to 3.0 GHz. The linear interpolation process produces a set of data that cannot be represented by a single equation but does give a reference file to which to compare the final curve-fit model. Results of the linear interpolation file for |S21| are plotted as a surface in Figure 11.1. Results of linear interpolation of /S21 phase angle data as shown in Table 11.2 are illustrated in Figure 11.2.

Develop a Model Equation for |S21|

A model for |S21| magnitude is developed first. There are two ways to approach the modeling of data in Table 11.1. Magnitude |S21| can be plotted as a function of frequency for each of the four current values given in Table 11.1. The second approach is to select equally spaced frequencies 0.5, 1.0, 1.5, 2.0, 2.5, and 3.0, GHz and plot |S21| as a function of current at the selected frequencies. This example uses the former approach of plotting |S21| as a function of frequency for the given current values. The resulting plots are shown in Figure 11.3. These are the primary curves for which the primary curve-fit equation set will be developed. Each curve will be modeled with four asymptotes. If the Curve Fit Utility software included with this book is used to develop equations

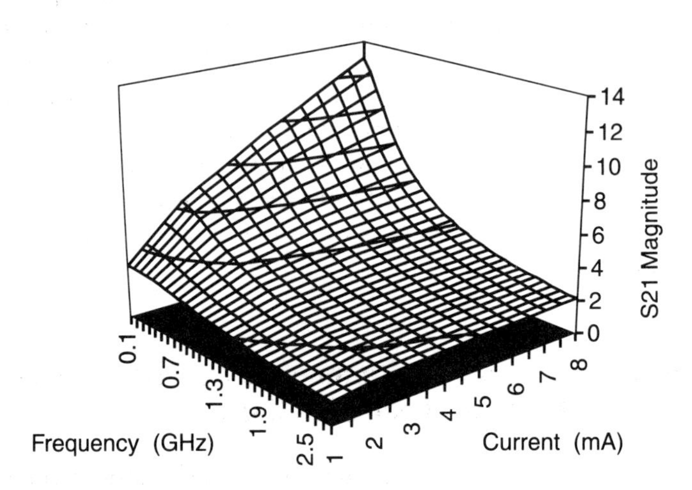

Figure 11.1 The INA-12063 |S21| magnitude linear interpolation file is plotted as a surface for reference.

Table 11.1
INA-12063 |S21| Values

Frequency (GHz)	1.5 mA	2.5 mA	5 mA	8 mA
0.1	4.26	6.33	9.56	12.97
0.2	4.2	6.19	9.18	12.17
0.3	4.11	5.98	8.65	11.1
0.4	4	5.74	8.07	10.06
0.5	3.83	5.37	7.34	8.89
0.6	3.69	5.07	6.75	8
0.7	3.49	4.73	6.18	7.2
0.8	3.32	4.43	5.68	6.56
0.9	3.18	4.18	5.26	6.02
1	3.03	3.93	4.88	5.53
1.1	2.89	3.71	4.55	5.08
1.2	2.72	3.47	4.24	4.72
1.3	2.64	3.31	3.99	4.4
1.4	2.54	3.15	3.76	4.18
1.5	2.43	2.98	3.56	3.86
1.6	2.33	2.84	3.37	3.71
1.7	2.23	2.72	3.21	3.48
1.8	2.15	2.6	3.05	3.34
1.9	2.08	2.51	2.93	3.14
2	1.99	2.4	2.81	3.01
2.1	1.93	2.31	2.67	2.89
2.2	1.83	2.2	2.55	2.77
2.3	1.82	2.15	2.47	2.68
2.4	1.72	2.05	2.37	2.59
2.5	1.7	2.01	2.31	2.51
2.6	1.65	1.95	2.24	2.42
2.7	1.6	1.89	2.17	2.32
2.8	1.54	1.81	2.08	2.25
2.9	1.5	1.75	2.02	2.17
3	1.49	1.75	2	2.18

Table 11.2
INA-12063 /S21 Values (Deg)

Frequency (GHz)	1.5 mA	2.5 mA	5 mA	8 mA
0.1	172	171	168	166
0.2	164	161	157	152
0.3	157	153	146	141
0.4	149	144	137	130
0.5	141	135	128	121
0.6	135	128	120	114
0.7	128	122	114	107
0.8	122	116	108	102
0.9	116	110	103	97
1	111	104	97	92
1.1	106	100	93	88
1.2	102	95	89	84
1.3	97	91	85	81
1.4	92	87	81	77
1.5	88	83	78	74
1.6	84	79	75	71
1.7	80	76	71	68
1.8	77	72	68	65
1.9	73	69	64	62
2	69	66	61	59
2.1	66	62	58	56
2.2	63	59	56	53
2.3	59	56	53	51
2.4	57	54	51	48
2.5	54	51	48	47
2.6	50	48	45	43
2.7	47	45	42	40
2.8	44	42	39	37
2.9	41	39	37	36
3	39	37	35	34

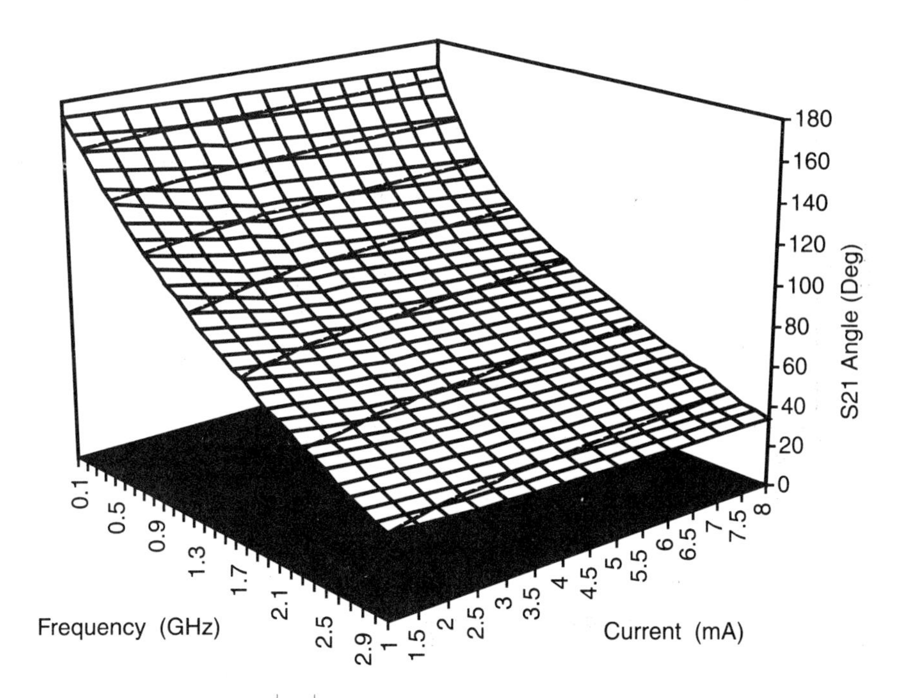

Figure 11.2 The INA-12063 |S21| phase angle linear interpolation file is plotted as a surface for reference.

for these plots the second asymptote **A2** will be selected as the reference. A reference line equation, one LHF and two RHFs will be used to describe each curve. This results in a primary set of equations that have elements

$$|S21|(f, i) = A(i) + B(i)f + C(i)\log_{10}\left[\frac{10^{\frac{f-f_1}{a_1}}}{1 + 10^{\frac{f-f_1}{a_1}}}\right]$$

$$+ D(i)\log_{10}\left[1 + 10^{\frac{f-f_2}{a_2}}\right] + E(i)\log_{10}\left[1 + 10^{\frac{f-f_3}{a_3}}\right] \quad (11.2)$$

An option is to define the left most asymptote **A1** as the reference and use three RHFs to fit the curve. This has the advantage of reducing the number keystrokes needed to enter a secondary equation (LHFs use more keystrokes than RHFs) but does not change the number of terms or coefficients needed. A method used to reduce the number of coefficients and the number of secondary equations needed is to assign constant values for transition range coefficients a_1, a_2, and a_3. This introduces small data tracking errors in transition regions.

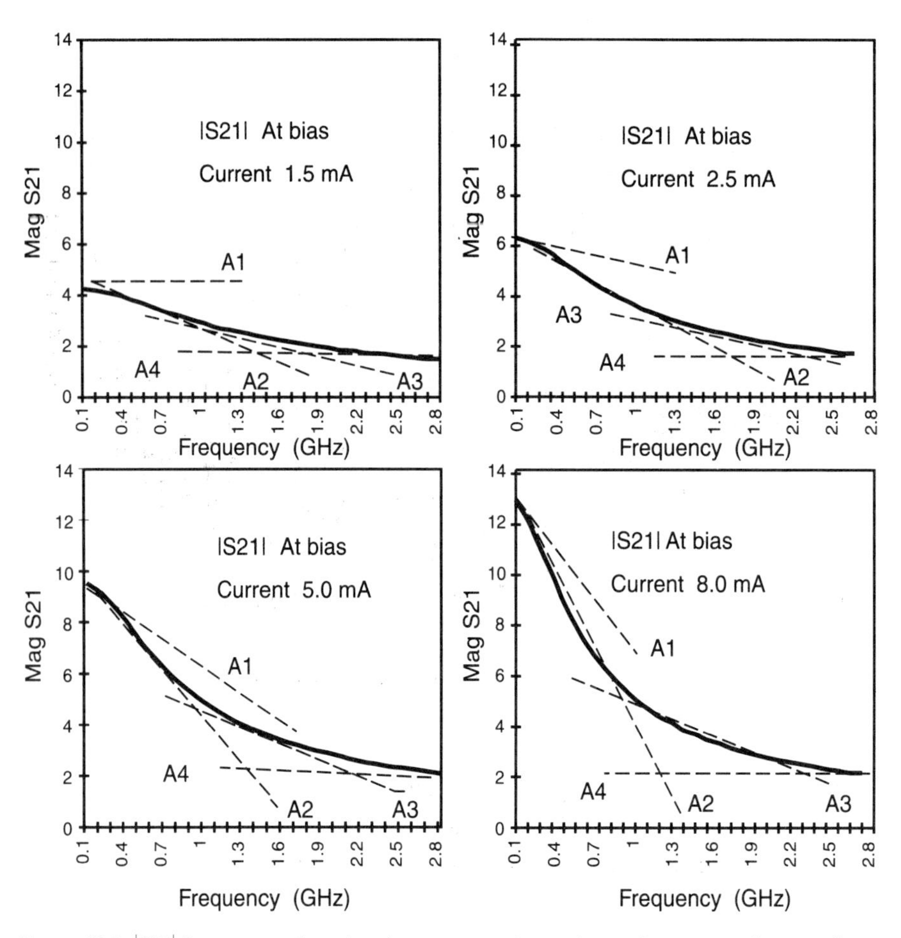

Figure 11.3 $|S21|$ Curves are plotted and asymptotes drawn for each current value as a function of frequency.

Curve-fit equations derived for the four current values are

$i = 1.5$ mA

$$|S21|(1.5) = 4.761 - 1.737f + 0.261 \log_{10}\left[\frac{10^{\frac{f-0.358}{0.2}}}{10^{\frac{f-0.358}{0.2}}}\right]$$

$$+ 0.659 \log_{10}\left[1 + 10^{\frac{f-1.41}{0.6}}\right] + 0.127 \log_{10}\left[1 + 10^{\frac{f-2.68}{0.2}}\right]$$

$i = 2.5$ mA

$$|S21|(2.5) = 7.25 - 3.54f + 0.41 \log_{10}\left[\frac{10^{\frac{f-0.383}{0.2}}}{1 + 10^{\frac{-0.383}{0.2}}}\right]$$

$$+ 1.49 \log_{10}\left[1 + 10^{\frac{f-1.10}{0.6}}\right] + 0.21 \log_{10}\left[1 + 10^{\frac{f-2.57}{0.2}}\right]$$

$i = 5.0$ mA

$$|S21|(5.0) = 10.41 - 6.08f + 0.514 \log_{10}\left[\frac{10^{\frac{f-0.217}{0.2}}}{1 + 10^{\frac{0.217}{0.2}}}\right]$$

$$+ 3.07 \log_{10}\left[1 + 10^{\frac{f-1.11}{0.6}}\right] + 0.193 \log_{10}\left[1 + 10^{\frac{f-2.79}{0.2}}\right]$$

$i = 8.0$ mA

$$|S21|(8.0) = 14.36 - 10.64f + 0.759 \log_{10}\left[\frac{10^{\frac{f-0.179}{0.2}}}{1 + 10^{\frac{-0.179}{0.2}}}\right]$$

$$+ 5.68 \log_{10}\left[1 + 10^{\frac{f-0.956}{0.6}}\right] + 0.234 \log_{10}\left[1 + 10^{\frac{f-2.66}{0.2}}\right]$$

Solutions for these equations as a function of frequency are compared in Figure 11.4 to plotted data from Table 11.1.

The coefficients $A(i)$, $B(i)$, $C(i)$, $D(i)$, $E(i)$, $f_1(i)$ $f_2(i)$ and $f_3(i)$ are now known. Curve-fit equations for the value of these coefficients as a function of current are the secondary set of equations. Table 11.3 organizes the primary equation coefficient values.

Each of the coefficients $A(i)$, $B(i)$, $C(i)$, $D(i)$, $E(i)$, $f_1(i)$ $f_2(i)$ and $f_3(i)$ in Table 11.3 can be modeled as a function of current using polynomial expansions or by using another set of RHF-LHF equations. Some of the coefficients are well behaved and can be approximated nicely with polynomial expressions. Others vary radically and cannot be modeled accurately with polynomial expansions. To illustrate the errors that develop, both approaches, polynomial expansion, and RHF-LHF models will be used in this example.

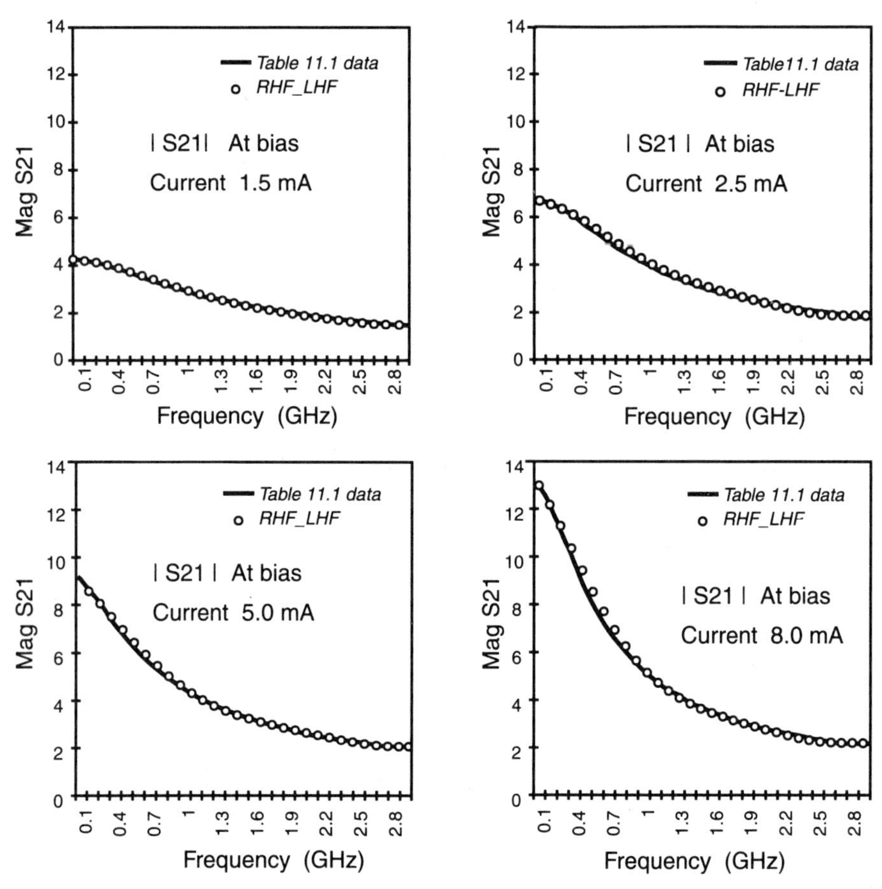

Figure 11.4 Curve-fit equation solutions are compared to original data.

Table 11.3
Primary Coefficients Are Tabulated as a Function of Current

Current	$A(i)$	$B(i)$	$C(i)$	$D(i)$	$E(i)$	$f_1(i)$	$f_2(i)$	$f_3(i)$
1.5	4.76	−1.74	0.261	0.659	0.127	0.358	1.41	2.68
2.5	7.25	−3.54	0.41	1.49	0.21	0.383	1.1	2.57
5.0	10.41	−6.08	0.514	3.07	0.193	0.217	1.11	2.79
8.0	14.36	−10.64	0.759	5.68	0.234	0.179	0.956	2.66

Primary Equation Coefficients Modeled by Polynomial Expansion

All of the polynomial expansions are performed about the bias current value of 4.0 mA. Equations for the first four coefficients are

$$\text{Coefficient } A(i) = 9.49 + 1.058(i - 4.0) - 0.1987(i - 4.0)^2 + 0.0596(i - 4.0)^3$$

$$\text{Coefficient } B(i) = -5.221 - 0.8956(i - 4.0) + 0.0787(i - 4.0)^2 - 0.0483(i - 4.0)^3$$

$$\text{Coefficient } C(i) = 0.496 + 0.0248(i - 4.0) - 0.0131(i - 4.0)^2 + 0.00583(i - 4.0)^3$$

$$\text{Coefficient } D(i) = 2.468 + 0.5999(i - 4.0) - 0.01089(i - 4.0)^2 + 0.0154(i - 4.0)^3$$

Figure 11.5 shows the first four coefficients $A(i)$, $B(i)$, $C(i)$, and $D(i)$, plotted with the polynomial expansions that best fit the curves.

Notice that the polynomial expansions for $A(i)$, $B(i)$, and $C(i)$, give reasonably good approximation to the curve, but the expansion for $D(i)$ has significant error. The effects of this and other similar errors will be noticed immediately when the final primary equation is plotted as a surface and compared to the linearly interpolated surface in Figure 11.1.

Coefficients $E(i)$, $f_1(i)$ $f_2(i)$ and $f_3(i)$ are plotted from data in Table 11.3 and illustrated in Figure 11.6. Also shown in the curves of Figure 11.6 are plots of the polynomial expansions derived for the secondary coefficient equation set.

The polynomial expansion approximations for coefficients $E(i)$, $f_1(i)$ $f_2(i)$ and $f_3(i)$ generate even greater error. Polynomial equations developed for coefficients $E(i)$, $f_1(i)$ $f_2(i)$ and $f_3(i)$ are

$$\text{Coefficient } E(i) = 0.2212 - 0.0206(i - 4.0) - 0.01199(i - 4.0)^2 + 0.00449(i - 4.0)^3$$

$$\text{Coefficient } f_1(i) = 1.036 + 0.05147(i - 4.0) + 0.04229(i - 4.0)^2 - 0.015(i - 4.0)^3$$

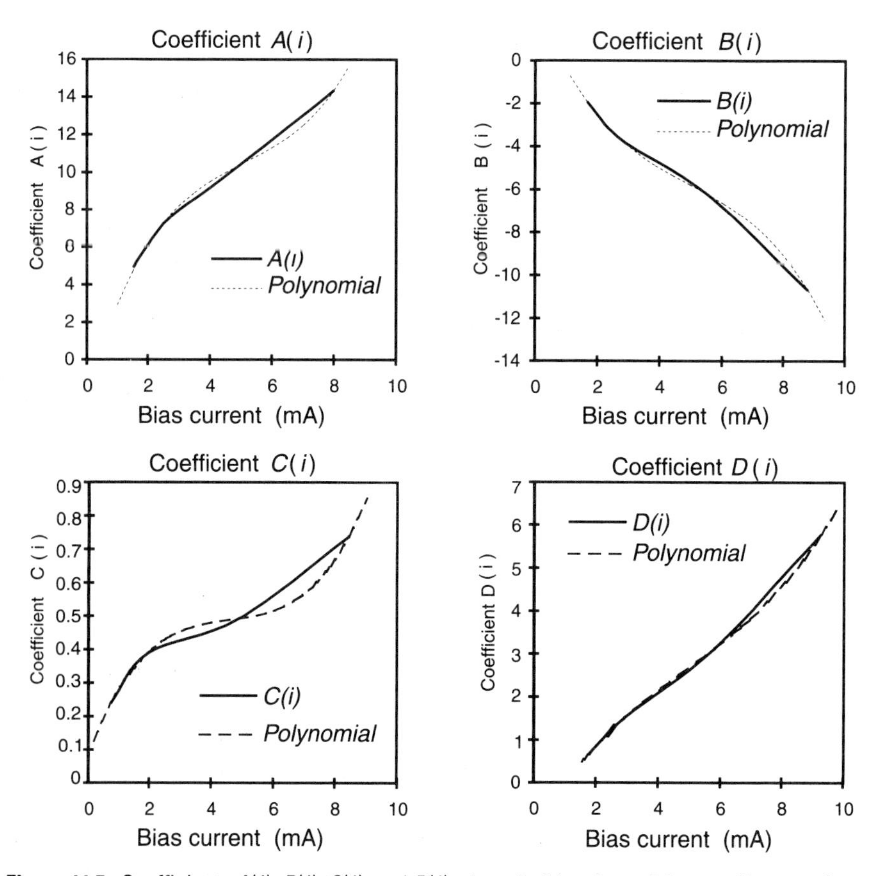

Figure 11.5 Coefficients $A(i)$, $B(i)$, $C(i)$, and $D(i)$, plotted with polynomial curve-fit expansions.

$$\text{Coefficient } f_2(i) = 0.3018 - 0.0808(i - 4.0) - 0.00955(i - 4.0)^2 + 0.00552(i - 4.0)^3$$

$$\text{Coefficient } f_3(i) = 2.6185 + 0.1458(i - 4.0) + 0.0467(i - 4.0)^2 - 0.0201(i - 4.0)^3$$

These secondary equations are substituted back into the primary equation (11.2) and data is generated for bias currents ranging from 1mA to 8.5 mA over the frequency range 0.1 GHz to 3.0 GHz. When that data is plotted as a surface the result shown in Figure 11.7 is obtained. Polynomial expansion errors are clearly evident in the surface plotted in Figure 11.7.

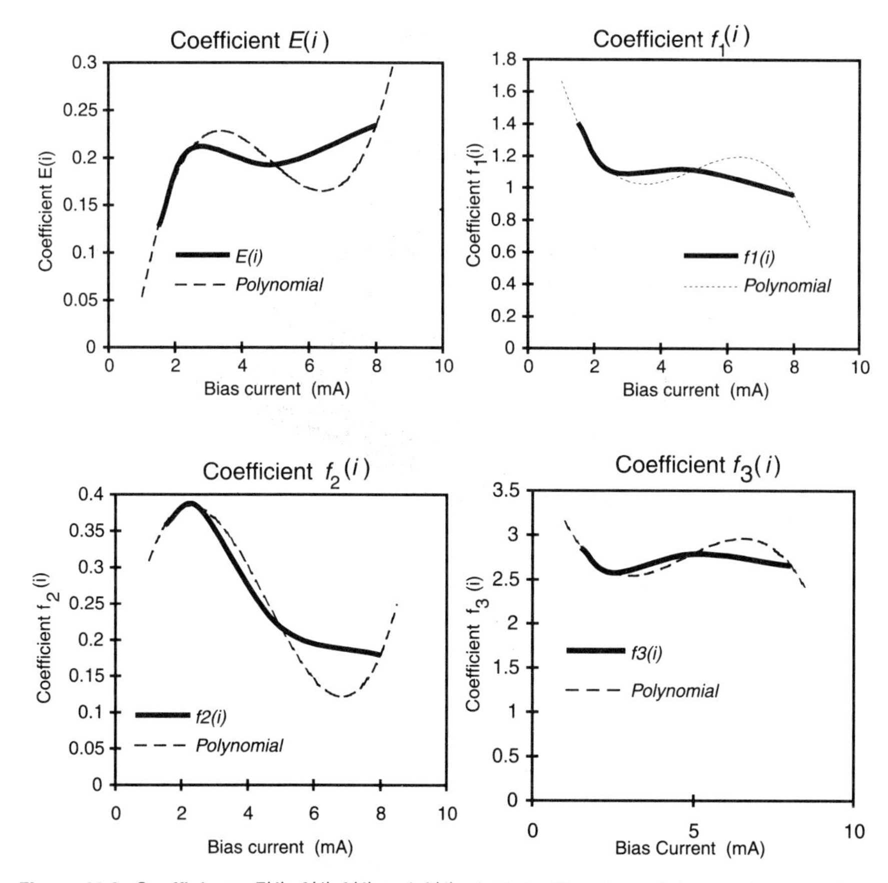

Figure 11.6 Coefficients $E(i)$, $f_1(i)$ $f_2(i)$ and $f_3(i)$, plotted with polynomial curve-fit expansions.

The linearly interpolated surface (Figure 11.1) is used as a reference, and its values are subtracted point by point from the surface shown in Figure 11.7. The result is an error surface showing the magnitude of error caused by polynomial expansion approximation of coefficient secondary equations. The error surface is plotted in Figure 11.8.

The polynomial expansion secondary equations have little or no error compared to the linearly interpolated surface as a function of frequency at bias current values of 1.5 mA, 2.5 mA, 5.0 mA, and 8.0 mA. However, percent error becomes as large as 50 percent at bias current values between the data points given in Table 11.1. The data points of Table 11.1 reveal nothing of how $|S21|$ behaves in the regions in between. It is known that polynomial expansions generate error when the function being fit is not well behaved. How much of the percent error between the linearly interpolated surface and the polynomial expansion secondary equation surface is real and how much is polynomial

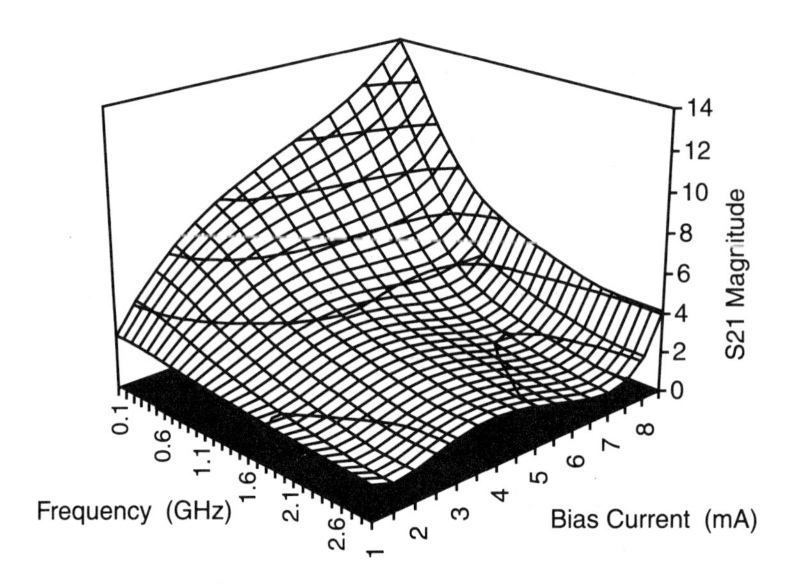

Figure 11.7 The model of |S21| generated by using polynomial expansions for secondary equations is plotted as a surface.

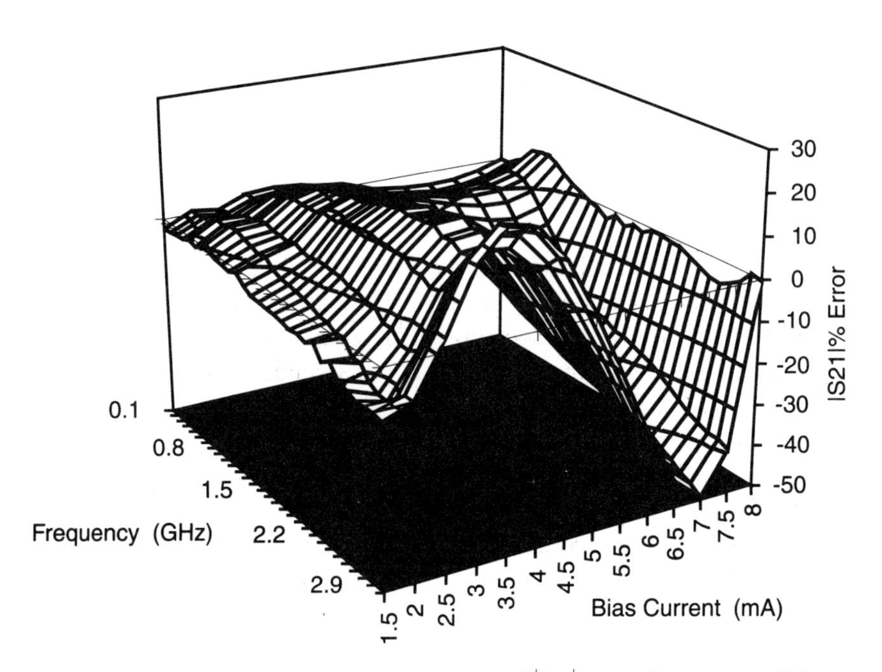

Figure 11.8 Percent error between linearly interpolated |S21| versus frequency and bias current surface and polynomial expansion coefficient surface.

equation error? To gain better understanding of the modeling dilemma, try modeling the secondary functions with RHF-LHF curve-fit equations instead of polynomial expansions.

Secondary Equation Coefficients Modeled by RHF-LHF Curve Fit

The secondary equation coefficients $A(i)$, $B(i)$, $C(i)$, $D(i)$, $E(i)$, $f_1(i)$ $f_2(i)$ and $f_3(i)$ are now modeled by the curve-fit technique introduced in Chapter 2. Each coefficient is modeled with an appropriate number of asymptotes. Model equations for coefficients $A(i)$, $B(i)$, $C(i)$, and $D(i)$ are

$$\text{Coefficient } A(i) = 3.2i - 0.578 \log_{10}\left[1 + 10^{\frac{i-2.13}{0.3}}\right]$$

$$\text{Coefficient } B(i) = -1.8 - 0.788i - 0.154 \log_{10}\left[1 + 10^{\frac{i-4.76}{0.2}}\right] - 0.313 \log_{10}\left[\frac{10^{\frac{i-2.35}{0.2}}}{1 + 10^{\frac{i-2.35}{0.2}}}\right]$$

$$\text{Coefficient } C(i) = 0.32 + 0.0363i + 0.0108 \log_{10}\left[1 + 10^{\frac{i-5.21}{0.2}}\right] + 0.0308 \log_{10}\left[\frac{10^{\frac{i-2.19}{0.2}}}{1 + 10^{\frac{i-2.19}{0.2}}}\right]$$

$$\text{Coefficient } D(i) = 0.1 + 0.587i + 0.0579 \log_{10}\left[1 + 10^{\frac{i-4.89}{0.2}}\right] + 0.579 \log_{10}\left[\frac{10^{\frac{i-2.13}{1.8}}}{1 + 10^{\frac{i-2.13}{1.8}}}\right].$$

Values for coefficients $A(i)$, $B(i)$, $C(i)$, and $D(i)$ taken from Table 11.3 are plotted in Figure 11.9 in comparison with results of the RHF-LHF curve-fit equations listed above. This is the same data the polynomial expansion equations represent. Compare the RHF-LHF curve-fit here with the polynomial curve fits in Figure 11.5.

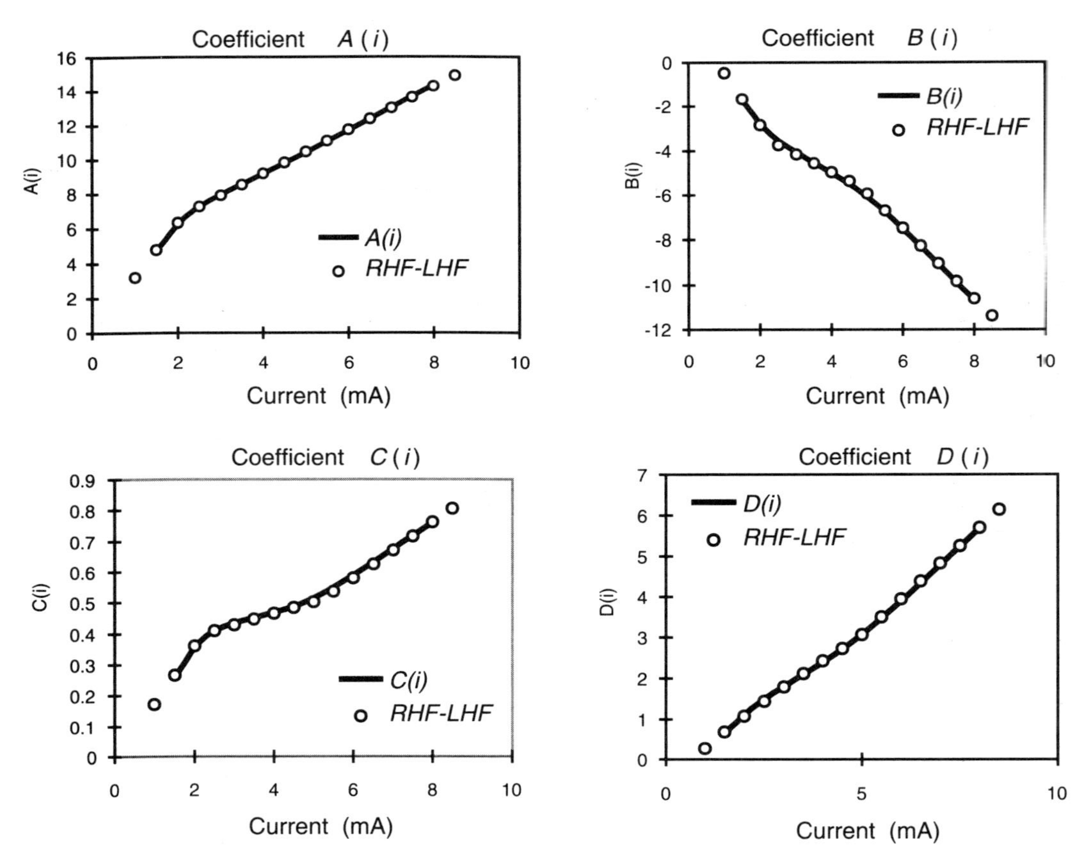

Figure 11.9 RHF-LHF curve-fit equations for coefficients *A*(*i*), *B*(*i*), *C*(*i*), and *D*(*i*) are plotted for comparison to Table 11.3 values.

The curve-fit functions match Table 11.3 values with little or no error. Curve-fit equations for primary equation coefficients $E(i)$, $f_1(i)$ $f_2(i)$ and $f_3(i)$ are now modeled. Each coefficient is modeled with an appropriate number of asymptotes. Model equations for coefficients $E(i)$, $f_1(i)$ $f_2(i)$ and $f_3(i)$ are

$$\text{Coefficient } E(i) = 0.25 - 0.014i + 0.0313 \log_{10}\left[1 + 10^{\frac{i-4.8}{1}}\right] + 0.072 \log_{10}\left[\frac{10^{\frac{i-2.05}{0.4}}}{1 + 10^{\frac{i-2.05}{0.4}}}\right]$$

$$\text{Coefficient } f_1(i) = 1.02 + 0.0213i - 0.016 \log_{10}\left[1 + 10^{\frac{i-5.12}{0.2}}\right] - 0.0885 \log_{10}\left[\frac{10^{\frac{i-2.26}{0.2}}}{1 + 10^{\frac{i-2.26}{0.2}}}\right]$$

$$\text{Coefficient } f_2(i) = 0.572 - 0.075i + 0.0679 \log_{10}\left[1 + 10^{\frac{i-4.96}{1}}\right] + 0.0788 \log_{10}\left[\frac{10^{\frac{i-2.3}{0.6}}}{1 + 10^{\frac{i-2.3}{0.6}}}\right]$$

$$\text{Coefficient } f_3(i) = 2.2 + 0.1273i - 0.145 \log_{10}\left[1 + 10^{\frac{i-4.93}{0.8}}\right] - 0.3329 \log_{10}\left[\frac{10^{\frac{i-2.34}{0.5}}}{1 + 10^{\frac{i-2.34}{0.5}}}\right].$$

Values for coefficients $E(i)$, $f_1(i)$ $f_2(i)$ and $f_3(i)$ taken from Table 11.3 are plotted in Figure 11.10 in comparison with results of the RHF-LHF curve-fit equations listed above. Compare the curve-fit obtained here with RHF-LHF models to those of polynomial expansion shown in Figure 11.6.

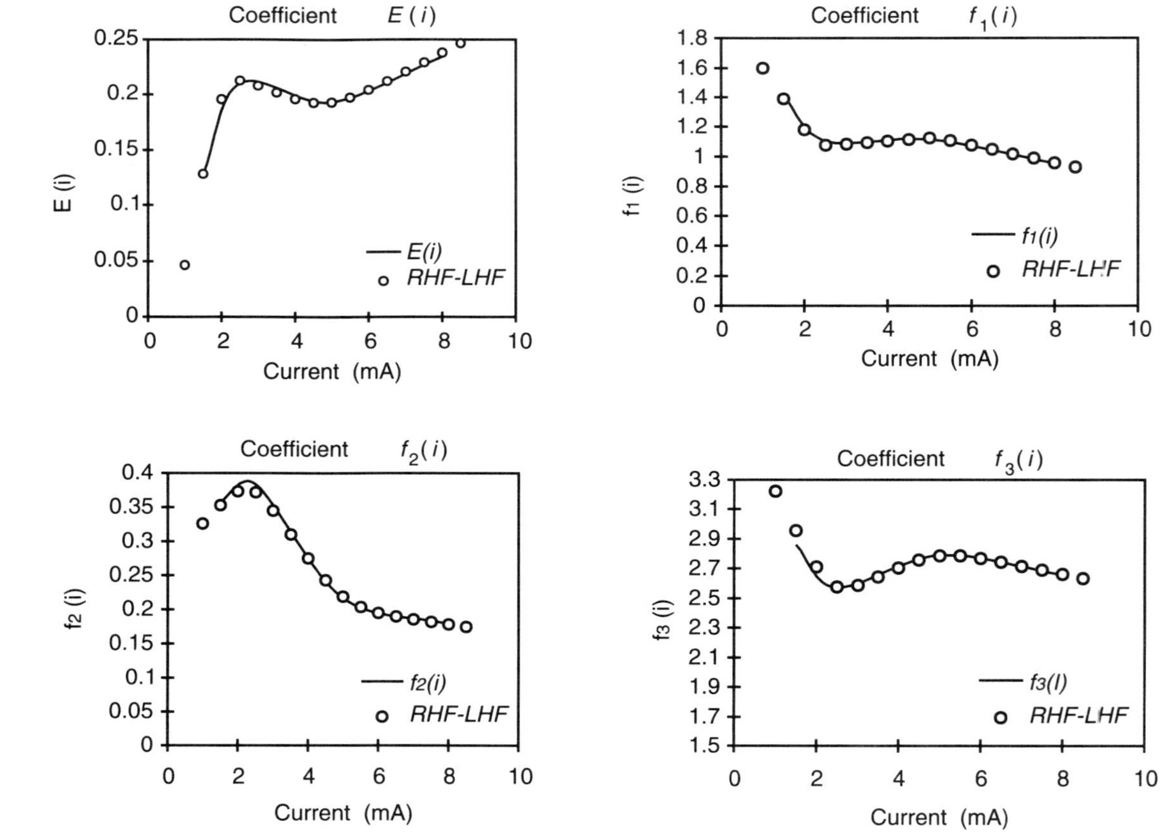

Figure 11.10 RHF-LHF Curve-fit equations for coefficients $E(i)$, $f_1(i)$ $f_2(i)$ and $f_3(i)$ are plotted for comparison to Table 11.3 values.

The secondary equations listed above that solve for primary equation coefficients by using RHF-LHF curve fits are substituted back into primary equation (11.2) and values for |S21| are calculated for bias current ranging from 1 mA to 8.5 mA and frequencies ranging from 0.1 GHz to 3.0 GHz. The results are plotted as a surface in Figure 11.11.

It is clear that the use of RHF-LHF curve-fit techniques for modeling of coefficients $A(i)$, $B(i)$, $C(i)$, $D(i)$, $E(i)$, $f_1(i)$ $f_2(i)$ and $f_3(i)$ results in less severe cyclical content in |S21| than the polynomial expansion method. The surface of Figure 11.11 is compared to the linearly interpolated surface Figure 11.1 and percent error is plotted as a surface in Figure 11.12. Maximum percent error of 35 percent using the RHF-LHF derived secondary equations is less than the 50 percent peak error of the polynomial expansion method shown in Figure 11.8. The most surprising feature of Figure 11.12 is the error along lines of constant current of 1.5 mA, 2.5 mA, 5.0 mA and 8.0 mA. The error should be minimum along these lines as it is in the polynomial expansion error plot Figure 11.10. In fact, the error is maximum along constant current lines of 1.5 mA, 2.5 mA and 5.0 mA for the RHF-LHF derived |S21|. A point to consider is the fact that the difference of two small numbers is taken when computing |S21| percent error. Percent error value is very sensitive to small absolute value differences when absolute values are small.

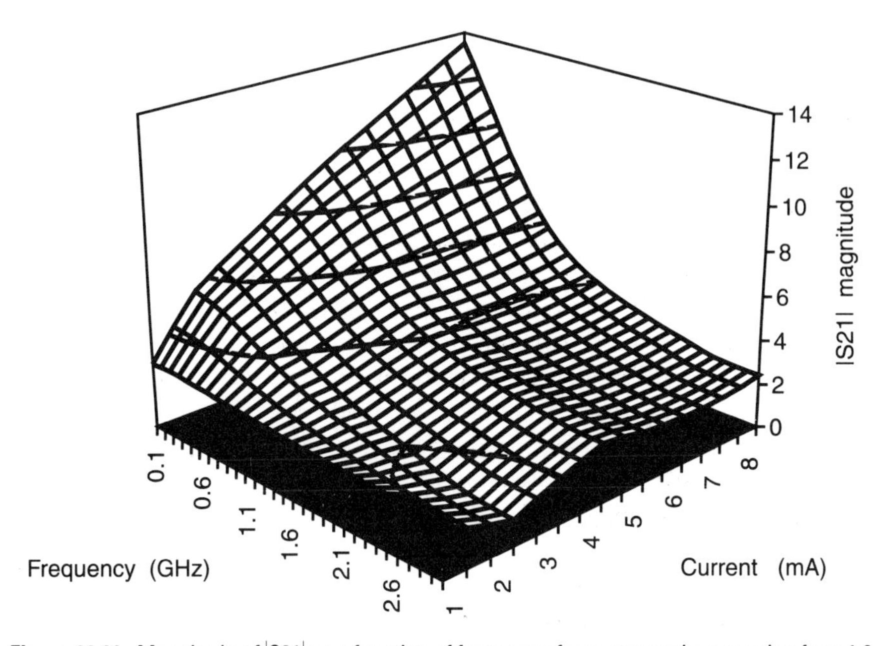

Figure 11.11 Magnitude of |S21| as a function of frequency for current values ranging from 1.0 mA to 8.0 mA derived using RHF-LHF curve fit.

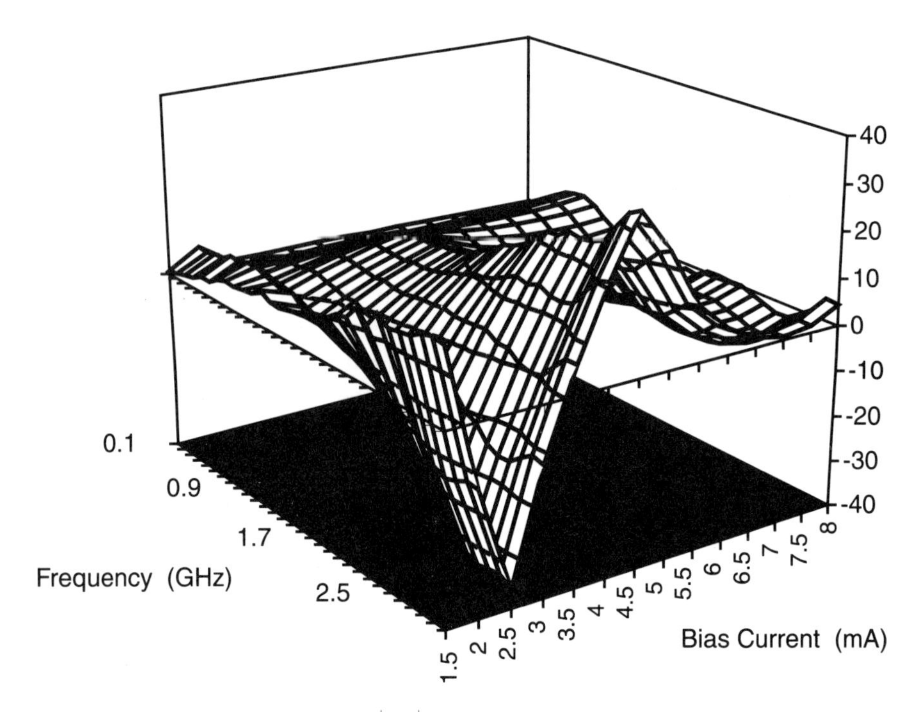

Figure 11.12 Percent error between |S21| solutions using linearly interpolated and RHF-LHF derived equations.

The Second Approach Results in Even Lower Percent Error

The approach used above to develop a closed-form equation for |S21| was to plot |S21| as a function of frequency for the four current values given. Then RHF-LHF curve fits were developed for those four plots such that a single primary equation could be written that represented the four curves. That primary equation had variable coefficients $A(i)$, $B(i)$, $C(i)$, $D(i)$, $E(i)$, $f_1(i)$ $f_2(i)$ and $f_3(i)$ for which RHF-LHF curve-fit secondary equations were developed. The resulting |S21| model produced 35 percent maximum error when compared to linearly interpolated data over the frequency and current ranges.

A second approach mentioned earlier is to plot |S21| as a function of current for equally spaced frequencies of 0.5 GHz, 1.0 GHz, 1.5 GHz, 2.0 GHz, 2.5 GHz, and 3.0 GHz (see Table 11.1). The resulting curves can be represented by a primary equation derived from two asymptotes having only four coefficients $G(f)$, $H(f)$, $I(f)$ and $i_a(f)$ that vary as a function of frequency. Secondary equations are developed for these four coefficients, and when substituted back into the primary equation, values for |S21| are computed over frequency and current. The result is a surface plot of |S21| as shown in Figure

11.13, which has less than 16 percent peak error when compared to the linearly interpolated plot of Figure 11.1.

A plot of percent error compared to the linearly interpolated surface is shown in Figure 11.14. Cyclic content in the surface derived by this approach is greater even though peak error is less.

A closed-form equation now exists for the calculation of $|S21|$ at any frequency in the range 0.1 GHz to 3.0 GHz and at any bias current 1.0 mA to 8.0 mA.

Modeling /S21 Phase Angle as a Function of Frequency and Bias Current

Parameter $\underline{/S21}$ phase angle listed in Table 11.2 is linearly interpolated and plotted as a surface in Figure 11.2. Data in each column of Table 11.2 are modeled with two asymptotes, resulting in equations consisting of a reference line and a RHF. Equations that represent $\underline{/S21}$ at bias currents 1.5 mA, 2.5 mA, 5.0 mA, and 8.0 mA are

$$\underline{/S21}(1.5) = 180 - 80f + 46.24 \log_{10}\left[1 + 10^{\frac{f-1.11}{0.9}}\right]$$

$$\underline{/S21}(2.5) = 179.66 - 96.59f + 56.86 \log_{10}\left[1 + 10^{\frac{f-0.902}{0.8}}\right]$$

$$\underline{/S21}(5.0) = 178.24 - 102.4f + 45.95 \log_{10}\left[1 + 10^{\frac{f-0.857}{0.6}}\right]$$

$$\underline{/S21}(8.0) = 177.7 - 126.9f + 49.84 \log_{10}\left[1 + 10^{\frac{f-0.6518}{0.5}}\right]$$

Results of plotting these equations in comparison to data from Table 11.2 are shown in Figure 11.15.

All of the equations for phase angle as a function of frequency at fixed bias currents are of the form

$$\underline{/S21}(i) = Y(i) - V(i)f + W(i) \log_{10}\left[1 + 10^{\frac{f-f(i)}{a(i)}}\right]. \tag{11.3}$$

This is the primary model equation for this particular set of phase angle data. Secondary equations $Y(i)$, $V(i)$, $W(i)$, $f(i)$, and $a(i)$ are developed using polynomial expansion and using RHF-LHF for comparison. Data used to develop polynomial expansion and RHF-LHF equations for the coefficients $Y(i)$, $V(i)$, $W(i)$, $f(i)$, and $a(i)$ are taken from the four equations above and listed in Table 11.4.

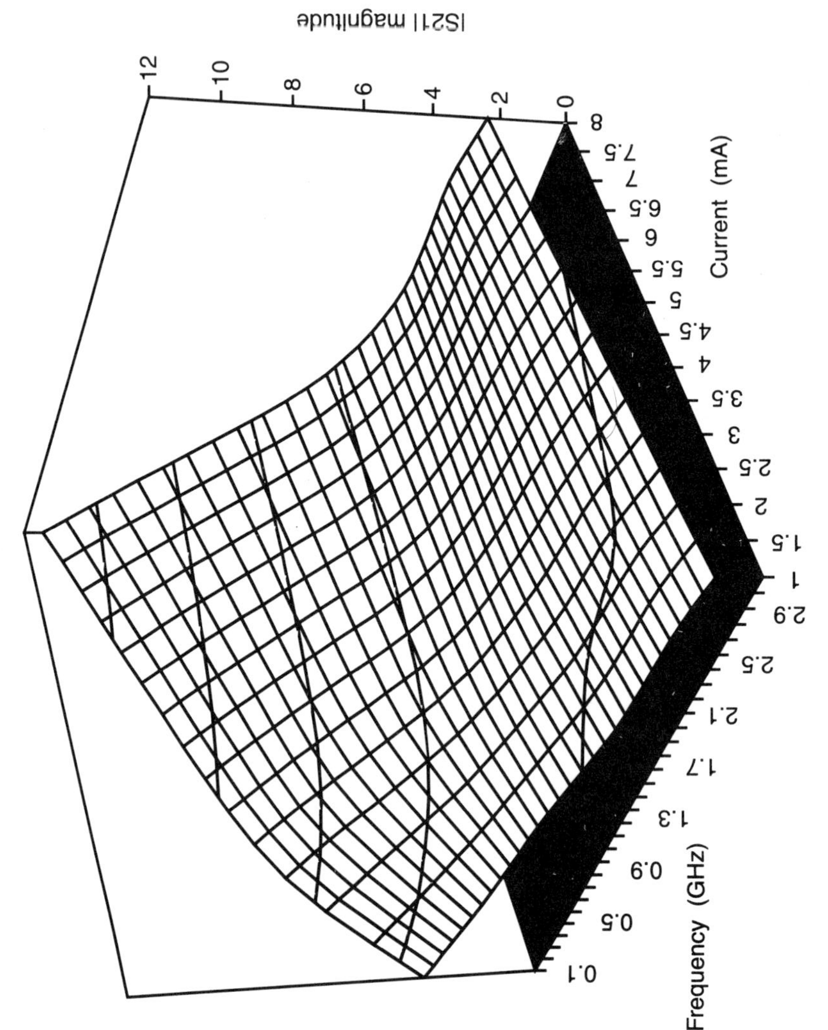

Figure 11.13 Magnitude of S21 as a function of current for frequency values ranging from 0.5 GHz to 3.0 GHz.

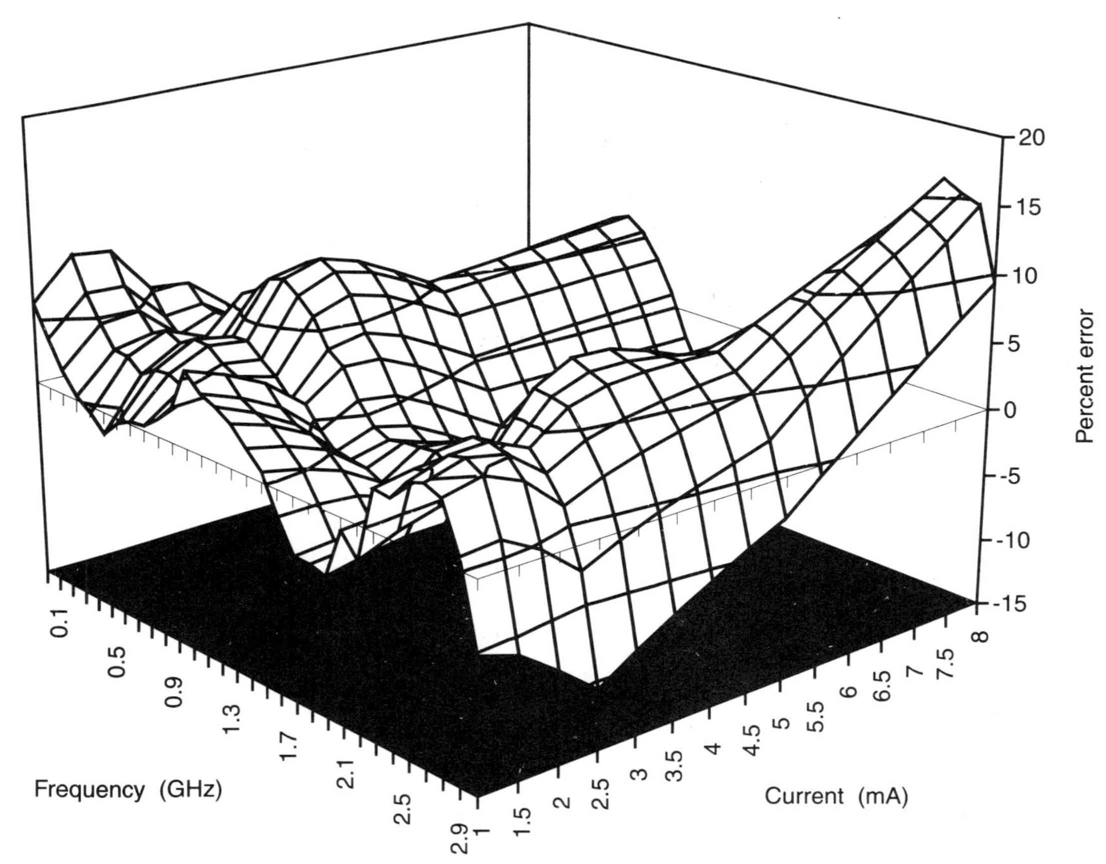

Figure 11.14 Percent error compared to linearly interpolated values.

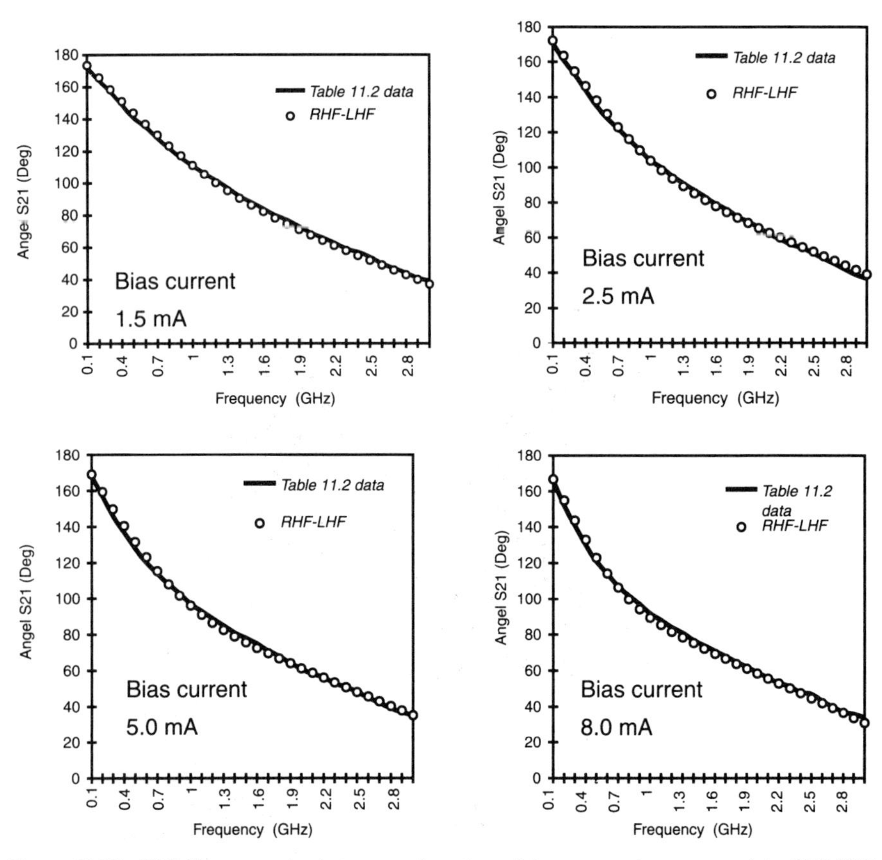

Figure 11.15 /$\underline{S21}$ Phase angle data as a function of frequency is compared to RHF-LHF model-generated data at constant bias current values.

Polynomial Expansion Coefficient Equations for Y(i), V(i), W(i), f(i), and a(i)

Polynomial expansion is performed about bias current value 4.0 mA. The resulting equations are

$$Y(i) = 178.82 - 0.6057(i-4) - 0.002797(i-4)^2 + 0.02078(i-4)^3$$

$$V(i) = -103.22 - 0.883(i-4) + 1.7044(i-4)^2 - 0.7905(i-4)^3$$

$$W(i) = 53.67 - 6.709(i-4) - 1.83(i-4)^2 + 0.8169(i-4)^3$$

$$f(i) = 0.8298 + 0.01207(i-4) + 0.0253(i-4)^2 - 0.00986(i-4)^3$$

$$a(i) = 0.6698 - 0.07724(i-4) + 0.00699(i-4)^2 + 0.000426(i-4)^3$$

Table 11.4
Data for Developing Phase Angle Coefficient Secondary Equations

Current	$Y(i)$	$V(i)$	$W(i)$	$f(i)$	$a(i)$
1.5	180	−80	46.24	1.112	0.9
2.5	179.66	−96.59	56.86	0.902	0.8
5	178.24	−102.4	45.95	0.8574	0.6
8	177.69	−126.9	49.84	0.6518	0.5

When these secondary equations for coefficients in the primary equation are substituted into (11.3), and data is generated for frequencies ranging from 0.1 GHz to 3.0 GHz and bias currents from 1.0 mA to 8.5 mA, the surface shown in Figure 11.16 is obtained.

The tendency for polynomial expansions to generate cyclical error is seen again in the surface plotted in Figure 11.16. Cyclical amplitude becomes exaggerated at 3.0 GHz where phase angle values are smallest. The exaggeration is clearly illustrated as percent error in Figure 11.17 where comparison is made to the linearly interpolated phase angle data from Figure 11.2.

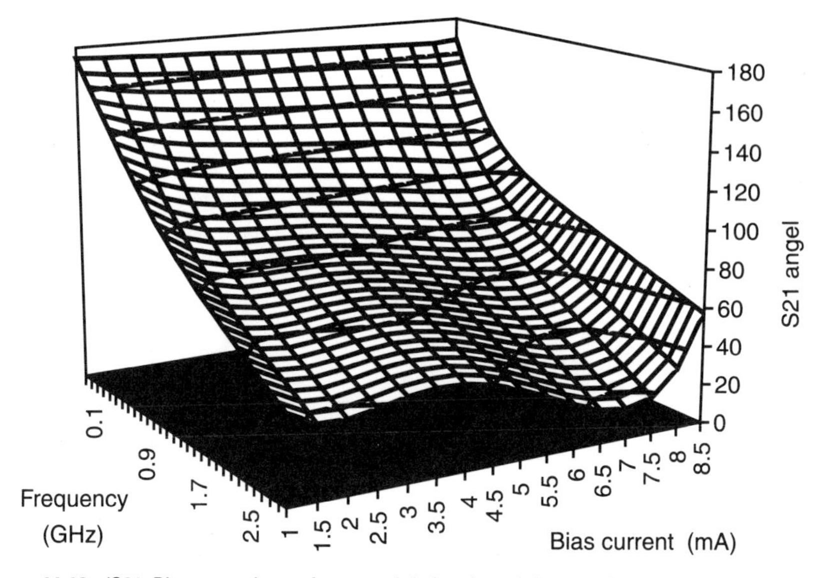

Figure 11.16 /S21 Phase angle surface model developed from polynomial expansion secondary equations.

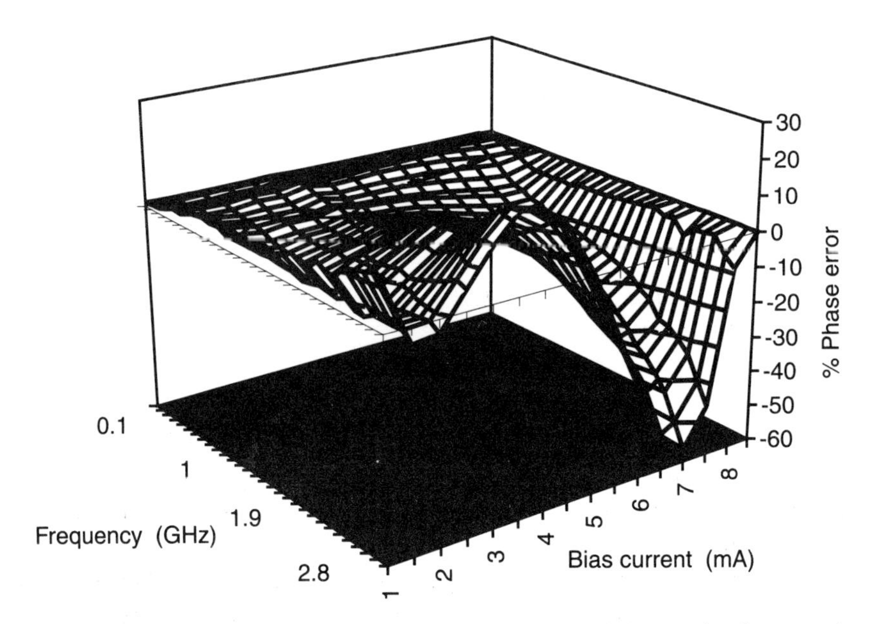

Figure 11.17 Phase angle percent error resulting from polynomial expansion for secondary equations.

RHF-LHF Models for Secondary Y(i), V(i), W(i), f(i), and a(i) Equations

The curve-fitting technique described in Chapter 2 is used to model $Y(i)$, $V(i)$, $W(i)$, $f(i)$, and $a(i)$ with Table 11.4 data. The five functions require from two to three asymptotes for reasonably accurate fit. Curve-fit equations obtained are

$$Y(i) = 181.2 - 0.617i + 0.1919 \log_{10}\left[1 + 10^{\frac{i-5.07}{0.4}}\right] + 0.0696 \log_{10}\left[\frac{10^{\frac{i-2.38}{0.2}}}{1 + 10^{\frac{i-2.38}{0.2}}}\right]$$

$$V(i) = -100 - 1.7 \log_{10}\left[1 + 10^{\frac{i-4.7}{0.2}}\right] - 4.24 \log_{10}\left[\frac{10^{\frac{i-2.31}{0.2}}}{1 + 10^{\frac{i-2.31}{0.2}}}\right]$$

$$W(i) = 73.72 - 5.965i + 3.236 \log_{10}\left[1 + 10^{\frac{i-4.91}{0.4}}\right] + 8.646 \log_{10}\left[\frac{10^{\frac{i-2.3}{0.4}}}{1 + 10^{\frac{i-2.3}{0.4}}}\right]$$

$$f(i) = 0.91 - 0.009i - 0.029 \log_{10}\left[1 + 10^{\frac{i-5.28}{0.4}}\right] - 0.1056 \log_{10}\left[\frac{10^{\frac{i-2.34}{0.4}}}{1 + 10^{\frac{i-2.34}{0.4}}}\right]$$

$$a(i) = 1.07 - 0.106i + 0.08 \log_{10}\left[1 + 10^{\frac{i-4.5}{1}}\right]$$

These secondary equations are substituted back into (11.3); this is used to calculate /S21 phase angle values for any frequency from 0.1 GHz to 3.0 GHz and any bias current from 1.0 mA to 8.0 mA. Data obtained forms the surface that is plotted in Figure 11.18. Some evidence of cyclic error is present even with RHF-LHF derived secondary equations. Percent error peaks as high as 25 percent when compared to linearly interpolated phase angle as shown in Figure 11.19. Percent error is greatest where phase shift values are smallest and the difference in small numbers magnifies the error.

A model with lower percent error could have been obtained if the primary equation had been created by first plotting phase angle as a function of current for several values of frequency. The primary equation would have had coefficients that varied as a function of frequency. Secondary equations for the coefficients would have been functions of frequency.

S-parameter Modeling Conclusions

A modeling technique has been shown that results in closed-form equations for S-parameter magnitude and phase angle. This example focused on S21 values. The remaining values S11, S12, and S22 can be modeled using the same technique. Model results are compared against linearly interpolated data. It is shown that there are multiple approaches to modeling parameter behavior which result in lower error when compared to linearly interpolated data. There is no knowledge of how data actually behaves in the interpolated regions. It

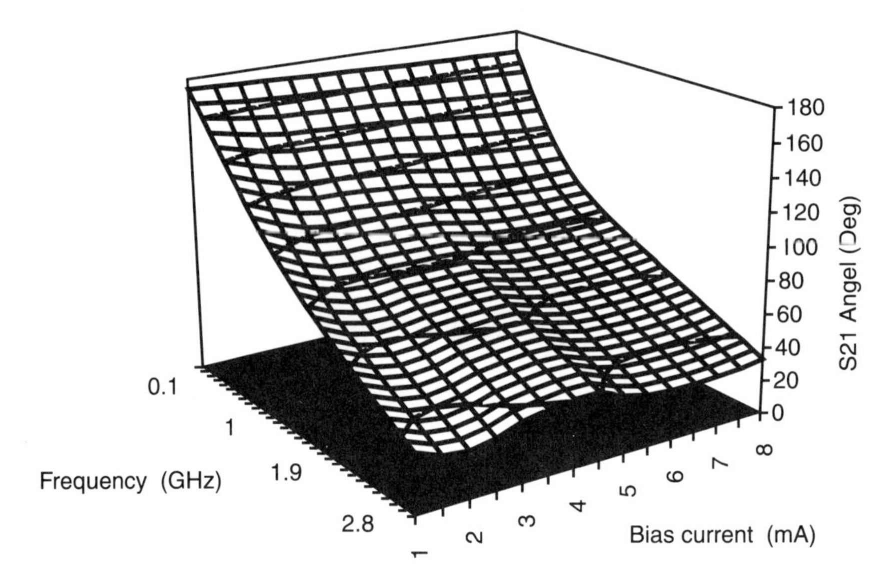

Figure 11.18 /S21 Phase angle as a function of frequency and bias current plotted as a surface.

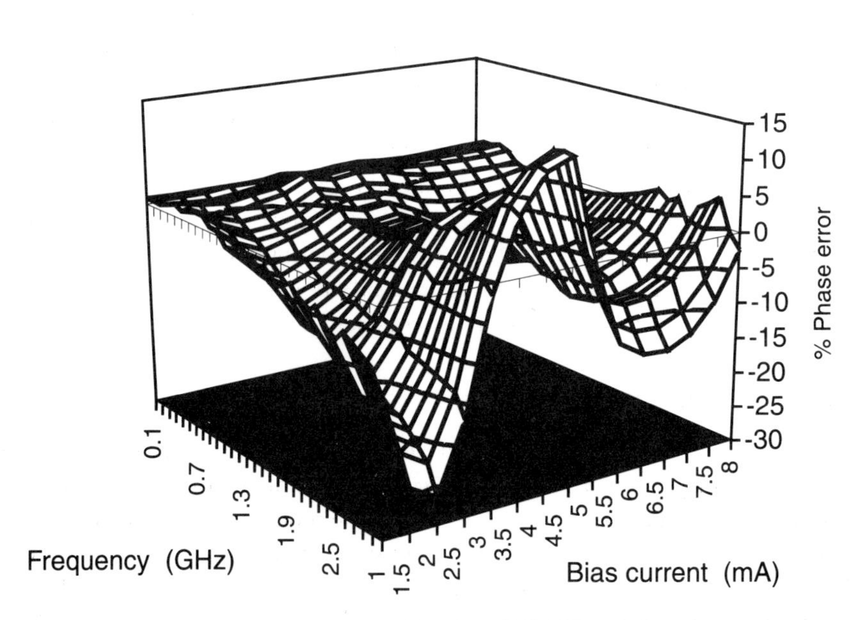

Figure 11.19 /S21 Percent phase error resulting from RHF-LHF modeling of secondary equations compared to linearly interpolated data.

could be linear as interpolated or it could be cyclic, similar to the modeled results. In this example, percent error is always derived in comparison to the linearly interpolated data. It is interesting to note that if percent error is derived by comparing polynomial expansion derived data to RHF-LHF derived data instead of comparing to linearly interpolated data, the values of percent error are of the same magnitude. Figure 11.20 illustrates percent phase shift error derived by the comparison of polynomial expansion to that of the RHF-LHF secondary equation model.

11.3 Modeling a Single Impulse in Time With a Closed-Form Equation

The RHF-LHF curve-fit methods described in Chapter 2 can be used to model measured data resulting in closed-form equations. In Chapter 6, the RHF-LHF equations were used to simulate Butterworth and Chebishev bandpass frequency response resulting in designed equations. They can be applied to a variety of unique modeling situations. For instance, consider the impulse function, a function that has zero value for all time leading up to the instant of occurrence. At the moment of occurrence the function increases super rapidly to some value A, remains at value A for an incredibly short time, then decays just as super rapidly to zero where it remains for all remaining time. The function is usually defined with three equations that apply in sequential segments of time.

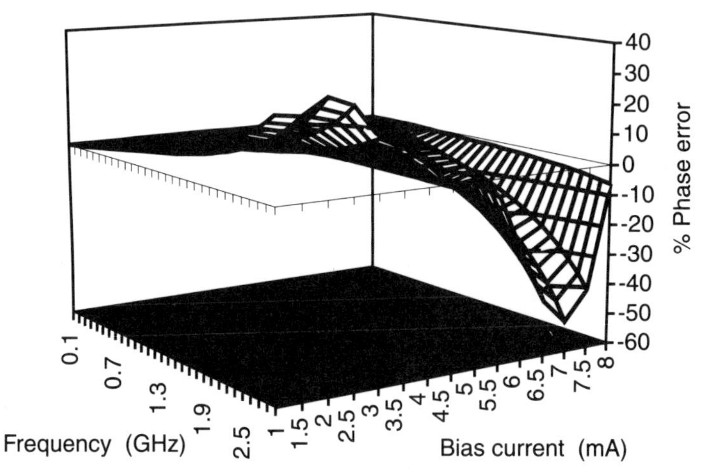

Figure 11.20 /S21 percent phase error derived by comparing polynomial expansion model to RHF-LHF model.

This type of function can easily be described by using five asymptotes and four RHFs. The first asymptote is a line described by the equation $y(t) = 0$ that is applied from $-\infty \leq t \leq 0$. At $t = 0$ the second asymptote is applied, rising from value 0 to value A in time interval Δt. At time Δt the third asymptote is applied and that asymptote is described by the equation $y(t) = A$. At time $t = 2\Delta t$, the fourth asymptote is applied going from value A to value 0 in time duration Δt. Finally at time $t = 3\Delta t$, the fifth asymptote described by the equation $y(t) = 0$ is applied. The slope change at time 0 is $+\frac{A}{\Delta t}$. Slope change at $t = \Delta t$ and at $t = 2\Delta t$ is $-\frac{A}{\Delta t}$. Slope change at $t = 3\Delta t$ is $+\frac{A}{\Delta t}$. Asymptote intersection points are $t = 0$, $t = \Delta t$, $t = 2\Delta t$, and, $t = 3\Delta t$. Transition range at each intersection is defined to be $a = \frac{\Delta t}{2}$. The product of slope change and transition range is $\frac{A}{2}$. Figure 11.21 illustrates the impulse function modeled with asymptotes.

The asymptotes are connected with right-hand functions to result in the closed-form equation

$$y(t) = \frac{A}{2}\left\{\log_{10}\left[1 + 10^{\frac{2t}{\Delta t}}\right] - \log_{10}\left[1 + 10^{\frac{2(t-2\Delta t)}{\Delta t}}\right] - \log_{10}\left[1 + 10^{\frac{(t-2\Delta t)}{\Delta t}}\right] + \log_{10}\left[1 + 10^{\frac{2(t-3\Delta t)}{\Delta t}}\right]\right\}$$

which reduces to

$$y(t) = \frac{A}{2}\log_{10}\left[\frac{\left(1 + 10^{\frac{2t}{\Delta t}}\right)\left(1 + 10^{\frac{2(t-3\Delta t)}{\Delta t}}\right)}{\left(1 + 10^{\frac{2(t-\Delta t)}{\Delta t}}\right)\left(1 + 10^{\frac{2(t-2\Delta t)}{\Delta t}}\right)}\right]. \tag{11.4}$$

Equation (11.4) is a unique function that can be used in a variety of ways. It can be used as a stand alone video pulse function or impulse function, and it can be used as a multiplier that causes a sine wave or cosine wave pulse of any

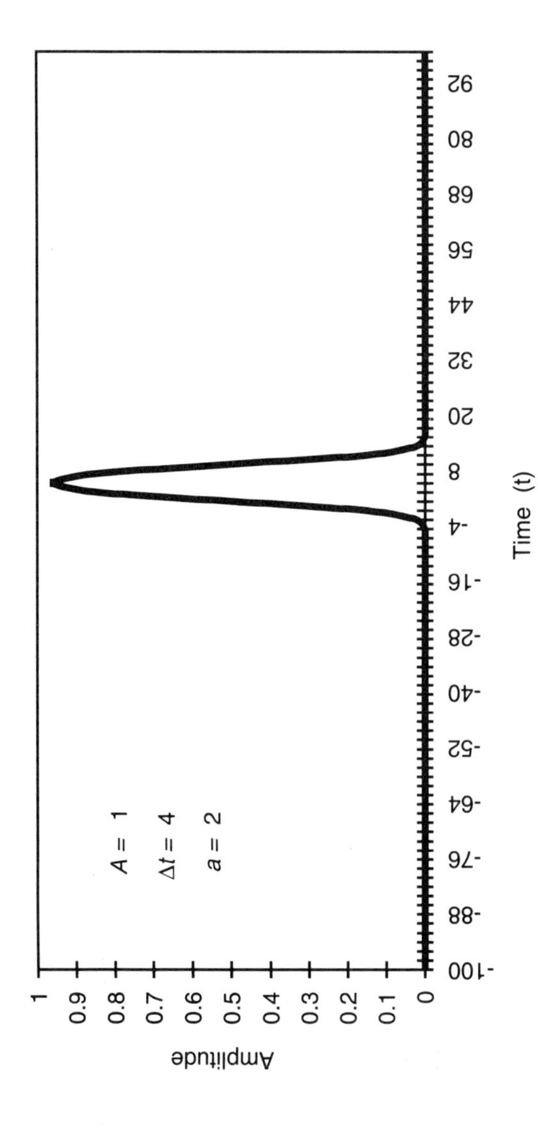

Figure 11.21 The impulse function is modeled by asymptotes and right-hand functions.

desired number of cycles, with any desired rise and fall time to occur once in all time.

11.4 Modeling a Sine Wave Burst in Time

Modeling a single sine wave or cosine wave burst of any number of cycles in time with any desired rise and fall time is as simple as modifying the intersection points in (11.4) to give the desired rise, fall, and duration to the burst then multiplying the modified (11.4) by $\sin(\varphi t)$ or $\cos(\varphi t)$. The behavioral model that applies is

$$y(t) = \frac{A}{2}\log_{10}\left[\frac{\left(1+10^{\frac{2(t+t_1)}{\Delta t}}\right)\left(1+10^{\frac{2(t-t_1)}{\Delta t}}\right)}{\left(1+10^{\frac{2(t+t_1-\Delta t)}{\Delta t}}\right)\left(1+10^{\frac{2(t-t_1+\Delta t)}{\Delta t}}\right)}\right]\sin(\varphi t) \quad (11.5)$$

where $-t_1$ is the moment in time before $t = 0$ that the sine wave begins to rise to amplitude A, time $t = -(t_1 - \Delta t)$ is when the sine wave reaches full amplitude A, time $t = (t_1 - \Delta t)$ is when the sine wave begins to fall to zero amplitude, and at time $t = t_1$ it reaches zero. Figure 11.22 illustrates the function applied to an RF signal burst of three cycles.

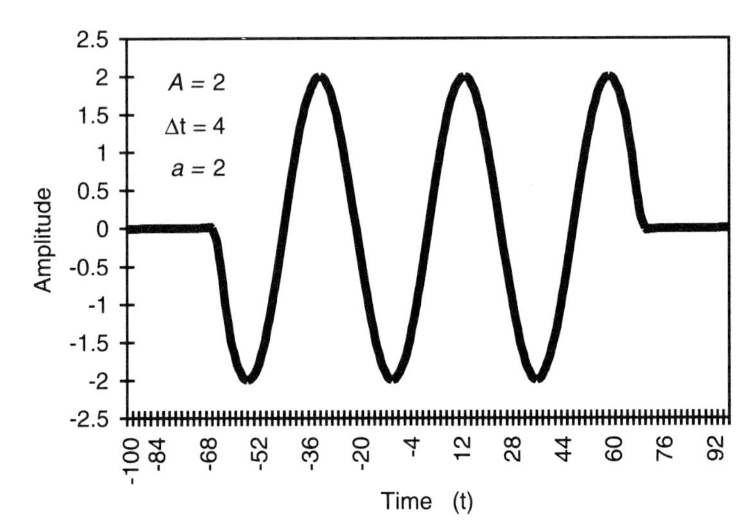

Figure 11.22 A burst of three sine wave cycles in time.

11.5 Modeling Junction Capacitance Under Forward or Reverse Bias with a Single Equation

Diode junction capacitance is presently modeled as diffusion capacitance, which predominates when the diode junction is forward biased, and depletion capacitance, which predominates when the junction is reverse biased. Two separate capacitance equations represent the different bias regions. Diffusion capacitance is defined as the rate of change of current per change in voltage across the junction multiplied by the diode transit time

$$C_s = \tau_d\left(\frac{\partial i_d}{\partial v}\right) \tag{11.6}$$

Depletion capacitance in the reverse bias region is derived from the space charge equation

$$C_d(v) = \frac{C_j(0)}{\left(1 - \frac{v}{v_j}\right)^M} \tag{11.7}$$

where $C_j(0)$ is junction capacitance at zero volts bias, M is doping profile constant, and v_j is internal junction potential. The value of M varies from $M = 0.33$ for linearly graded junctions to $M = 0.5$ for abrupt junctions. The value of v_j ranges from $v_j = 0.5$ volt to $v_j = 1.0$ volt depending on the semiconductor material, the doping profile, and the temperature. When bias voltage v approaches v_j, the depletion capacitance equation value increases rapidly and goes to infinity at $v = v_j$.

Measured junction capacitance value does not wildly ramp off to infinity as bias voltage v approaches v_j and goes into the forward bias region. Measured capacitance value tends to increase linearly along a continuous slope that is established at some negative bias voltage v_f just below v_j. Conventional modeling of capacitance in the forward bias region is based on determination of the first derivative of (11.7) with respect to bias voltage

$$\frac{\partial C_j(0)}{\partial v} = \frac{MC_j(0)}{v_j}\left(1 - \frac{v}{v_j}\right)^{-(M+1)} \tag{11.8}$$

The value of slope of a tangent to the function $C_j(v)$ at bias voltage v_f is

$$\frac{\partial C_{\mathrm{j}}(v_{\mathrm{f}})}{\partial v} = \frac{MC_{\mathrm{j}}(0)}{v_{\mathrm{j}}}\left(1 - \frac{v_{\mathrm{f}}}{v_{\mathrm{j}}}\right)^{-(M+1)} \tag{11.9}$$

An equation for a line tangent to $C_{\mathrm{j}}(v)$ at v_{f} is derived by solving the line equation $C(v) = av + b$ where a is the slope value from (11.9). The value b is determined by entering values into the line equation at $v = v_{\mathrm{f}}$

$$C_{\mathrm{j}}(0)\left(1 - \frac{v_{\mathrm{f}}}{v_{\mathrm{j}}}\right)^{-M} = \frac{MC_{\mathrm{j}}(0)v_{\mathrm{j}}}{v_{\mathrm{j}}}\left(1 - \frac{v_{\mathrm{f}}}{v_{\mathrm{j}}}\right)^{-(M+1)} + b$$

and solving for the constant b

$$b = C_{\mathrm{j}}(0)\left(1 - \frac{v_{\mathrm{f}}}{v_{\mathrm{j}}}\right)^{-M}\left[1 - M\left(1 - \frac{v_{\mathrm{f}}}{v_{\mathrm{j}}}\right)^{-1}\right] \tag{11.10}$$

The line equation is then

$$Cj(v) = C_{\mathrm{j}}(0)\left(1 - \frac{v_{\mathrm{f}}}{v_{\mathrm{j}}}\right)^{-M}\left[1 - M\left(1 - \frac{v_{\mathrm{f}}}{v_{\mathrm{j}}}\right)^{-1}\right] + \frac{MC_{\mathrm{j}}(0)v}{v_{\mathrm{j}}}\left(1 - \frac{v_{\mathrm{f}}}{v_{\mathrm{j}}}\right)^{-(M+1)}$$

which simplifies to

$$C_{\mathrm{j}}(v) = C_{\mathrm{j}}(0)\left(1 - \frac{v_{\mathrm{f}}}{v_{\mathrm{j}}}\right)^{-M}\left(1 - M\frac{(v_{\mathrm{f}} - v)}{(v_{\mathrm{j}} - v_{\mathrm{f}})}\right) \tag{11.11}$$

Computer-aided design programs like SPICE use (11.7) for instantaneous junction voltage in the range $-\infty < v < v_{\mathrm{f}}$ and switch to (11.11) for instantaneous junction voltage in the range $v_{\mathrm{f}} < v < +\infty$. This is awkward and often leads to errors in modeling, particularly where measured data shows junction capacitance peaking at a maximum value then folding back as shown in Figure 11.23.

Develop a Single Equation for Junction Capacitance

A single equation can be written to model junction capacitance over the depletion region, the linearly projected region where $v > v_{\mathrm{f}}$, and the region where capacitance value peaks and folds over. The model technique involves fitting four asymptotes to the entire curve, basing values for intersections, slopes, and transition ranges on parameters $C_{\mathrm{j}}(0)$, M, v_{f}, v_{j}, and v_{p}. Coefficient v_{p} is defined as the voltage where measured capacitance value peaks and rolls over. The first asymptote **A1** approximates capacitance at negative voltages $v \ll 0$ *volts*.

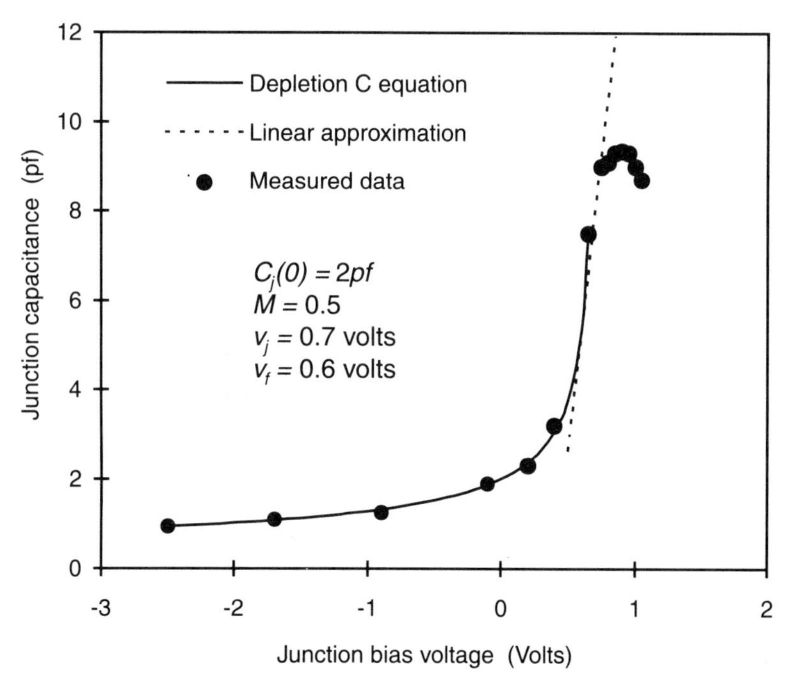

Figure 11.23 Junction capacitance calculated with two separate equations fails to model measured data accurately.

The second asymptote **A2** and transition range value a_1 approximates capacitance in the range $-v_j < v < 0$. The third asymptote **A3** is the linear approximation defined by (11.11) for the voltage range $v_f < v < +\infty$. Transition range value a_2 approximates capacitance in the range $0 < v < v_j$. The fourth asymptote **A4** is used to model the peak and rollover in capacitance at voltage v_p as voltage increases above v_j. Transition range a_3 approximates capacitance value in the region of v_p. The asymptotes and transition ranges are shown in Figure 11.24.

Determine an equation for the first asymptote, which is arbitrarily assigned to be tangent to the depletion capacitance curve at $v = -5v_j$. Slope of asymptote **A1** is

$$S1 = \frac{C_j(0)M}{v_j}\left(1 + \frac{5v_j}{v_j}\right)^{-(M+1)} = \frac{C_j(0)M}{v_j 6^{(M+1)}} \quad (11.12)$$

Line equation constant b_1 is determined by solving $f(v) = S1v + b_1$ at $v = -5v_j$ and obtaining

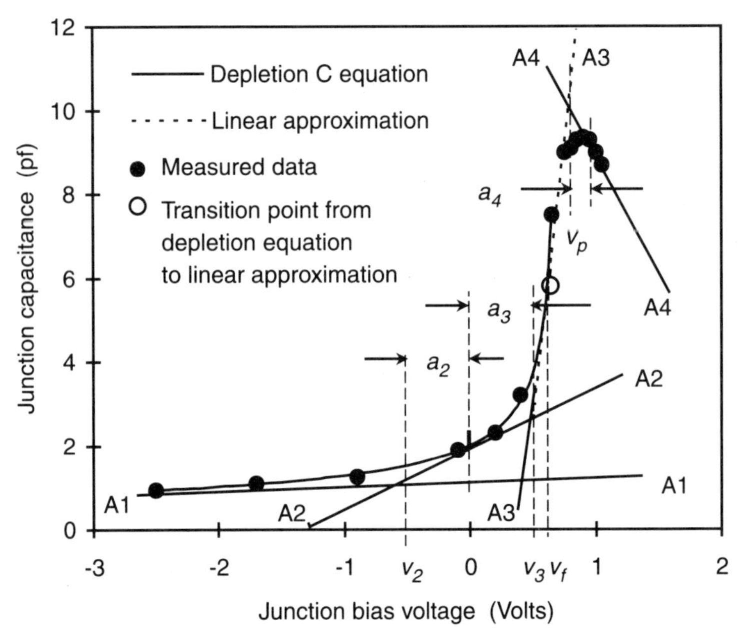

Figure 11.24 Four asymptotes are used to model forward and reverse biased junction capacitance.

$$b_1 = C_j(0)\left[\frac{1}{6^M} + \frac{5M}{6^{(M+1)}}\right]$$

The equation for asymptote **A1** is

$$A1(v) = \frac{C_j(0)}{6^M}\left[1 + \frac{5M}{6} + \frac{M}{6}\frac{v}{v_j}\right] \tag{11.13}$$

Let asymptote **A2** be tangent to the depletion capacitance curve at $v = 0$. Slope of asymptote **A2** is

$$S2 = \frac{C_j(0)M}{v_j}\left(1 + \frac{0}{v_j}\right)^{-(M+1)} = \frac{C_j(0)M}{v_j} \tag{11.14}$$

Line equation constant b_2 is determined by solving $f(v) = S2v + b_2$ at $v = -5v_j$ and obtaining

$$b_2 = C_j(0)$$

The equation for asymptote **A2** is

$$A2(v) = Cj(0)\left(1 + M\frac{v}{v_j}\right) \tag{11.15}$$

Asymptote **A3** is tangent to the depletion capacitance curve at voltage $v = v_f$. The (11.9) repeated here is the slope of **A3**,

$$S3 = \frac{MC_j(0)}{v_j}\left(1 - \frac{v_f}{v_j}\right)^{-(M+1)} \tag{11.16}$$

and (11.11) repeated here is the equation for **A3**

$$A3(v) = C_j(0)\left(1 - \frac{v_f}{v_j}\right)^{-M}\left(1 - M\frac{(v_f - v)}{(v_j - v_f)}\right) \tag{11.17}$$

These three asymptotes are directly related to the depletion capacitance equation by the coefficient values that appear in them. The fourth asymptote **A4** is related to measured capacitance data and the slope along which measured capacitance diminishes as bias voltage continues to increase into the forward bias region. A voltage v_p is defined as the voltage at which measured capacitance value peaks as shown in Figure 11.23. The slope **S4** of asymptote **A4** is determined by fitting a line to the measured data. Typical values of **S4** range from -5 to -25 depending on the value of $C_j(0)$.

Intersection points of asymptotes **A1&A2**, and **A2&A3**, are determined by setting equations **A1**(v) = **A2**(v), and **A3**(v) = **A4**(v) and solving for variable v. The intersection of **A3&A4** is defined to be at $v = v_p$.

Intersection of **A1&A2**

$$v_2 = \frac{v_j}{M}\left(\frac{6 + 5M - 6^{(M+1)}}{1 - 6^{(M+1)}}\right) \tag{11.18}$$

Intersection of **A2&A3**

$$v_3 = \frac{v_j}{M}\left[\frac{1 - \left(1 - \frac{v_f}{v_j}\right)^{-M} + M\frac{v_f}{v_j}\left(1 - \frac{v_f}{v_j}\right)^{-(M+1)}}{\left(1 - \frac{v_f}{v_j}\right)^{-(M+1)} - 1}\right] \tag{11.19}$$

Transition range a_2 from asymptote **A1** to **A2** is defined to be the range from 0 volts to the intersection of **A1**&**A2** at voltage v_2. The value of $a_2 = |v_2|$, the absolute value of v_2,

$$a_2 = \left| \frac{v_j}{M} \left(\frac{6 + 5M - 6^{(M+1)}}{1 - 6^{(M+1)}} \right) \right| \tag{11.20}$$

Transition range a_3 from asymptote **A2** to **A3** is defined to be half of the range from 0 volts to the intersection voltage v_3.

$$a_3 = \frac{v_j}{2M} \left[\frac{1 - \left(1 - \frac{v_f}{v_j}\right)^{-M} + M\frac{v_f}{v_j}\left(1 - \frac{v_f}{v_j}\right)^{-(M+1)}}{\left(1 - \frac{v_f}{v_j}\right)^{-(M+1)} - 1} \right] \tag{11.21}$$

Transition range a_4 value about the intersection of **A3**&**A4** at voltage v_p is selected to match the trace of measured capacitance value as it peaks and rolls over. A typical value is $a_4 = 0.2$ volt.

A model equation for junction capacitance over the full range of bias voltage from reverse bias into forward bias and into capacitance value roll over is

$$C(v) = \frac{C_j(0)}{6^M}\left[1 + \frac{5M}{6} + \frac{M}{6}\frac{v}{v_j}\right] + (S2 - S1)a_2 \log_{10}\left[1 + 10^{\frac{(v-v_2)}{a_2}}\right]$$

$$+ (S3 - S2)a_3 \log_{10}\left[1 + 10^{\frac{(v-v_3)}{a_3}}\right] + (S4 - S3)a_4 \log_{10}\left[1 + 10^{\frac{(v-v_p)}{a_4}}\right] \tag{11.22}$$

where values S1, S2, S3, S4, a_2, a_3, a_4, v_2, and v_3 are from the equations developed above. The model equation (11.22) is shown plotted in Figure 11.25 in comparison with measured data where $C_j(0) = 2.0$ pf, $M = 0.5$, $v_j = 0.7$ volts, $v_f = 0.6$ volts, $v_p = 0.8$ volts, and slope $S4 = -6$ pf/volt.

11.6 Summary

Modeling techniques described in Chapter 2 have been applied to the characterization of S-parameters as a function of frequency and bias current. An example is given illustrating the modeling process by developing a primary equation from

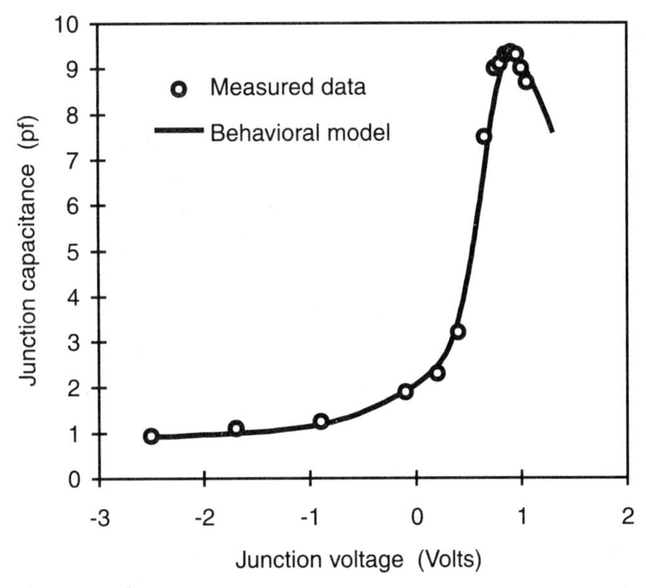

Figure 11.25 The junction capacitance behavioral model equation is plotted for comparison with measured data.

current data plotted as a function of frequency. Comparisons are made between results obtained when secondary equations are polynomial expansions and when they are RHF-LHF models. The resulting closed-form equations are plotted as three-dimensional surfaces and compared to a surface derived by linear interpolation of values from initial S-parameter data. Comparison of the results obtained using linear interpolation, polynomial expansion, and RHF-LHF curve fits show as much as 50 percent difference in $|S21|$ parameter values particularly where $|S21|$ parameter value is small. A second approach is used developing a primary model equation from data plotted as a function of current for several frequencies to show that alternate approaches result in lower peak error but higher cyclic content. Secondary equations developed from RHF-LHF curve fits are used. Sixteen percent peak error is obtained compared to linearly interpolated data.

The curve-fit technique of Chapter 2 is applied to the modeling of an impulse function and to a sine wave burst of three cycles, each function occurring once in all time. A closed-form equation is derived for each of the two functions. The functions are shown as examples of the utility of the modeling technique and hint at applications yet to be applied.

Finally, the curve-fit technique of Chapter 2 is applied to the development of a single behavioral model equation representing junction capacitance as a function of junction bias from reverse bias into the forward bias voltage range.

11.7 Problems

Problem 11.1

Use the data in Table 11.1 and develop a primary equation for |S21| magnitude as a function of current at frequencies 0.5 GHz, 1.0 GHz, 1.5 GHz, 2.0 GHz, 2.5 GHz, and 3.0 GHz. Develop a table of primary equation coefficient values $G(f)$, $H(f)$, $I(f)$, and $i(f)$ for the six RHF-LHF primary equations. Find RHF-LHF curve-fit secondary equations for $G(f)$, $H(f)$, $I(f)$, and $i(f)$. Compute |S21| values for all frequency and current values and create a surface plot like Figure 11.13. Develop a linearly interpolated data set from Table 11.1 and compare the modeled data to the linearly interpolated data. Plot percent error and compare to Figure 11.14.

Problem 11.2

Use the data in Table 11.2 and develop a primary equation for /S21 phase angle as a function of current at frequencies 0.5 GHz, 1.0 GHz, 1.5 GHz, 2.0 GHz, 2.5 GHz, and 3.0 GHz. Develop a table of primary equation coefficient values $P(f)$, $Q(f)$, $R(f)$, and $i(f)$ for the six RHF-LHF primary equations. Find RHF-LHF curve-fit secondary equations for $P(f)$, $Q(f)$, $R(f)$, and $i(f)$. Compute /S21 phase angle values for all frequency and current values and create a surface plot like that in Figure 11.18. Develop a linearly interpolated dataset from Table 11.2 and compare the modeled data to the linearly interpolated data. Plot percent error and compare to Figure 11.19.

Appendix A
Answers to Problems

Chapter 1

Problem 1.1

$$\Delta f = 0.24896 - 0.34107(T-25) - 0.0004076(T-25)^2 + 0.0001079(T-25)^3$$

Chapter 2

Problem 2.1

$$f(t) = \frac{A\pi}{16}\log_{10}\left[\frac{\left(1+10^{\frac{16t}{\pi}}\right)\left(1+10^{\frac{(4t-6\pi)}{\pi}}\right)^8}{\left(1+10^{\frac{(16t-32\pi)}{\pi}}\right)\left(1+10^{\frac{(4t-2\pi)}{\pi}}\right)^8}\right]$$

Problem 2.2

$$f(x,U) = A(U) + B(U)x + C(U)\log_{10}\left[1+10^{\frac{x-x_1(U)}{2}}\right] + D(U)\log_{10}\left[\frac{10^{\frac{x-x_2(U)}{2}}}{1+10^{\frac{x-x_2(U)}{2}}}\right]$$

Where

$$A(U) = 5.292 - 7.382(U - 1.9) + 4.389(U - 1.9)^2$$

$$B(U) = -0.00754 + 0.773(U - 1.9) - 0.1831(U - 1.9)^2$$

$$C(U) = 7.9375 + 2.9046(U - 1.9) - 9.2402(U - 1.9)^2$$

$$D(U) = -1.5987 - 3.4753(U - 1.9) + 2.8514(U - 1.9)^2$$

$$x_1(U) = 8.234 - 0.3375(U - 1.9) - 1.9691(U - 1.9)^2$$

$$x_2(U) = 3.5477 - 0.442(U - 1.9) - 0.512(U - 1.9)^2$$

Chapter 4

Problem 4.1

Compression coefficient values for transistors biased at 50 percent i_{dss} or i_{max}

Bipolar	$K = 6.15$
Square Law MESFET	$K = 5.43$
Step-Doped MESFET	$K = 6.8$
Typical MESFET	$K = 6.1$

Problem 4.2

Bipolar	$0.541\ i_d(\text{max})$
Square Law MESFET	$0.538\ i_{dss}$
Step-Doped MESFET	$0.486\ i_{dss}$
Typical MESFET	$0.526\ i_{dss}$

Problem 4.3

4.0 dB per dB increase in P_{in}

Problem 4.4

Square Law MESFET Class A

Assume v_k = 1 volt

$K = 5.43$

P_{sat} = +22.5 dBm

P_{OIP} = +30.62 dBm

Bias current i = 50.8 mA

NF = 2.75 dB

Problem 4.5

Step-Doped MESFET Class A

Assume v_k = 1 volt

$K = 6.8$

P_{sat} = +22.68 dBm

P_{OIP} = +29.63 dBm

Bias current i = 52.9 mA

NF = 2.80 dB

Problem 4.6

Bipolar Class A

Assume v_k = 1 volt

$K = 6.15$

P_{sat} = +22.1 dBm

$P_{OIP} = +29.6$ dBm

Bias current $i = 46.0$ mA

$NF = 2.65$ dB

Chapter 8

Problem 8.1

Assume $v_k = 1$ volt

MMIC $P_{sat} = +23.38$ dBm

MMIC $NF = 2.27$ dB

Std Dev of current $\sigma_i = 0.016$ A

Std Dev of Gain $\sigma_G = 1.0$ dB

Std Dev of NF $\sigma_{NF} = 0.12$ dB

Problem 8.2

With certainty, 90.25 percent of the amplifiers in the population are compressed more than 2.0 dB when driven with +15 dBm. Likewise, 97.55 percent of the amplifiers have $P_{out} > +24$ dBm when driven at +15 dBm.

Problem 8.3

The amplifier's output third-order intercept point mean value is +34.17 dBm with standard deviation of 0.23 dB. The amplifier's 1 dB compression point mean value is +24.54 dBm with standard deviation of 0.23 dB.

Chapter 9

Problem 9.1

$P_{out} = +28.31$ dBm

Compression = 5.69 dB

Phase Shift = 28.45 Degrees

Chapter 11

Problem 11.1

The primary equation expresses |S21| as a function of current and is developed using two asymptotes. Coefficients of the primary equation are functions of frequency.

$$|S21| = G(f) + H(f)i + I(f)\log_{10}\left[1 + 10^{\frac{i-i(f)}{1}}\right]$$

Secondary equations are developed using RHF and LHF

$$G(f) = 3.286 - 1.0286f + 0.2009\log_{10}\left[1 + 10^{\frac{f-1.97}{0.2}}\right]$$

$$- 0.122\log_{10}\left[1 + 10^{\frac{f-2.62}{0.2}}\right] + 0.569\log_{10}\left[\frac{10^{\frac{f-1.33}{0.3}}}{1 + 10^{\frac{f-1.33}{0.3}}}\right]$$

$$H(f) = 0.8 - 0.2f + 0.0684\log_{10}\left[1 + 10^{\frac{f-2.24}{0.3}}\right]$$

$$- 0.667\log_{10}\left[\frac{10^{\frac{f-1.19}{0.4}}}{1 + 10^{\frac{f-1.19}{0.4}}}\right]$$

$$I(f) = -0.42 + 0.0256\log_{10}\left[1 + 10^{\frac{f-1.867}{0.1}}\right]$$

$$- 0.022\log_{10}\left[1 + 10^{\frac{f-2.58}{0.1}}\right] + 0.483\log_{10}\left[\frac{10^{\frac{f-1.25}{0.4}}}{1 + 10^{\frac{f-1.25}{0.4}}}\right]$$

$$i(f) = 3.93 - 0.96f + 0.296 \log_{10}\left[1 + 10^{\frac{f-1.497}{0.1}}\right] - 1.8 \log_{10}\left[1 + 10^{\frac{f-2.23}{0.3}}\right] + 0.1699 \log_{10}\left[\frac{10^{\frac{f-0.97}{0.1}}}{1 + 10^{\frac{f-0.97}{0.1}}}\right]$$

Appendix B
Computing Input Third-Order Intercept of Cascaded Amplifier Stages

Recall Properties of Logarithms.

An equality that occurs in the following development of input third-order intercept IIP of multiple cascaded stages is reviewed. The logarithm of an argument value x is a number a to which a base number B is raised such that

$$\log_{\mathrm{B}}(x) = a$$

and

$$x = B^{a}$$

By simple substitution, it is true that

$$x = B^{\log_{\mathrm{B}}(x)}$$

Also recall that the addition of logarithms is equivalent to the multiplication of the logarithm arguments

$$\log_{\mathrm{B}}[x] + \log_{\mathrm{B}}[y] = \log_{\mathrm{B}}[xy] = \log_{\mathrm{B}}[B^{\log_{B}(x)+\log_{B}(y)}]$$

These identities are used in the following development.

Consider the IIP of Two Cascaded Stages

Joint intercept point JIP of two cascaded stages is defined in Chapter 9, Section 9.2.1, (9.1). Equation (9.1) is reproduced here in terms of its logarithm equivalent (not its decibel equivalent):

$$JIP = -\log_{10}[10^{-OIP_1} + 10^{-IIP_2}] \tag{B.1}$$

Recall from (9.3) that cascaded input third-order intercept IIP_c is related to joint intercept point by gain in amplifier stage 1

$$IIP_c = JIP + G_1.$$

Substituting JIP from equation (B.1) into the equation for IIP_c, obtain

$$IIP_c = -\log_{10}[10^{-OIP_1} + 10^{-IIP_2}] + G_1 = \log_{10}[10^{G_1}] - \log_{10}[10^{-OIP_1} + 10^{-IIP_2}] \tag{B.2}$$

Recognize that $OIP_1 = IIP_1 + G_1$ and substitute this identity into equation (B.2) for OIP_1. Combine logarithms on the right side of equality (B.2) to obtain

$$IIP_c = -\log_{10}[10^{-IIP_1} + 10^{-IIP_2+G_1}] \tag{B.3}$$

an equation that gives input third-order intercept of two cascaded amplifier stages in terms of input third-order intercept of each stage and gain of the first stage. The units of IIP_c, IIP_1, and IIP_2, can be decibels with respect to a milliwatt or a watt as long as there is consistency throughout. Gain G_1 can be in terms of decibels.

Add a Third Cascaded Stage

Suppose the second stage of the example above is actually two stages. There exists a second stage gain G_2 and input third-order intercept IIP_2, and a third stage gain G_3 and a third stage input third-order intercept IIP_3. A joint intercept point between stages two and three, JIP_B, also exists such that input third-order intercept value looking into the cascaded second and third stages is $IIP_B = JIP_B + G_2$. Joint intercept third-order intercept value

$$JIP_B = -\log_{10}[10^{-OIP_2} + 10^{-IIP_3}] = -\log_{10}[10^{-IIP_2-G_2} + 10^{-IIP_3}] \tag{B.4}$$

Substitute this back into the equality for IIP_B and obtain

$$IIP_B = -\log_{10}[10^{-IIP_2} + 10^{-IIP_3+G_2}] \tag{B.5}$$

Now substitute (B.5) into (B.3) as the value for input third-order intercept of stage 2, IIP2, and recall the properties of logarithms to obtain the equation for input third-order intercept of three cascaded amplifier stages

$$IIP_c = -\log_{10}[10^{-IIP_1} + 10^{-IIP_2+G_1} + 10^{-IIP_3+G_1+G_2}] \tag{B.6}$$

Input Third-order Intercept for N Stages

The extension of (B.6) to N cascaded amplifier stages is obvious

$$IIP_c =$$

$$-\log_{10}[10^{-IIP_1} + 10^{-IIP_2+G_1} + 10^{-IIP_3+G_1+G_2} + \cdots + 10^{-IIP_n+G_1+G_2+\cdots+G_{n-1}}] \tag{B.7}$$

Appendix C
Noise Figure Degradation Due to Cascading Circuit Elements

Noise figure of n cascaded circuit elements, NF_c, is calculated using the familiar noise figure equation

$$NF_c = 10 * \log_{10}\left[F_1 + \frac{F_2 - 1}{g_1} + \frac{F_3 - 1}{g_1 g_2} \circ\circ\circ + \frac{F_n - 1}{g_1 g_2 g_3 \circ\circ\circ g_{n-1}}\right] \quad \text{(C.1)}$$

where noise factor of each circuit element

$$F_n = 10^{\frac{NF_n}{10}} \quad \text{(C.2)}$$

is determined from noise figure NF_n in dB of that element, and small signal gain

$$g_n = 10^{\frac{G_n}{10}} \quad \text{(C.3)}$$

is determined from gain G_n in dB of that element.

Consider two circuit elements in cascade where the cascaded noise figure is

$$NF_c = 10 * \log_{10}\left[F_1 + \frac{F_2 - 1}{g_1}\right] \quad \text{(C.4)}$$

Modify the logarithm argument by factoring noise factor F_1 as follows

$$NF_c = 10 * \log_{10}\left[F_1\left(1 + \frac{F_2 - 1}{g_1 F_1}\right)\right] \tag{C.5}$$

The product of two factors within the logarithm argument as shown in equation (C.5) can be rewritten

$$NF_c = 10 * \log_{10}[F_1] + 10 * \log_{10}\left[\left(1 + \frac{F_2 - 1}{g_1 F_1}\right)\right]$$

$$= NF_1 + 10 * \log_{10}\left[\left(1 + \frac{F_2 - 1}{g_1 F_1}\right)\right]$$

by recognizing that $NF_1 = 10 * \log_{10}[F_1]$.

The change in noise figure of circuit element 1, ΔNF_1, caused by cascading it with circuit element 2 is determined by subtracting the value of NF_1 from both sides of the equation above obtaining

$$\Delta NF_1 = (NF_c - NF_1) = 10 * \log_{10}\left[\left(1 + \frac{F_2 - 1}{g_1 F_1}\right)\right] \tag{C.6}$$

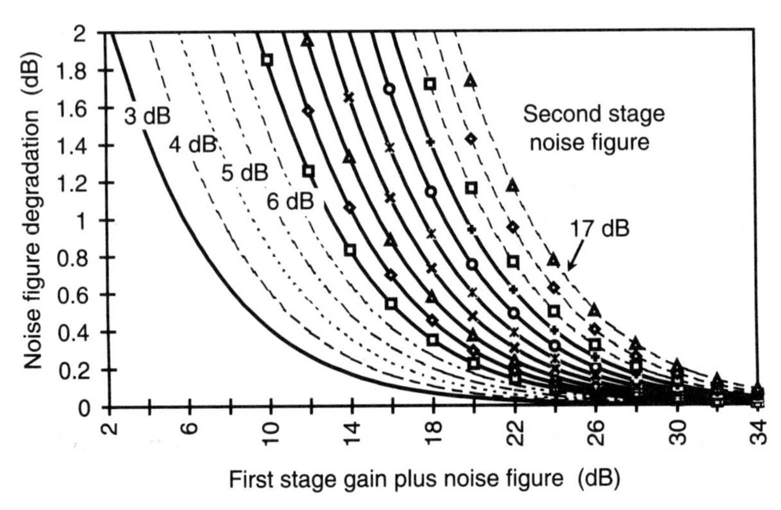

Figure C.1 Cascaded circuit noise figure degradation as a function of circuit element 1 gain plus noise figure and circuit element 2 noise figure.

Change in circuit element 1 noise figure is a function of noise factor of circuit element 2, F_2, and the product of circuit element 1 gain and circuit element 1 noise factor g_1F_1. This can be restated in terms of decibels. Noise figure of circuit element 1 is a function of circuit element noise figure NF_2 and of the sum of circuit element 1 gain and noise figure ($G_1 + NF_1$) in decibels. Values of noise figure degradation are easily calculated by entering equation (C.6) into a spreadsheet and plotting to results as illustrated in Figure C.1.

Appendix D
List of Symbols

A	Amperes
A_i	Arbitrary constants
A1, A2, A3–AN	Asymptote identification
B_{data}	Binary data command word
B_i	Arbitrary constants
$C(b)$	Secondary equation, a function of bias coefficient
C_i	Arbitrary constants
$C_j(0)$	Diode junction depletion capacitance at 0 volts
D_k	Arbitrary constants
$D(b)$	Secondary equation, a function of bias coefficient
$E(b)$	Secondary equation, a function of bias coefficient
F	Noise factor
$F(b)$	Secondary equation, a function of bias coefficient
G	Amplifier or amplifier assembly gain (dB)
$G(b)$	Secondary equation, a function of bias coefficient
$G_{bp}(f)$	Average gain in a frequency bandwidth
G_{cas}	Cascaded amplifier stage total small signal gain (dB)
G_{MMIC}	Monolithic integrated circuit total small signal gain (dB)
G_{ss}	Transistor or amplifier stage small signal gain (dB)
$G_{ss}(\max)$	Population maximum expected small signal gain (dB)
$G_{ss}(\text{mean})$	Population mean small signal gain value (dB)
$G_{ss}(\min)$	Population minimum expected small signal gain (dB)
H	Arbitrary constant
$H(b)$	Secondary equation, a function of bias coefficient
I_c	Collector current
I_{ds}	Drain to source current (amperes)
IF	Intermediate frequency
I_s	Junction reverse or saturation current

IIP	Amplifier stage input third-order intercept point (dBm)
IIP_{cas}	Cascaded amplifier stage input third-order intercept point (dBm)
I_{max}	Bipolar transistor saturation current
JIP	Joint third-order intercept point (dBm)
$J(b)$	Secondary equation, a function of bias coefficient
K	Compression coefficient
$L(b)$	Secondary equation, a function of bias coefficient
(LHF)	Left-hand function
LO	Local oscillator frequency
M	Cascaded power amplifier stage *PAE* optimization factor
M	Diode junction doping profile
$M(b)$	Secondary equation, a function of bias coefficient
N	Number of asymptotes needed
N	Ratio of saturated power output of two cascaded amplifier stages
NF	Transistor or amplifier stage noise figure (dB)
NF_{cas}	Cascaded amplifier stage noise figure (dB)
OIP	Amplifier stage output third-order intercept point (dBm)
OIP_{cas}	Cascaded amplifier stage output third-order intercept point (dBm)
PAE	Power-added efficiency
P_{IIP}	Input third-order intercept point (dBm)
P_{in}	Power input (dBm)
$P_{in}(1dB)$	Amplifier power input in dBm at 1 dB compression
P_{K}	Amplifier power output at beginning of compression
P_{OIP}	Output third-order intercept point (dBm)
P_{out}	Power output (dBm)
P_{sat}	Saturated power output (dBm)
$P_{sat}(max)$	Population maximum expected saturated power output (dBm)
$P_{sat}(mean)$	Population mean saturated power output value (dBm)
$P_{sat}(min)$	Population minimum expected saturated power output (dBm)
$P1(b)$	Constant used to determine current as function of P_{in}
$P2(b)$	Constant used to determine current as function of P_{in}
P_{1dB}	One dB compressed power output
P_{3rdOIP}	Third-order intercept point (dBm)

Q	Ripple amplitude correction factor
R	Bandpass ripple amplitude (dB)
(RHF)	Right-hand function
R_{in}	Input impedance
R_l	Load line resistance
S_j	Asymptote slope value
T	Temperature degrees Kelvin
U	Secondary function equation variable
$U_G(f)$	Amplifier gain sensitivity to temperature as a function of frequency
$U_{NF}(f)$	Noise figure sensitivity to temperature as a function of frequency
$U_{sat}(f)$	Amplifier saturated power output sensitivity to temperature as a function of frequency
$U1, U2, U3, -UN$	Secondary functions
V_{dc}	Applied DC voltage
V_{ce}	Collector to emitter voltage
V_{cmd}	Analog command Voltage
V_k	Knee voltage
a	Arbitrary constant
a_i	Right-hand function transition range
b	Arbitrary constant
b	Bias coefficient
b_{avg}	Average bias coefficient
b_i	Left-hand function transition range
b_q	Quiescent bias coefficient
c	Arbitrary constant
c_i	Arbitrary constants
d_i	Arbitrary constants
e	Natural base (2.7182818)
e_{c1}	Carrier signal 1 amplitude in volts
e_{c2}	Carrier signal 2 amplitude in volts
e_{im}	Intermodulation signal amplitude in volts
e_m	Modulating signal amplitude in volts
f	Frequency
f_c	Bandpass geometric center frequency
f_{hi}	Bandpass upper 3dB point
f_{lo}	Bandpass lower 3dB point
f_t	Transistor unity gain frequency
f_0	Transistor frequency at 3dB gain reduction
g	Transistor or amplifier stage gain power ratio

g_m	Transistor transconductance
g_{ss}	Transistor or amplifier stage small signal gain power ratio
i_{avg}	Average bias current
i_{ds}	Drain-to-source current (amperes)
i_{dss}	MESFET saturation current at $V_{gs} = 0$ volts
iip	Amplifier stage input third-order intercept point (Watts)
i_{max}	Maximum saturated transistor current
i_{peak}	Peak current along a load line
i_q	Quiescent bias current
jip	Joint third-order intercept point (Watts)
k	Boltzman's constant
m	Arbitrary constant
n	Number of poles in a bandpass network
oip	Amplifier stage output third-order intercept point (Watts)
p_{in}	Power input (Watts)
p_{out}	Power output (Watts)
p_{sat}	Saturated power output (Watts)
q	Electron charge value
t	Time as a variable
t_i	Specific points in time
u	Arbitrary dependent variable
v_{ga}	Voltage at point of asymptote tangency
v_{ds}	Drain-to-source voltage
v_f	Voltage at point of asymptote tangency
v_{gs}	Gate-to-source voltage
v_j	Junction voltage
v_k	Transistor knee voltage
v_p	Voltage where junction capacitance peaks
v_{po}	Transistor pinch-off voltage
x	Arbitrary independent variable
x_0	Weibull probability density function location factor
y	Arbitrary dependent variable
z	Arbitrary independent variable
α	Arbitrary constant
α	Weibull probability density function scale factor
β	Base current to collector current amplification ratio
β	Weibull probability density function shape factor
β	Arbitrary constant
φ	Compression phase shift

φ_{in}	Input signal phase angle
φ_{out}	Output signal phase angle
λ	Arbitrary constant
θ	AM to PM conversion sensitivity
σ	Standard deviation
σ_{cas}	Standard deviation of a cascaded amplifier stage parameter
3rd *OIP*	Third-order intercept point
ΔfBW	Bandwidth between 3dB points
ΔS	Difference in asymptote slope values
$\%BW$	Percent bandwidth

About the Author

Mr. Turlington graduated from Baltimore Polytechnic Institute in February, 1956. He immediately began his career with the Westinghouse Electric Corporation's Air Arm Division near Baltimore, where he was awarded a position in the prestigious Westinghouse-Johns Hopkins Scholarship Program, earning a Bachelor of Science degree in engineering in June, 1961. Continuing engineering studies at Johns Hopkins and in the Westinghouse School of Applied Engineering Science, he earned recognition for completing an Advanced Engineering Technologies Design Program in May, 1972.

During his forty-two year employment with the world-renowned Westinghouse Defense Electronics Center near Baltimore, he worked on numerous ground-based, airborne and space-borne radar programs. He is credited with numerous achievement awards including The First Space Shuttle Flight Award for work on the shuttle's master timing unit, and the SEASAT Synthetic Aperture Radar Team Award for the high quality radar imagery obtained from NASA's first space-borne synthetic aperture radar system. Most recently he was awarded the George Westinghouse Signature Award for his work on an electronic countermeasures solid state active aperture transmitter. He has published papers on injection locking of microwave oscillators, linear control circuitry for crystal oscillator temperature controlled ovens, the solid state L-Band Seasat Imaging Radar Transmitter, and active aperture transmit-receive modules. He holds patents on low-noise parametric varactor diode crystal oscillators and transmit receive modules. He remained with the Baltimore Defense Electronics Center for two years after it was acquired by Northrop Grumman before retiring in April, 1998. Presently, he lectures for Besser Associates on behavioral modeling. Mr. Turlington has developed this methodology of Behavioral Modeling over the last ten years and has applied it to the understanding and characterization of transmit receive modules, frequency selective limiters, and numerous nonlinear circuit elements and devices. This book is a compendium of those basic behavioral modeling ideas as they are applied to nonlinear microwave systems, subsystems, circuits, and transistors.

Index

Recent Titles in the Artech House Microwave Library

Advanced Automated Smith Chart Software and User's Manual, Version 3.0, Leonard M. Schwab

Behavioral Modeling of Nonlinear RF and Microwave Devices, Thomas R. Turlington

C/NL2 for Windows: Linear and Nonlinear Microwave Circuit Analysis and Optimization, Software and User's Manual, Stephen A. Maas and Arthur Nichols

Computer-Aided Analysis of Nonlinear Microwave Circuits, Paulo J. C. Rodrigues

Design of FET Frequency Multipliers and Harmonic Oscillators, Edmar Camargo

Design of RF and Microwave Amplifiers and Oscillators, Pieter L. D. Abrie

Feedforward Linear Power Amplifiers, Nick Pothecary

FINCAD: Fin-Line Analysis and Synthesis for Millimeter-Wave Application Software and User's Manual, S. Raghu Kumar, Anita Sarraf, and R. Sathyavageeswaran

Generalized Filter Design by Computer Optimization, Djuradj Budimir

GSPICE for Windows, Sigcad Ltd.

Introduction to Microelectromechanical (MEM) Microwave Systems, Hector J. De Los Santos

LINPAR for Windows: Matrix Parameters for Multiconductor Transmission Lines, Software and User's Manual, Version 2.0, Antonije Djordjevic *et al.*

Microwave Engineers' Handbook, Two Volumes, Theodore Saad, editor

Microwave Filters, Impedance-Matching Networks, and Coupling Structures, George L. Matthaei, Leo Young, and E.M.T. Jones

Microwave Mixers, Second Edition, Stephen Maas

Microwaves and Wireless Simplified, Thomas S. Laverghetta

Microwave Radio Transmission Design Guide, Trevor Manning

MULTLIN for Windows: Circuit-Analysis Models for Multiconductor Transmssion Lines, Software and User's Manual, Antonije R. Djordjevic *et al.*

The RF and Microwave Circuit Design Handbook, Stephen A. Maas

RF Design Guide: Systems, Circuits, and Equations, Peter Vizmuller

RF and Microwave Coupled-Line Circuits, Rajesh Mongia, Inder Bahl, and Prakash Bhartia

RF Power Amplifiers for Wireless Communications, Steve C. Cripps

RF Systems, Components, and Circuits Handbook, Ferril Losee

SPURPLOT: Mixer Spurious-Response Analysis with Tunable Filtering, Software and User's Manual, Version 2.0, Robert Kyle

TRANSLIN: Transmission Line Analysis and Design, Software and User's Manual, Paolo Delmastro

TRAVIS Pro: Transmission Line Visualization Software and User's Manual, Professional Version, Robert G. Kaires and Barton T. Hickman

For further information on these and other Artech House titles, including previously considered out-of-print books now available through our In-Print-Forever® (IPF®) program, contact:

Artech House
685 Canton Street
Norwood, MA 02062
Phone: 781-769-9750
Fax: 781-769-6334
e-mail: artech@artechhouse.com

Artech House
46 Gillingham Street
London SW1V 1AH UK
Phone: +44 (0)20-7596-8750
Fax: +44 (0)20-7630-0166
e-mail: artech-uk@artechhouse.com

Find us on the World Wide Web at: www.artechhouse.com